Computational Fluid Flow and Heat Transfer

The text provides insight into the different mathematical tools and techniques that can be applied to the analysis and numerical computations of flow models. It further discusses important topics such as the heat transfer effect on boundary layer flow, modeling of flows through porous media, anisotropic polytrophic gas model, and thermal instability in viscoelastic fluids.

This book:

- Discusses modeling of Rayleigh-Taylor instability in nanofluid layer and thermal instability in viscoelastic fluids.
- Covers open FOAM simulation of free surface problems, and anisotropic polytrophic gas model.
- Highlights the Sensitivity Analysis in Aerospace Engineering, MHD Flow of a Micropolar Hybrid Nanofluid, and IoT-Enabled Monitoring for Natural Convection.
- Presents thermal behavior of nanofluid in complex geometries and heat transfer effect on Boundary layer flow.
- Explains natural convection heat transfer in non-Newtonian fluids and homotropy series solution of the boundary layer flow.
- Illustrates modeling of flows through porous media and investigates Shock-driven Richtmyer-Meshkov instability.

It is primarily written for senior undergraduate, graduate students, and academic researchers in the fields of Applied Sciences, Mechanical Engineering, Manufacturing Engineering, Production Engineering, Industrial engineering, Automotive engineering, and Aerospace engineering.

Advances in Manufacturing, Design and Computational Intelligence Techniques

Series Editor-Ashwani Kumar – *Senior Lecturer, Mechanical Engineering, at Technical Education Department, Uttar Pradesh, Kanpur, India*

The book series editor is inviting edited, reference, and text book proposal submission in the book series. The main objective of this book series is to provide researchers a platform to present state of the art innovations, research related to advanced materials applications, cutting edge manufacturing techniques, innovative design and computational intelligence methods used for solving nonlinear problems of engineering. The series includes a comprehensive range of topics and its application in engineering areas such as additive manufacturing, nanomanufacturing, micromachining, biodegradable composites, material synthesis and processing, energy materials, polymers and soft matter, nonlinear dynamics, dynamics of complex systems, MEMS, green and sustainable technologies, vibration control, AI in power station, analog-digital hybrid modulation, advancement in inverter technology, adaptive piezoelectric energy harvesting circuit, contactless energy transfer system, energy efficient motors, bioinformatics, computer aided inspection planning, hybrid electrical vehicle, autonomous vehicle, object identification, machine intelligence, deep learning, control-robotics-automation, knowledge-based simulation, biomedical imaging, image processing, and visualization. This book series compiled all aspects of manufacturing, design and computational intelligence techniques from fundamental principles to current advanced concepts.

Hybrid Metal Additive Manufacturing: Technology and Applications
Edited by Parnika Shrivastava, Anil Dhanola, and Kishor Kumar Gajrani

Thermal Energy Systems: Design, Computational Techniques, and Applications
Edited by Ashwani Kumar, Varun Pratap Singh, Chandan Swaroop Meena, and Nitesh Dutt

Computational Fluid Flow and Heat Transfer: Advances, Design, Control, and Applications
Edited by Mukesh Kumar Awasthi, Ashwani Kumar, Nitesh Dutt, and Satyvir Singh

https://www.routledge.com/Advances-in-Manufacturing-Design-and-Computational-Intelligence-Techniques/book-series/CRCAIMDCIT?publishedFilter=alltitles&pd=published,forthcoming&pg=1&pp=12&so=pub&view=list?publishedFilter=alltitles&pd=published,forthcoming&pg=1&pp=12&so=pub&view=list

Computational Fluid Flow and Heat Transfer
Advances, Design, Control, and Applications

Edited by
Mukesh Kumar Awasthi
Ashwani Kumar
Nitesh Dutt
Satyvir Singh

CRC Press
Taylor & Francis Group
Boca Raton London New York

CRC Press is an imprint of the
Taylor & Francis Group, an **informa** business

First edition published 2024
by CRC Press
2385 NW Executive Center Drive, Suite 320, Boca Raton FL 33431

and by CRC Press
4 Park Square, Milton Park, Abingdon, Oxon, OX14 4RN

CRC Press is an imprint of Taylor & Francis Group, LLC

ISBN: 978-1-032-60318-6 (hbk)
ISBN: 978-1-032-73713-3 (pbk)
ISBN: 978-1-003-46517-1 (ebk)

DOI: 10.1201/9781003465171

Typeset in Sabon
by SPi Technologies India Pvt Ltd (Straive)

Contents

Aim and scope

The book *Computational Fluid Flow and Heat Transfer: Advances, Design, Control, and Applications* aims to provide a reference for the applications of flow problems with or without heat transfer using analytical/numerical/experimental techniques in various unique industry problems in the era of Industry 4.0, and it will offer a sound background with adequate applications. Apart from the diversified industrial problems we are also aiming to discuss the flow modeling with real time data in the current scenario. This book is intended to be multi-domain applications; this helps researchers and scholars to think beyond one specific field and work on cohort domain in mathematics, engineering, and computer science. This book will help researchers to think/use artificial intelligence and machine learning to solve the flow problems in complex geometries. The book can be an idea tickling for audience working in the domain of modeling and simulations.

In recent years, scientists have been trying to solve practical flow problems using mathematical models, and they are also getting success. As computing technologies have grown tremendously, the possibilities of solving complex problems have also increased. Most of the flow systems are very complex and it takes a lot of time and money to find their solutions. To design an engineering flow system and understand its processes, it is very important to understand and analyze all the parameters used in that system and mathematical models are very helpful in understanding and analyzing such parameters. This book focuses on the modeling of flow phenomenon in complex geometries.

The flow of fluids is very important in confined geometries. In case of cylindrical configuration, there are many applications of capillary such as water jet, film boiling, and film-wise condensate Ink jet printer is an excellent example of combination of capillary flow and electric field. The fluids, depending upon their nature, make a significant impact on our day-to-day life. Non-Newtonian fluids like polymers and viscoelastic fluids need special attention from the researchers. The exact solution of the Navier-Stokes equation is not known till date, and therefore, numerical methods are very helpful for the computation of various properties of fluid flows. In this book, we are looking to add the nature-inspired flow problems and their experimental/analytical/numerical solutions.

Instability of fluid flow is responsible for the turbulence of fluid flow. There are a large number of applications of fluid's instability in engineering and the natural environment. The instability phenomenon is also found in the fields of astrophysics, applied mathematics, geophysics, biology, physics, oceanography, etc. On applying some perturbation in the fluid system which may be in motion or stable, its flow properties will be different from the original state. The system is called stable if a small perturbation imposed on the system dies and the system comes back to its original state, while the system is called unstable if perturbation increases with time and the properties of the system change forever. A particular emphasis will be on the recent modeling and computational trends to investigate stability of multiphase flows.

Heat transfer occurs whenever there is inhomogeneous temperature distribution within the system or between the system and its surrounding environment. There are four basic modes of heat transfer: conduction, convection, phase change, and radiation. The process by which heat diffuses through a solid or a stationary fluid is termed as heat conduction. The transfer of heat from a wetted surface is assisted by the motion of the fluid, which is known as heat convection. However, when the fluid undergoes a phase change at or near the wetted surface, it is called phase-change heat transfer. The thermal radiation is by electromagnetic wave propagation. There is no medium needed for thermal radiation, although the electromagnetic waves can be transferred through gases. This book will include the studies related to heat transfer on industrial applications.

The relative migration of species in a mixture caused by concentration gradients is explicitly referred to as mass transfer. A net mass transfer from a region of high concentration to a region of low concentration results from the molecule's random mobility. The diffusion of chemical contaminants from point sources, whether natural or manufactured, into lakes, rivers, and oceans are two typical instances of mass transfer processes. Besides the modeling and simulation of heat transfer effect, the book will highlight the mathematical modeling and computation of the problems arising in engineering systems, with special emphasis on the mass transfer in various geometries. The book has 14 chapters, and the applicability of this book covers a wide range of industries such as fluid dynamics, heat transfer, production sector, design engineers, research and development engineers, automotive industry, aviation sector, electronics industry, civil engineers, and researchers in heat transfer sectors, and it will help the audience in conducting research in these industries.

Editors
Dr. Mukesh Kumar Awasthi
Dr. Ashwani Kumar
Dr. Nitesh Dutt
Dr. Satyvir Singh

Preface

The contemporary world is moving fast into the knowledge and innovation of science and technology. The fluid flow phenomena play a very crucial role in understanding various complex processes arising in natural and engineering systems. That's why, nowadays, one of the prime agendas of scientific research is to strengthen the researchers in the areas, such as mathematical modeling, simulation, and numerical techniques. This book provides a comprehensive resource for readers to make them familiar with recent research developments in the mathematical modeling of the real-world problems arising in diverse areas of science and engineering. The applications covered in the book will equip the readers with the fundamental understanding of how to identify appropriate situations for modeling and how to apply mathematical symbols and ideas to create flow models. It will provide an insight into different mathematical tools and techniques that can be applied for the analysis and numerical computations of the flow models.

The book *Computational Fluid Flow and Heat Transfer: Advances, Design, Control, and Applications* aims to provide a reference for application of flow problems with or without heat transfer using analytical/numerical/experimental techniques. Chapters included in the book illustrate new methodologies, model, techniques, and applications along with description of modeling, optimization and experimental works. **Chapter 1** covers the introduction to advance computation tools applied in fluid flow and heat transfer analysis. **Chapters 2 and 3** talk about the natural convection in Newtonian fluids and non-Newtonian fluids. **Chapter 4** covers how forced convection works. In continuation **Chapter 5** covers heat transfer effect on boundary layer flow in aerospace applications. **Chapters 6 and 7** provide information on the sensitivity analysis in aerospace applications. **Chapter 8** examines the different modeling and optimization to improve the flows through porous media. The chapter discusses theoretical and industrial applications of porous media fluid flows analysis. **Chapters 9–13** highlight different applications and thermal behavior of nanofluids. These chapters are crucial since applications of nanofluids are increasing in heat transfer enhancement techniques. Chapter 12 focuses on analyzing the MHD flow of a micropolar hybrid nanofluid between two porous disks. The flow nature is significantly influenced by the micro rotation of the hybrid nanoparticles. To provide clarity, a comparison between the analytical and numerical results is stated. In this study, TiO_2-Cu is chosen as the hybrid nanoparticle mixed with a water-based fluid. In continuation, Chapter 13 presents a comprehensive investigation of the Rayleigh-Taylor (RT) instability occurring at the edge of a planar configuration, where a viscous fluid and a nanofluid are involved. The setup consists of a lower region containing the viscous fluid and an upper region containing the nanofluid. The growth rate of perturbations is found to follow a quadratic pattern in the dispersion relationship.

Chapter 14 deals with the impact study of shock wave strengths on the hydrodynamic instability at elliptical bubble. The elliptical bubble is composed of helium gas, which is surrounded by nitrogen gas. Three different shock wave strengths are considered: Ms = 1.12, 1.25, and 1.5.

The book *Computational Fluid Flow and Heat Transfer: Advances, Design, Control, and Applications* is dedicated to the application of fluid and heat transfer that makes it ideal for usage in interdisciplinary undergraduate and postgraduate curricula, and makes it useful for teaching and research purposes. Additionally, the organization of the 14 chapters in this book facilitates self-study by independent pupils. It is also beneficial for engineers, economists, and decision-makers since it helps them comprehend the background, current, and prospective future of the subject under discussion.

Editors
Dr. Mukesh Kumar Awasthi
Dr. Ashwani Kumar
Dr. Nitesh Dutt
Dr. Satyvir Singh

Acknowledgments

We express our gratitude to CRC Press (Taylor & Francis Group) and the editorial team for their suggestions and support during completion of this book. We are grateful to all authors for submitting their quality work and reviewers for their illuminating views on each book chapter presented in the book *Computational Fluid Flow and Heat Transfer: Advances, Design, Control, and Applications*.

This book is dedicated to all engineers, researchers, and academicians.

Contributors

Mukesh Kumar Awasthi
Department of Mathematics
Babasaheb Bhimarao Ambedkar
  University
Lucknow, India

B.S. Bhadauria
Department of Mathematics
Babasaheb Bhimrao Ambedkar
  University
Lucknow, India

Awani Bhushan
Department of Mechanical
  Engineering
School of Mechanical Engineering
Chennai, India

G. S. Charana
Indian Institute of Engineering
  Science and Technology
Howrah, India

Nitesh Dutt
Department of Mechanical
  Engineering
COER University
Roorkee, Uttarakhand, India

K. Gnanaprasanna
Division of Mathematics, School of
  Advanced Science
Vellore Institute of Technology
Chennai, India

Reshu Gupta
Applied Science Cluster
  (Mathematics), UPES
Dehradun, India

Robert G. Hardin
Department of Biological and
  Agricultural Engineering
Texas A&M University
College Station, TX

Ankush Hedau
Department of Hydro and
  Renewable Energy
Indian Institute of Technology
  Roorkee
Uttarakhand, India

Ismail
Babasaheb Bhimrao Ambedkar
  University
Lucknow, India

Maria King
Department of Biological and
  Agricultural Engineering
Texas A&M University
College Station, TX

Anish Kumar
Department of Mathematics
Babasaheb Bhimrao Ambedkar
  University
Lucknow, India

Ashwani Kumar
Department of Mechanical
  Engineering
Technical Education Department
  Uttar Pradesh
Kanpur, India

Sunil Kumar
Department of Biological and
  Agricultural Engineering
Texas A&M University
College Station, TX

S.N. Rai
Department of Mathematics
Babasaheb Bhimrao Ambedkar
  University
Lucknow, India

L. Prince Raj
Indian Institute of Engineering
  Science and Technology
Howrah, India

Seetu Rana
Department of Mathematics
Government College Hisar
Hisar, India

Bidesh Sengupta
School of Computer Science and
  Engineering
Nanyang Technological University
Singapore

Shilpee
Babasaheb Bhimrao Ambedkar
  University
Lucknow, India

Pankaj Shukla
Division of Mathematics
School of Advanced Science
Vellore Institute of Technology
Chennai, India

P. M. Mohamed Abubacker Siddique
Indian Institute of Engineering
  Science and Technology
Howrah, India

Abhishek Kumar Singh
Division of Mathematics
School of Advanced Science
Vellore Institute of Technology
Chennai, India

Ankit Rajkumar Singh
Mechanical Engineering
  Department
Indian Institute of Technology
  Bombay
Mumbai, India

Satyvir Singh
Applied and Computational
  Mathematics
RWTH Aachen University
Aachen, Germany

Varun Pratap Singh
Department of Mechanical
  Engineering
School of Advanced Engineering,
  UPES
Dehradun, India

Anand Kumar Solanki
Department of Mechanical
  Engineering
Gayatri Vidya Parishad College
  of Engineering
Visakhapatnam, India

M. Sridharan
Department of Mechanical
  Engineering
SRM Institute of Science and
  Technology
Tiruchirappalli, India

Anurag Srivastava
Department of Mathematics
Babasaheb Bhimrao Ambedkar
    University
Lucknow, India

Mrityunjai Verma
Department of Mechanical
    Engineering
School of Advanced Engineering,
    UPES
Dehradun, India

About the editors

Dr. Mukesh Kumar Awasthi did his PhD on the topic "Viscous Correction for the Potential Flow Analysis of Capillary and Kelvin-Helmholtz instability." He is an Assistant Professor in the Department of Mathematics at Babasaheb Bhimrao Ambedkar University, Lucknow. Dr. Awasthi specialized in the mathematical modeling of flow problems. He has taught courses such as Fluid Mechanics, Discrete Mathematics, Partial differential equations, Abstract Algebra, Mathematical Methods, and Measure theory to postgraduate students. He has excellent knowledge in the mathematical modeling of flow problems, and he can solve these problems analytically as well as numerically. He has a good grasp of subjects such as viscous potential flow, electro-hydrodynamics, magneto-hydrodynamics, heat, and mass transfer. He has excellent communication skills and leadership qualities. Dr. Awasthi has qualified for the National Eligibility Test (NET) conducted on all India levels in 2008 by the Council of Scientific and Industrial Research (CSIR) and got Junior Research Fellowship (JRF) and Senior Research Fellowship (SRF) for doing research. He has published 115 plus research publications (journal articles/books/book chapters/conference articles) in Elsevier, Taylor & Francis, Springer, Emerald, World Scientific, and many other national and international journals and conferences. Also, he has published eight books. He has attended many symposia, workshops, and conferences in mathematics as well as fluid mechanics. He has won the Research Awards four times consecutively from 2013–2016 from the University of Petroleum and Energy Studies, Dehradun, India. He has also received the start-up research fund for his project "Nonlinear study of the interface in multilayer fluid system" from UGC, New Delhi. Recently in 2023 Dr. Awasthi name has been included in world's top 2% researcher/ scientist list jointly issued by Stanford University USA and Scopus Database.

Dr. Ashwani Kumar earned a PhD in Mechanical Engineering in the area of Mechanical Vibration and Design. He is currently a Senior Lecturer, Mechanical Engineering (Gazetted Officer Group B) at the Technical Education Department Uttar Pradesh (under Government of Uttar Pradesh) Kanpur, India since December 2013. He was an Assistant Professor in Department of Mechanical Engineering, Graphic Era University Dehradun India (NIRF

Ranking 55) from July 2010 to November 2013. He has more than 13 years of research, academic, and administrative experience including Coordinator for AICTE-Extension of Approval, Nodal officer for PMKVY-TI Scheme (Government of India), internal coordinator-CDTP scheme (Government of Uttar Pradesh), Industry Academia relation officer, Assistant Centre Superintendent (ACS)-Institute Examination Cell, Zonal Officer to conduct Joint Entrance Examination (JEE-Diploma), Sector Magistrate for Lok Sabha-Vidhan Sabha Election, among others. He is Series Editor of four book series (1) Advances in Manufacturing, Design and Computational Intelligence Techniques, (2) Renewable and Sustainable Energy Developments, (3) Smart Innovations and Technological Advancements in Mechanical and Materials Engineering, and (4) Computational Intelligence and Biomedical Engineering. He is guest editor of a special issue titled Sustainable Buildings, Resilient Cities and Infrastructure Systems (Buildings ISSN: 2075-5309, I.F. 3.8). He is Editor-in-Chief for International Journal of Materials, Manufacturing and Sustainable Technologies (IJMMST, ISSN: 2583-6625), and Editor of International Journal of Energy Resources Applications (IJERA, ISSN: 2583-6617). He is guest editor and editorial board member of eight international journals and acts as review board member of 20 prestigious (Indexed in SCI/SCIE/Scopus) international journals with high impact factor, i.e. Applied Acoustics, Measurement, JESTEC, AJSE, SV-JME, and LAJSS. In addition, he has published 100+ research articles in journals, book chapters, and conferences. He has authored/co-authored cum edited 30+ books on mechanical, materials, and energy engineering. He has published two patents. He is associated with International Conferences as Invited Speaker/Session Chair/Advisory Board/Review Board Member/Program Committee Member. He has delivered many invited talks in webinar, FDP, and Workshops. He has been awarded Best Teacher for excellence in academic and research. He has successfully guided 15 BTech, MTech, and PhD theses. He is an external doctoral committee member of S.R.M. University, New Delhi. He is currently involved in the research areas of AI & ML in Mechanical Engineering, Smart Materials & Manufacturing Techniques, Thermal Energy Storage, Building Efficiency, Renewable Energy Harvesting, Sustainable Transportation and Heavy Vehicle Dynamics. His Orcid is 0000-0003-4099-935X, Google Scholar web link is https://scholar.google.com/citations?hl=en&user=KOILpEkAAAAJ and research gate web link is https://www.researchgate.net/profile/Ashwani-Kumar-45.

Dr. Nitesh Dutt is Head of Department and Associate Professor in the Department of Mechanical Engineering, COER University Roorkee, Uttarakhand, India. He has more than seven years of teaching experience. He has a bachelor's degree in Mechanical Engineering, and Masters and PhD from IIT Roorkee. He has published more than 11 research articles in international journals and conferences. His main areas of research are

Renewable Energy, Green Energy Harvesting, Nuclear Engineering, Heat and Mass Transfer, Thermodynamics, Fluid Mechanics, Refrigeration, and Air Conditioning, Computational Fluid Dynamics (CFD).

Dr. Satyvir Singh is a postdoc research associate in Applied and Computational Mathematics at RWTH Aachen University, Germany. Prior to that, he worked as a Research Fellow in Nanyang Technological University, Singapore. He completed his PhD in the School of Mechanical and Aerospace Engineering from Gyeongsang National University, South Korea. He earned his MTech in the Department of Mathematics from IIT Madras, India. Dr. Singh has a vast area of research that includes computational Fluid dynamics, high-order numerical methods, multi-component flows, kinetic theory of gases and quantum physics, and computational biology. He has qualified for the CSIR-NET-JRF exam with A.I.R.-38. He has published 30 plus research publications (journal articles/book chapters/conference articles) in Elsevier, AIP, APS, Springer, and many other international journals. He has also attended many international conferences and presented his research work in the USA, the UK, South Korea, Singapore, Germany, China, Japan, and India.

Chapter 1

Introduction to mathematical and computational methods

Mukesh Kumar Awasthi

Babasaheb Bhimarao Ambedkar University, Lucknow, India

Nitesh Dutt

COER University, Roorkee, Uttarakhand, India

Ashwani Kumar

Technical Education Department Uttar Pradesh Kanpur, India

Ankush Hedau

Indian Institute of Technology Roorkee, Uttarakhand, India

1.1 INTRODUCTION

In the field of fluid flow and heat transfer analysis, computational tools play a pivotal role in simulating and analyzing complex engineering systems. These tools leverage advanced computational algorithms and numerical techniques to solve the governing equations of fluid dynamics and heat transfer, providing engineers and researchers with invaluable insights into the behavior of fluids and thermal processes. The development of computational tools for fluid flow and heat transfer analysis has revolutionized the way engineers approach the design, optimization, and analysis of various engineering systems. Traditionally, experimental methods and theoretical calculations were relied upon to understand fluid flow and heat transfer phenomena. However, these approaches often had limitations in terms of cost, time, and accuracy. The advent of computational tools has mitigated these limitations and opened up new avenues for studying and optimizing fluid flow and heat transfer processes.

One of the key advantages of computational tools is their ability to model and simulate complex fluid flow and heat transfer scenarios that are often challenging or impossible to replicate in experimental setups. By discretizing the governing equations into a computational domain and employing numerical methods, computational tools can simulate the behavior of fluids and heat transfer phenomena with high accuracy and efficiency. This enables engineers to explore a wide range of design parameters, boundary conditions, and operating scenarios, facilitating the optimization of engineering

DOI: 10.1201/9781003465171-1

1

systems in a virtual environment. Another significant advantage of computational tools is their capability to provide detailed insights into the flow and thermal fields, enabling the visualization and analysis of complex fluid dynamics and heat transfer phenomena. Engineers can obtain velocity profiles, pressure distributions, temperature gradients, and other essential parameters within the fluid domain. This detailed information aids in understanding the underlying physics of the system, identifying flow patterns, detecting areas of high heat transfer or fluid resistance, and optimizing the design to enhance performance and efficiency.

Furthermore, computational tools offer the flexibility to simulate and analyze a diverse range of fluid flow and heat transfer problems across different industries and applications. Whether it is the aerodynamics of aircraft, the cooling of electronic devices, the flow in industrial processes, or the hydrodynamics of offshore structures, computational tools can be tailored to address specific engineering challenges. This versatility has made them indispensable tools in numerous fields, including aerospace, automotive, energy, environmental engineering, and many more. In recent years, the advancement in computational power, numerical algorithms, and modeling techniques has further expanded the capabilities of computational tools for fluid flow and heat transfer analysis. High-performance computing clusters, coupled with parallel processing, enable simulations of larger and more complex systems in a reasonable timeframe. Advanced turbulence models, multiphysics simulations, and coupled fluid-structure interaction analyses have improved the fidelity and accuracy of simulations, making them even more reliable for real-world applications.

1.2 ADVANCED COMPUTATIONAL TOOLS FOR FLUID FLOW ANALYSIS

Fluid flow analysis is a fundamental aspect of many engineering disciplines, including aerospace, automotive, chemical, and environmental engineering. The accurate prediction and understanding of fluid behavior are crucial for optimizing designs, improving performance, and ensuring safe and efficient operation of various systems. In recent years, the development of advanced computational tools has significantly enhanced our capabilities to simulate and analyze fluid flow phenomena. This chapter explores the state-of-the-art computational tools used for fluid flow analysis, focusing on their underlying principles, numerical techniques, and modeling strategies.

1.2.1 Governing equations and numerical methods

The governing equations that describe fluid flow, namely the Navier-Stokes equations, form the foundation for computational fluid dynamics (CFD) simulations. Computational tools employ numerical methods such as finite volume, finite element, or finite difference methods to discretize and solve these equations. These methods divide the computational domain into smaller control volumes or elements, enabling the approximation and solution of the flow field at discrete points. This numerical approach allows engineers and researchers to obtain solutions for complex flow scenarios that are difficult or impossible to solve analytically. Governing equations of the flow of a compressible Newtonian fluid are as follows:

$$\text{Continuity } \frac{\partial \rho}{\partial t} + div(\rho q) = 0 \tag{1.1}$$

$$x - \text{momentum } \frac{\partial(\rho u)}{\partial t} + div(\rho u q) = -\frac{\partial p}{\partial x} + div(\mu \nabla u) + S_x \tag{1.2a}$$

$$y - \text{momentum} \frac{\partial(\rho v)}{\partial t} + div(\rho v q) = -\frac{\partial p}{\partial y} + div(\mu \nabla y) + S_y \tag{1.2b}$$

$$z - \text{momentum} \frac{\partial(\rho w)}{\partial t} + div(\rho w q) = -\frac{\partial p}{\partial z} + div(\mu \nabla w) + S_z \tag{1.2c}$$

$$\text{Energy } \frac{\partial(\rho i)}{\partial t} + div(\rho i q) = -p\nabla \cdot q + div(k\nabla T) + \Phi + S_i \tag{1.3}$$

$$\text{Equation of state } p = (\rho, T) \quad \text{and} \quad i = i(\rho, T) \tag{1.4}$$

$$\text{For perfect gas } p = \rho RT \quad \text{and} \quad i = C_v T \tag{1.5}$$

Here ρ represents density, $q = (u, v, w)$ is the fluid velocity, p denotes pressure, μ is the viscosity of the fluid, $S = (S_x, S_y, S_z)$ is the external force.

1.2.2 Turbulence modeling

Turbulent flows, characterized by chaotic and unpredictable behavior, are prevalent in many engineering applications. Accurately simulating turbulent flows is a challenging task due to their complex nature. Advanced turbulence models, such as Reynolds-Averaged Navier-Stokes (RANS), Large Eddy Simulation (LES), or Direct Numerical Simulation (DNS), are incorporated into computational tools to capture the turbulent behavior and its impact on the flow field. These models utilize statistical or resolved approaches to simulate the effects of turbulence, enabling engineers to study turbulence phenomena, predict flow patterns, and estimate quantities of interest, such as drag, lift, or mixing efficiency.

1.2.3 Multiphase flow simulations

In numerous engineering applications, fluids exist in multiple phases, such as gas-liquid, liquid-liquid, or solid-liquid systems. Computational tools have evolved to handle multiphase flow simulations, enabling engineers to study complex interactions between different phases. These tools incorporate appropriate models and numerical techniques to capture interphase phenomena, such as phase change, particle dispersion, or surface tension effects. Multiphase flow simulations find applications in various industries, including petroleum, chemical, and pharmaceutical, where understanding the behavior of mixtures or suspensions is crucial for process optimization and system design.

1.2.4 High-performance computing and parallel processing

The computational demands of fluid flow simulations, particularly for large-scale and complex systems, necessitate the use of high-performance computing (HPC) resources. Computational tools leverage parallel processing techniques and distributed computing architectures to accelerate simulations and reduce computational time. This allows engineers and researchers to analyze larger domains, refine spatial and temporal resolutions, and obtain results within reasonable timeframes. HPC capabilities are essential for addressing real-world engineering challenges, where accuracy and efficiency are paramount.

1.2.5 Validation and verification

Validating and verifying computational results against experimental or analytical data are a critical step in ensuring the reliability and accuracy of computational tools. Validation involves comparing simulation results with experimental measurements, while verification assesses the correctness and convergence of the numerical solution. Computational tools employ various techniques, such as grid refinement studies, sensitivity analysis, or benchmark comparisons, to validate and verify their results. These procedures instill confidence in the computational tools and their ability to accurately predict fluid flow behavior.

1.2.6 Convection and radiation modeling

In addition to heat conduction, convective and radiative heat transfer mechanisms are often present in practical engineering systems. Computational tools incorporate advanced modeling techniques to simulate and analyze convective and radiative heat transfer. Convective heat transfer is characterized by the transfer of heat between a solid surface and a moving fluid, such as air or water. Computational tools utilize empirical correlations or CFD-based simulations to predict convective heat transfer coefficients. Radiative heat transfer involves the transfer of thermal energy through electromagnetic waves. Computational tools employ radiation models, such as the discrete ordinates or the Monte Carlo method, to accurately simulate radiative heat transfer in participating media.

1.2.7 Phase change and boiling

Phase change phenomena, such as boiling or condensation, significantly impact heat transfer processes in many applications, including power plants, refrigeration systems, or nuclear reactors. Computational tools incorporate specific models to simulate phase change phenomena accurately. Boiling models, for instance, capture the formation and dynamics of vapor bubbles, their growth, departure, and interaction with the surrounding fluid. These models enable engineers to analyze heat transfer enhancement during boiling and optimize heat exchanger designs in boiling applications.

1.2.8 Multiphysics coupling

Heat transfer phenomena are often coupled with other physical phenomena, such as fluid flow, structural mechanics, or electromagnetic fields. Computational tools have evolved to handle multiphysics simulations, where the interaction between different physical domains is considered. This enables engineers to study complex systems where heat transfer is influenced by other factors or vice versa. Multiphysics simulations find applications in various fields, including electronics cooling, combustion analysis, or electromagnetic heating processes.

1.2.9 High-performance computing and parallel processing

Computational heat transfer simulations can be computationally intensive, particularly for large-scale and complex systems. To address this, advanced computational tools leverage high-performance computing (HPC) resources and parallel processing techniques. These tools distribute the computational workload across multiple processors or computing nodes, enabling faster simulations and reduced computational time. HPC capabilities are essential for tackling real-world engineering challenges, where accuracy and efficiency are critical.

1.2.10 Visualization and post-processing

Computational tools provide powerful visualization and post-processing capabilities to interpret and analyze heat transfer simulation results. Engineers can visualize temperature distributions, heat flux profiles, thermal gradients, or heat transfer coefficients. Post-processing tools enable the extraction of quantitative data, such as heat transfer rates, temperature differences, or thermal resistance, facilitating the interpretation and comparison of different design scenarios. Visualization and post-processing techniques enhance the understanding of heat transfer phenomena, aid in design optimization, and support decision-making processes.

1.3 CFD SOFTWARE APPLICATION IN DIFFERENT INDUSTRIES

There are several CFD software available in the market. Some of the most popular ones are given in Table 1.1.

Software for computational fluid dynamics (CFD) is widely utilized in many industries for a variety of purposes (Table 1.2). Following are some examples of the markets and uses for CFD software, along with pertinent citations from academic studies: In general, a large number of sectors and applications, including aircraft, automotive, shipbuilding, turbo machinery, electronics, energy, and medical, use CFD software. Aerodynamic analysis, thermal analysis, hydrodynamic analysis, and design optimization are only a few of the uses ofthe software. The uses ofCFD analysis software are essentially limitless.

Table 1.1 Software names and applications

Sr. No.	Software name	Applications
1.	CONVERGE CFD Software [1], ANSYS Fluids [2], Autodesk CFD [3], SimScale[4], OpenFOAM[5], Cadence CFD Simulation Software [6]	• Used for simulating complex fluid flow problems. • Widely used in the automotive, aerospace, and energy industries. • Cited in numerous scientific publications.
2.	FLACS [7]	• Used for complex, high-hazard risk and safety assessments. • Simulates the dispersion of hazardous materials, fire, and explosions, ensuring in-depth safety studies.
3.	Solidworks Simulation Software [8]	• Used for design optimization. • Used for comprehensive testing like natural frequencies, heat transfer characteristics, and buckling instabilities.

Table 1.2 CFD software's application in different industries

Sr. No.	Field	Applications
1.	Aerospace [9]	Aerodynamic analysis of aircraft and spacecraft
		Design optimization of airfoils and wings
		Thermal analysis of engines and heat shields
2.	Automotive [10]	Aerodynamic analysis of vehicles
		Engine cooling and thermal management
		Exhaust system design
3.	Shipbuilding [11, 12]	Hydrodynamic analysis of ships and
		Offshore structures
		Propeller design and optimization
		Maneuvering and stability analysis
4.	Turbomachinery [13]	Design and optimization of turbines, compressors, and pumps [5]
		Flow analysis in turbomachinery components [5]
5.	Electronics [9]	Thermal analysis of electronic components and systems
		Cooling system design and optimization
6.	Energy [14]	Analysis of wind turbines and wind
		Thermal analysis of power plants and nuclear reactors
7.	Medical [15, 16]	Analysis of blood flow and drug delivery in the human body
		Design and optimization of medical devices
8.	Mechanical engineering field [17–23], [24–32], [33–37]	Heat transfer problems as in renewablesolar air heater,
		Combustion problems
		Boiling
		Multiphase flow, Instability Analysis
		Refrigeration
		Air conditioning
		Turbines
		Turbomachinery

1.4 STEPS INVOLVED IN CFD SIMULATION

Prior to anything else, the problem must be stated, and within this definition, the objectives must be made crystal apparent. The geometry of the issue, whether steady-state or transient conditions are present, and the characteristics of the materials (including fluids and solids) must all be stated, for instance. Additionally, boundary conditions must be precisely established.

Step involvement on simulation software

1. Create the geometry (3D or 2D) of the system to be analyzed using Boolean operations if necessary.
2. Discretize the domain in small parts to generate the grid or mesh.

3. Prepare the nomenclature of the domain and boundary conditions. (Nomenclature can also be done before the meshing.)
4. Select the solver for simulation.
5. Apply boundary conditions and assign properties to the materials. Use solver scheme and run the simulation, until convergence is achieved.
6. Post-process the results: analyze and visualize the simulation results. Evaluate the accuracy and validity of the simulation.

1.5 COMPARISON OF MATLAB CODE RESULTS WITH THE CFD RESULTS

In this section, a problem is formulated for steady-state conduction in a steel rod (diameter is 1 cm and length is 10 cm), density (8030 kg/m³), specific heat (502.48 J/kg/k), thermal conductivity (16.27 w/m/k). One end of the rod is maintained at 300°C and the other end is at 30°C. The problem is solved using the finite volume method by implementing MATLAB code and is also simulated using CFD simulation in ANSYS. The problem is simplified with the assumption that heat flows solely along the length of the cylinder, making it one-dimensional. The following equation is utilized:

$$\frac{d}{dx}\left(k\frac{dT}{dx}\right) = 0. \tag{1.6}$$

1.5.1 Computational solution using Matlab

We can partition the rod into ten identical control volumes, as depicted in Figure 1.1 with $\delta x = 0.01$ m.

As there is no source term in the equation (1.6), the nodes 2–9, the temperatures are available on the nodes to the east (E) and west (W) [38], and therefore, the discretization of equation (1.6) using control volume method can be given as:

$$\left(\frac{k_E}{\delta x_E}A_E + \frac{k_W}{\delta x_W}A_W\right)T_n = \left(\frac{k_W}{\delta x_W}A_W\right)T_W + \left(\frac{k_E}{\delta x_E}A_E\right)T_E \tag{1.7}$$

As $k_E = k_W = k$, $\delta x_E = \delta x_W = \delta x$ and $A_E = A_W = A$, equation (1.7) can be written as

$$\left(a_E + a_W\right)T_n = \left(a_W\right)T_W + \left(a_E\right)T_E \tag{1.8}$$

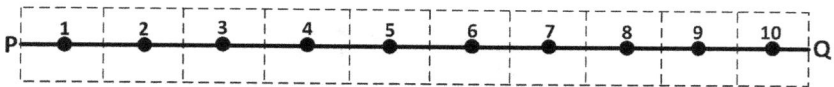

Figure 1.1 Discretization of the domain.

Here,

$$a_E = \frac{k_E}{\delta x_E} A_E, \quad a_W = \frac{k_W}{\delta x_W} A_W$$

The nodes 1 and 10 require special attention as they are boundary nodes. Integrating equation (1.6) over the control volume surrounding point 1 gives us

$$kA\left(\frac{T_E - T_n}{\delta x}\right) - kA\left(\frac{T_n - T_P}{\delta x / 2}\right) = 0 \tag{1.9}$$

This expression shows that the flux through control volume boundary P has been approximated by assuming a linear relationship between temperatures at boundary points P and node 1.

Rearranging equation (1.9), we get

$$\left(\frac{k}{\delta x} A + \frac{2k}{\delta x} A\right) T_n = 0 \cdot T_W + \left(\frac{2k}{\delta x} A\right) T_P + \left(\frac{k}{\delta x} A\right) T_E \tag{1.10}$$

From this equation, it can be easily identified that the fixed temperature boundary condition enters the calculation as a source term $S_u + S_n T_n$, where $S_n = -\frac{2kA}{\delta x}$, $S_u = \frac{2kA}{\delta x} T_P$.

Hence, equation (1.10) takes the form as:

$$a_n T_n = a_W T_W + a_E T_E + S_u \tag{1.11}$$

with

$$a_W = 0; a_n = a_W + a_E - S_n; \; S_u = \frac{2kA}{\delta x} T_P.$$

Again, integrating equation (1.6) over the control volume surrounding node 10 gives us

$$kA\left(\frac{T_Q - T_n}{\delta x / 2}\right) - kA\left(\frac{T_n - T_W}{\delta x}\right) = 0 \tag{1.12}$$

Rearranging equation (1.12), we get

$$\left(\frac{k}{\delta x} A + \frac{2k}{\delta x} A\right) T_n = 0 \cdot T_E + \left(\frac{2k}{\delta x} A\right) T_Q + \left(\frac{k}{\delta x} A\right) T_W \tag{1.13}$$

Hence, equation (1.13) takes the form as:

$$a_n T_n = a_W T_W + a_E T_E + S_u \qquad (1.14)$$

With

$$a_E = 0; a_n = a_W + a_E - S_n; \; S_u = \frac{2kA}{\delta x} T_Q.$$

The discretization of each node is given in equations (1.8), (1.11), and (1.14). Substitution of numerical values gives us $\frac{k}{\delta x} A = 0.040675\pi$. The coefficients of each discretized equation are given in Table 1.3.

The set of algebraic equations can be written as:

$$\left.\begin{aligned}
0.122025\pi T_1 &= 0.040675\pi T_2 + 0.08135\pi T_P \\
0.08135\pi T_2 &= 0.040675\pi T_3 + 0.040675\pi T_1 \\
0.08135\pi T_3 &= 0.040675\pi T_4 + 0.040675\pi T_2 \\
0.08135\pi T_4 &= 0.040675\pi T_5 + 0.040675\pi T_3 \\
0.08135\pi T_5 &= 0.040675\pi T_6 + 0.040675\pi T_4 \\
0.08135\pi T_6 &= 0.040675\pi T_7 + 0.040675\pi T_5 \\
0.08135\pi T_7 &= 0.040675\pi T_8 + 0.040675\pi T_6 \\
0.08135\pi T_8 &= 0.040675\pi T_9 + 0.040675\pi T_7 \\
0.08135\pi T_9 &= 0.040675\pi T_{10} + 0.040675\pi T_8 \\
0.122025\pi T_{10} &= 0.040675\pi T_9 + 0.08135\pi T_Q
\end{aligned}\right\} \qquad (1.15)$$

Table 1.3 Coefficient values for various nodes

Node	a_W	a_E	S_u	S_n	$a_n = a_W + a_E - S_n$
1	0	0.040675π	$0.08135\pi T_P$	-0.08135π	0.122025π
2	0.040675π	0.040675π	0	0	0.08135π
3	0.040675π	0.040675π	0	0	0.08135π
4	0.040675π	0.040675π	0	0	0.08135π
5	0.040675π	0.040675π	0	0	0.08135π
6	0.040675π	0.040675π	0	0	0.08135π
7	0.040675π	0.040675π	0	0	0.08135π
8	0.040675π	0.040675π	0	0	0.08135π
9	0.040675π	0.040675π	0	0	0.08135π
10	0.040675π	0	$0.08135\pi T_Q$	-0.08135π	0.122025π

The system (1.15) in matrix form can be written as:

$$
\begin{bmatrix}
0.122025\pi & -0.040675\pi & 0 & 0 & 0 & 0 & 0 & 0 & 0 & 0 \\
-0.040675\pi & 0.08135\pi & -0.040675\pi & 0 & 0 & 0 & 0 & 0 & 0 & 0 \\
0 & -0.040675\pi & 0.08135\pi & -0.040675\pi & 0 & 0 & 0 & 0 & 0 & 0 \\
0 & 0 & -0.040675\pi & 0.08135\pi & -0.040675\pi & 0 & 0 & 0 & 0 & 0 \\
0 & 0 & 0 & -0.040675\pi & 0.08135\pi & -0.040675\pi & 0 & 0 & 0 & 0 \\
0 & 0 & 0 & 0 & -0.040675\pi & 0.08135\pi & -0.040675\pi & 0 & 0 & 0 \\
0 & 0 & 0 & 0 & 0 & -0.040675\pi & 0.08135\pi & -0.040675\pi & 0 & 0 \\
0 & 0 & 0 & 0 & 0 & 0 & -0.040675\pi & 0.08135\pi & -0.040675\pi & 0 \\
0 & 0 & 0 & 0 & 0 & 0 & 0 & -0.040675\pi & 0.08135\pi & -0.040675\pi \\
0 & 0 & 0 & 0 & 0 & 0 & 0 & 0 & -0.040675\pi & 0.122025\pi
\end{bmatrix}
\times
\begin{bmatrix}
T_1 \\ T_2 \\ T_3 \\ T_4 \\ T_5 \\ T_6 \\ T_7 \\ T_8 \\ T_9 \\ T_{10}
\end{bmatrix}
=
\begin{bmatrix}
0.08135\pi T_P \\ 0 \\ 0 \\ 0 \\ 0 \\ 0 \\ 0 \\ 0 \\ 0 \\ 0.08135\pi T_Q
\end{bmatrix}
$$

$$(1.16)$$

Taking $T_p = 300$, $T_Q = 30$ and solving the above system, we have

$$
\begin{bmatrix} T_1 \\ T_2 \\ T_3 \\ T_4 \\ T_5 \\ T_6 \\ T_7 \\ T_8 \\ T_9 \\ T_{10} \end{bmatrix} = \begin{bmatrix} 286.5 \\ 259.5 \\ 232.5 \\ 205.5 \\ 178.5 \\ 151.5 \\ 124.5 \\ 97.5 \\ 70.5 \\ 43.5 \end{bmatrix}
$$

(1.17)

Figure 1.2 represents the temperature values generated at the nodes of the control volume with variation in the length of the rod. It shows the linear profile of temperature along the length of the rod.

Analytical Solution: Consider the differential given equation

$$
\frac{d}{dx}\left(k \frac{dT}{dx} \right) = 0.
$$

(1.18)

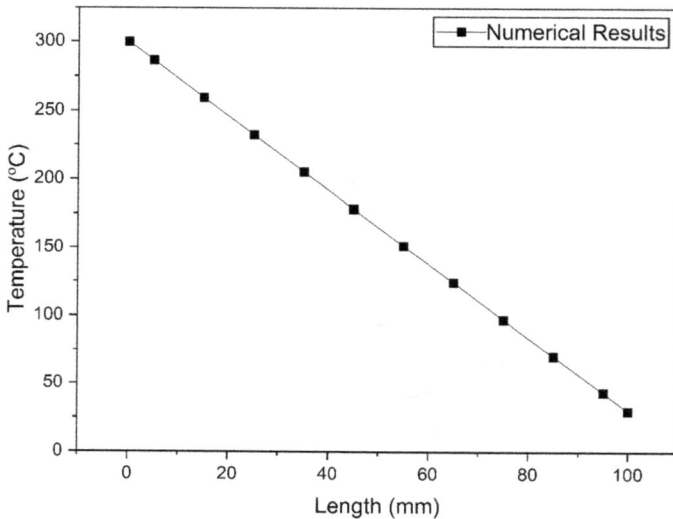

Figure 1.2 Numerical results using CFD approach.

Integrating

$$k\frac{dT}{dx} = A.$$ (1.19)

where A is a constant.

As the thermal conductivity k is given as constant, equation (1.19) can be easily integrated as:

$$T = \frac{A}{k}x + B.$$ (1.20)

Here B is also a constant.

To compute A and B, we use the boundary conditions. If point P is considered at $x = 0$, point Q lies at $x = 0.1$. Therefore, $T = 300$ at $x = 0$ and $T = 30$ at $x = 0.1$. Using these conditions, equation (1.20) can be written as:

$$\left. \begin{array}{l} 300 = \dfrac{A}{k}(0) + B, \\ 30 = \dfrac{A}{k}(0.1) + B. \end{array} \right\}$$ (1.21)

Solving equation (1.21), we get $A = -2700k$ and $B = 300$. Hence, $T = -2700x + 300$.

One can easily write a MATLAB code as

```
clear all
clc
x = 0 : 0.01 : 0.1;
T = -2700*x + 300;
plot(x, T)
```

Figure 1.3 shows the temperature values along the length of the rod. Temperature values are created at the location of the nodes.

1.6 COMPUTATIONAL SOLUTION USING ANSYS

In this section, the above problem is also solved through software ANSYS Fluent. Domain is discretized in small domains called mesh. Boundary conditions of 300°C are applied at the left face of the rod and 30°C is applied at the right face of the rod. It is assumed that heat transfer takes place only along the length of the rod. Hence boundary condition of heat flux (zero) is

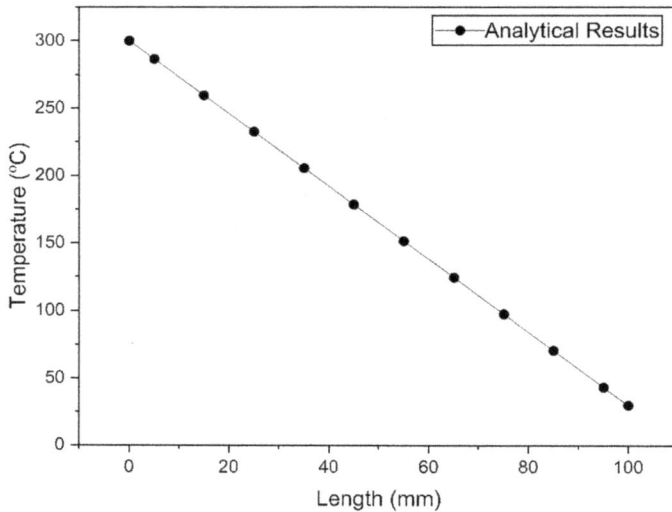

Figure 1.3 Temperature vs length using Matlab code.

Table 1.4 Grid independence results

Parameter	Average			Temperature (°C)
Mesh elements	Orthogonal Quality	Aspect Ratio	Skewness	Temperature at node 6 (55 mm from the hot end of the rod)
18909	0.97	1.96	0.163	151.6
5184	0.98	2.02	0.167	151.6
501830	0.98	1.54	0.10	151.6
965150	0.99	1.63	0.08	151.6

applied along the curved surface of the rod. The properties of the steel rod are imposed in the geometry domain. Simulation is carried out until convergence is achieved. Grid independence results are shown in Table 1.4. It has been observed that in all cases the temperature value at node 6 (55 mm from the hot end of the rod) is the same. The reason for this behavior is explained as follows. It is clear that the only conduction equation used for the analysis is linear in nature. Due to single conduction equation and linear nature of the equation, there is no difference in results observed by varying the grid size. Figure 1.4 shows the mesh generation of the domain (rod).

Figure 1.5 represents the results in the terms of temperature contours (K) along the length of the rod. The results of the temperature vs length plot of

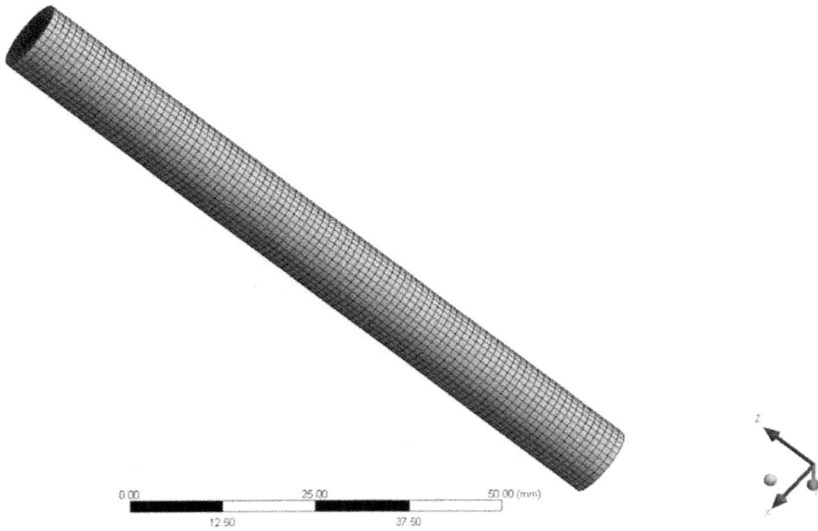

Figure 1.4 Meshing of the domain.

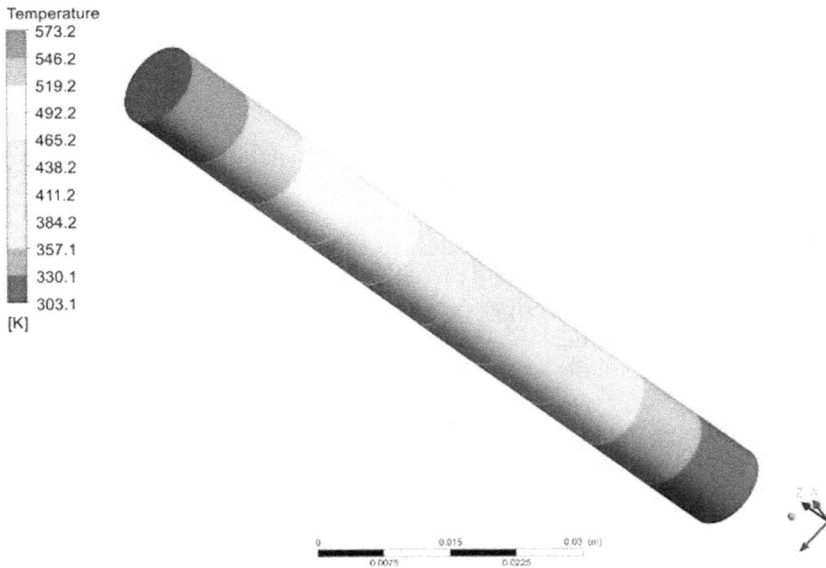

Figure 1.5 Temperature contours of CFD domain.

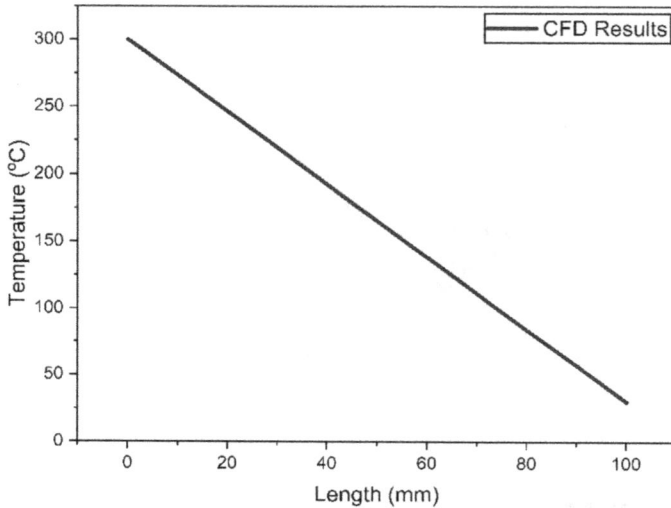

Figure 1.6 Temperature vs length using CFD simulation.

the CFD simulation shown in figure reveal the same trend as that obtained for the numerical and analytical approach as discussed above (Figure 1.6).

Overall analysis reveals that for the above particular problem, the results obtained using CFD (numerical approach), analytical analysis and CFD software (ANSYS Fluent) are the same, which justifies the strength of the CFD approach, especially that CFD software is remarkable and is the future of the scientific community.

1.7 BENEFITS AND LIMITATIONS OF CFD

CFD simulations benefit engineers and scientists in a variety of ways. Notably, because these simulations can be carried out effectively on a computer, they result in significant cost savings by eliminating the need for time-consuming and expensive physical experimentation and testing. Additionally, CFD models are exceptionally accurate, offering nuanced and precise insights into flow dynamics, especially in complex flow systems. This improved precision encourages a deeper comprehension of the system's physical phenomena and flow characteristics, enabling engineers and scientists to make wise design choices. Before any physical prototypes are built, thorough testing and design optimization are possible thanks to virtual prototyping, a crucial component of CFD simulations [39].

CFD simulations include benefits and drawbacks that need be taken into account. On the plus side, they significantly reduce expenses by removing the requirement for actual experimentation, but it's important to note that these simulations are frequently processor-intensive, which could result in increased computational time and modeling costs. The intricacy of CFD simulations also necessitates expertise in order to decide when, where, and how they should be used [40]. Another drawback is that they are only marginally useful in the conceptual design stage, with most of their applications in the detailed design and troubleshooting phases in sectors like oil and gas. The related costs, which may include the need for additional software and hardware, can also be a disadvantage and present financial difficulties. In conclusion, CFD offers indisputable benefits in terms of financial savings, accuracy, thorough comprehension, virtual prototyping, and safety evaluations. It should be understood, nonetheless, that CFD models have several disadvantages, such as computational requirements, complexity, limited applications in some design phases, and cost considerations.

1.8 FUTURE TRENDS AND CHALLENGES

The field of computational heat transfer analysis continues to evolve, driven by advances in numerical algorithms, computing resources, and modeling techniques. Future trends include the integration of machine learning and artificial intelligence algorithms for enhanced predictive capabilities, the development of reduced-order modeling techniques for faster simulations, and the incorporation of uncertainty quantification methodologies. However, challenges remain, such as accurately capturing transient heat transfer phenomena, modeling complex geometries, and improving the efficiency and robustness of computational tools.

1.9 CONCLUSION

Advanced computational tools have transformed the way we study and analyze fluid flow phenomena and heat transfer. In this chapter, a problem of heat conduction is solved usingthe numerical CFD approach analytically and also usingthe CFD software ANSYS Fluent. In all cases, same results of the temperature values are obtained along the length of the rod. Thisrepresents that CFD (numerical approach) is useful to predict the behavior of the system. For most of the complicated problems like irregular geometries, different boundary conditions, and multiphase flow, it is a challenging task to solve it numerically. However, using CFD software, problems can be solved easily and results can be obtained up to the remarkable point. By solving the governing equations and employing numerical methods, these tools provide engineers and researchers with powerful means to simulate, visualize, and analyze complex flow scenarios. They enable the optimization of engineering designs, the exploration of different flow conditions the prediction of flow-related quantities, and the prediction of heat transfer characteristics.

With ongoing advancements and future developments, computational tools will continue to play a pivotal role in advancing our understanding of fluid flow and heat transfer, leading to innovative engineering solutions and improved system performance.

REFERENCES

[1] "Convergence at a glance. https://convergecfd.com/
[2] "Ansys Fluids Computational Fluid Dynamics (CFD) Simulation Software. https://www.ansys.com/en-in/products/fluids
[3] "Autodesk CFD Software. https://geekflare.com/best-cfd-analysis-software/
[4] "SimScale CFD Software.
[5] "Open FOAM", [Online]. Available: https://geekflare.com/best-cfd-analysis-software/
[6] "Cadence CFD Software. https://www.cadence.com/en_US/home/tools/system-analysis/computational-fluid-dynamics.html
[7] "FLACS-CFD explosion, fire & dispersion modelling software", [Online]. Available: https://www.gexcon.com/software/flacs-cfd/
[8] "SOLIDWORKS Simulation Software. https://www.solidworks.com/product/solidworks-simulation
[9] "Applications & Limitations of the CFD Aerospace Industry. https://imaginationeering.com/applications-limitations-of-the-cfd-aerospace-industry/
[10] "Reducing Carbon Footprint In Automotive By Modeling And Simulations. https://www.knowledgeridge.com/c/ExpertsViewsDetails/728
[11] "CFD Simulations. https://ulstein.com/ship-design/cfd-simulations
[12] D. Kim and T. Tezdogan, "CFD-based Hydrodynamic Analyses of Ship Course Keeping Control and Turning Performance in Irregular Waves". *Ocean Eng.* 248 (2022) 110808, doi: 10.1016/J.OCEANENG.2022.110808
[13] "13 Key Applications of CFD Simulation and Modeling. https://blog.spatial.com/cfd-modeling-applications
[14] A. Iranzo, "CFD Applications in Energy Engineering Research and Simulation: An Introduction to Published Reviews," *Processes* 7 no. 12 (2019) 1–17, doi: 10.3390/pr7120883
[15] L. Reid, "An Introduction to Biomedical Computational Fluid Dynamics," *Adv. Exp. Med. Biol.* 1334 (2021) 205–222. doi: 10.1007/978-3-030-76951-2_10
[16] A. Prasad, A. Kumar, and M. Gupta. *Advanced Materials and Manufacturing Techniques in Biomedical Applications*, 1st ed. Wiley, 2023, https://doi.org/10.1002/9781394166985
[17] W. Quitiaquez, J. Estupinán-Campos, C. Nieto-Londoño, C. A. Isaza-Roldán, P. Quitiaquez, and F. Toapanta-Ramos, "CFD Analysis of Heat Transfer Enhancement in a Flat-Plate Solar Collector with Different Geometric Variations in the Superficial Section," *Int. J. Adv. Sci. Eng. Inf. Technol.* 11, no. 5 (2021) 2039–2045, doi: 10.18517/IJASEIT.11.5.15288
[18] A. Kumar, V. P. Singh, C. S. Meena, and N. Dutt, *Thermal Energy Systems: Design, Computational Techniques, and Applications*, 1st ed. CRC Press, 2023.
[19] V. P. Singh et al., "Recent Developments and Advancements in Solar Air Heaters: A Detailed Review," *Sustainability* 14 (2022) 1–57. doi: 10.3390/su141912149

[20] N. Dutt, A. J. Hedau, A. Kumar, M. K. Awasthi, V. P. Singh, and G. Dwivedi, "Thermo-hydraulic Performance of Solar Air Heater Having Discrete D-shaped Ribs As Artificial Roughness," *Environ. Sci. Pollut. Res.* (2023). doi: 10.1007/s11356-023-28247-9

[21] N. Dutt, R. Kumar, and K. Murugesan, "Experimental Performance Analysis of Multi-Parabolic Flat Plate Solar CollectorBT - Emerging Trends in Mechanical and Industrial Engineering: Select Proceedings of ICETMIE 2022," X. Li, M. M. Rashidi, R. S. Lather, and R. Raman, Eds., Singapore: Springer Nature Singapore, 2023, pp. 159–172. doi: 10.1007/978-981-19-6945-4_12

[22] Varun Pratap Singh, Chandan Swaroop Meena, Ashwani Kumar, Nitesh Dutt, "Double Pass Solar Air Heater: A Review," *Int. J. Energy Resour. Appl.* 1, no. 2 22–43.

[23] J. Hammond, N. Pepper, F. Montomoli, and V. Michelassi, "Machine Learning Methods in CFD for Turbomachinery: A Review," *Int. J. Turbomach. Propuls. Power* 7, no. 2 2022. doi: 10.3390/ijtpp7020016

[24] M. K. Awasthi, N. Dutt, A. Kumar, and S. Kumar, Electrohydrodynamic Capillary Instability of Rivlin–Ericksen Viscoelastic Fluid Film with Mass and Heat Transfer. *Heat Transfer* (2023) 1–19. doi:10.1002/htj.22944

[25] P. K. Kushwaha, N. K. Sharma, A. Kumar, and C. S. Meena, "Recent Advancements in Augmentation of Solar Water Heaters Using Nanocomposites with PCM: Past, Present, and Future". *Buildings* 13 (2023) 79. https://doi.org/10.3390/buildings13010079

[26] R. Kumar, A. Kumar, L. Kant, A. Prasad, S. Bhoi, C. S. Meena, V. P. Singh, and A. Ghosh. Experimental and RSM-based Process-Parameters Optimisation for Turning Operation of EN36B Steel. *Materials* 16 (2023) 339. https://doi.org/10.3390/ma16010339

[27] V. P. Singh, S. Jain, A. Karn, A. Kumar. Experimental Assessment of Variation in Open Area Ratio on Thermohydraulic Performance of Parallel Flow Solar Air Heater". *Arab. J. Sci. Eng.* (2022). https://doi.org/10.1007/s13369-022-07525-7

[28] C. S. Meena, A. N. Prajapati, A. Kumar, and M. Kumar. Utilization of Solar Energy for Water Heating Application to Improve Building Energy Efficiency: An Experimental Study. *Buildings* 12 (2022) 2166. https://doi.org/10.3390/buildings12122166

[29] S. Bhoi, A. Kumar, A. Prasad, C. S. Meena, R. B. Sarkar, B. Mahto, and A. Ghosh, Performance Evaluation of Different Coating Materials in Delamination for Micro- Milling Applications on High-Speed Steel Substrate. *Micromachines* 13 (2022) 1277. https://doi.org/10.3390/mi13081277

[30] V. Verma, C. S. Meena, S. Thangavel, A. Kumar, T. Choudhary, G. Dwivedi, "Ground and Solar Assisted Heat Pump Systems for Space Heating and Cooling Applications in the Northern Region of India – A Study On Energy and CO2 Saving Potential". *Sustai. Energy Technol. Assess.* 59 (2023) 103405, ISSN2213-1388. https://doi.org/10.1016/j.seta.2023.103405

[31] G. Pant, C. S. Meena, A. Saxena, A. Kumar, V. P. Singh, N. Dutt (2023). Study the Temperature Variation in Alternate Coils of Insulated Condenser Cum Storage Tank: Experimental Study. In: Sikarwar, B.S., Sharma, S.K., Jain, A., Singh, K.M. (eds) *Advances in Fluid and Thermal Engineering.* FLAME 2022. Lecture Notes in Mechanical Engineering. Springer, Singapore. https://doi.org/10.1007/978-981-99-2382-3_52

[32] A. Saxena, A. N. Prajapati, G. Pant, C. S. Meena, A. Kumar, V. P. Singh (2023). Water Consumption Optimization of Hybrid Heat Pump Water Heating System. In: Shukla, A.K., Sharma, B.P., Arabkoohsar, A., Kumar, P. (eds) *Recent Advances in Mechanical Engineering. FLAME 2022*. Lecture Notes in Mechanical Engineering. Springer, Singapore. https://doi.org/10.1007/978-981-99-1894-2_61

[33] M. K. Awasthi, R. Asthana and G. S. Agrawal "Viscous Potential Flow Analysis of Nonlinear Rayleigh-Taylor Instability with Heat and Mass Transfer", *Microgravity Sci. Technol.* 24 (2012) 351–363.

[34] M. K. Awasthi, "Rayleigh–Taylor Instability of Swirling Annular Layer with Mass Transfer" *ASME-J. Fluid Eng.* 141 (2019) 071202 (5 pages).

[35] M. K. Awasthi "Nonlinear analysis of Rayleigh-Taylor instability of cylindrical flow with heat and mass transfer" *ASME J. Fluid Eng.* 135 (2013) 061205 (7 pages).

[36] M. K. Awasthi, Z. Uddin and R. Asthana, "Temporal Instability of a Power-law Viscoelastic Nanofluid Layer", *Eur. Phys. J. ST* 230 (2021) 1427–1434.

[37] M. K. Awasthi, Dharamendra and D. Yadav, "Temporal Instability of Nanofluid Layer in a Circular Cylindrical Cavity", *Eur. Phys. J. ST* 231 (2022) 2773–2779.

[38] H. K. Versteeg and W. Malalasekera, "An Introduction to Computational Fluid Dynamics," 1995.

[39] "Advantages and disadvantages of Computational Fluid Dynamics. https://www.quadco.engineering/en/know-how/cfd-advantages-and-disadvantages.htm

[40] "The Benefits and Limitations of Computational Fluid Dynamics in the Offshore Oil and Gas Industry. https://www.peritusint.com/benefits-and-limitations-of-computational-fluid-dynamics/

Chapter 2

IoT-enabled monitoring for natural convection solar dryers

M. Sridharan

SRM Institute of Science and Technology, Tiruchirappalli, India

2.1 INTRODUCTION

Recent researchers working with various branches of science and engineering are driving their study toward integration of Internet of Things. Industry 4.0, an historic period framed with an ultimate mission of automating entire system of trade setting. The following are the advantages of automating any industrial process system: reduced human efforts in monitoring real-time process, maintaining or tracking quality is easy, flexible operations, any bias inside the process from the required can be rectified instantly, easy to make effective decisions, etc. Table 2.1 presents some highlighted applications of Internet of Things–based performance monitoring in commercial and non-commercial environments. Implementation of such automatic data monitoring in any industrial environment is called smart factory. Such a smart factory is characterized by its flexibility, efficiency, accuracy, ergonomics, and adaptability. It is the integration of value process, customers, and business partners in business. Its fundamental technical foundation is laid by Internet of Things (IoT) system.

Numerous research works are conducted with an objective of increasing the performance of solar dryers. Such research works are conducted on the basis of two categories, namely, active and passive drying. Active or forced convection solar dryer and another type is passive or natural convection solar dryer. In an active type of solar dryer, the external source of hot air will be given as an additional element for drying the product. In a passive type of solar dryer, open sun drying process will be carried out without any additional external sources. From the literature, it is observed that solar dryers have wide applications in many industries like paper industries, food processing industries, pharmaceuticals, wood, textiles, and agro-based industries, among which the food processing industries mainly demands for drying substances such as carrot, mango, green peas, spices, and fish.

Among the different classifications of dryers, this study focused on an automatic process monitoring of conventional passive or natural convection solar dryer. The performance of any solar dryer system completely depends on solar irradiance, ambient temperature, and absorber plate temperature.

DOI: 10.1201/9781003465171-2

Table 2.1 Applications of Internet of Things–based performance monitoring

Applications	Field of research	Purpose of IoT	References
Double pipe heat exchanger	Thermal engineering	Process monitoring	Sridharan et al. [1]
Weather forecasting	Meteorology Engineering	Process monitoring	Fowdur et al. [2]
Electro synthesis system	Electro-Chemical Science	Process controlling	Radmannia et al. [3]
Smart Homes	Electrical and Electronics Engineering	Monitoring and controlling	Mocrii et al. [4]
Human action prediction	Biological Science	Monitoring and controlling	Machado et al. [5]
Smart metal forming	Manufacturing Engineering	Process monitoring	Yang [6]
Plant leaf disease detection	Biological Science	Process monitoring	Mallikarjuna et al. [7]
Automatic washing system	Electrical and Electronics Engineering	Monitoring and controlling	Hadipour et al. [8]
Solar Dryer	Engineering	Application in enhance drying in mango and fish	M. Chandrasekar and others [9, 11–12]
Solar dryer system	Solar energy	Comparative process monitoring	Current study

In this work an attempt has been made to study the performance monitoring rate of IoT-based system with the manually monitored mean. Deviations in such comparative measurements throughout the experiment are proposed.

2.1.1 Novelty

The major novelty of this technical brief includes the following: (1) First-ever IoT-based process monitoring and control system for any box-type solar energy conversion devices. (2) First-ever IoT-based process monitoring and control system for solar dryer. (3) First-ever IoT-based system with data transfer accuracy of 97%. (4) Instantaneous display of process temperature variations to the user with less time delay (average of 10s).

2.2 MATERIALS AND METHODS

2.2.1 Experimental setup

The solar dryer experimental rig [14] consisted of a 2 mm thick stainless steel 610 × 670 mm² absorber plate and was covered with a transparent

Table 2.2 Uncertainties in measuring instrument

Measured parameter	Instruments used	Range	Error
Absorber plate temperature	Temperature gun/RTD	−50°C to + 110°C	± 0.1°C
Mass flow rate	Flow sensor	0.50–6.5 lit/min	± 1.42%
Irradiance	Pyranometer	0–2000 W/m²	± 0.62%
Initial/Final weight of the sample	Digital weighing machine	0–600 g	± 0.1 g

5 mm thick toughened glass cover. The gap between the glass cover and absorber plate was maintained at 100 mm. The outer shell of the drying chamber is made of plywood. The drying chamber was provided with sufficient insulation to reduce thermal losses to the surroundings of the system. The schematic view of the in-house fabricated natural convection solar dryer is shown in Figure 2.5. A calibrated digital-type thermometer having 0.1°C resolution, ± 1°C accuracy, and 10 seconds response time was used for measuring variable absorber plate temperatures. A weighing balance (digital type) with a readability of 0.01 g was used to determine the mass of the product (carrot) under study before, during, and after drying process. The solar irradiance data is measured using a Gantner instruments pyranometer having an operation range of 0–2000 W/m² and an uncertainty of ± 0.67% on the full scale. Results of uncertainty analysis of instruments used are given in Table 2.2. The experiments were conducted on clear sunshine days for four continuous days (07/01/2019 to 10/01/2019) in the location of the solar thermal laboratory, Saranathan College of Engineering, Tiruchirappalli (10.7905° N, 78.7047° E), Tamilnadu, India [13]. Tables 2.4 and 2.5 consist of the average values of such results. The solar dryer in this study is kept facing north-south direction in which the module is sloping toward south. The tilt angle is 19° with reference to the horizontal. Figure 2.5 shows the real-time experimental setup integrated with IoT components.

2.2.2 Performance analysis

The efficiency of the dryer is obtained by [10] using the following formula given in Eq. (2.1).

$$\eta_{dryer} = \frac{\dfrac{M_W \times h_{fg}}{\Delta t}}{A_d \times I} \tag{2.1}$$

2.3 UNCERTAINTY ANALYSIS

Quantities used to measure absorber plate temperature and ambient temperature are subjected to uncertainties. Such uncertainties are due to errors in measurement. Quantities like absorber plate temperature and ambient temperature are measured and probabilities of error in each of them are calculated, which in turn are used to estimate uncertainties related with experimental data.

General form for uncertainty in the result will be given by Eq. (2.2).

$$\sigma_x = \sqrt{\left(\frac{\delta X}{\delta x_1}\right)(\sigma_{x1})^2 + \left(\frac{\delta X}{\delta x_2}\right)(\sigma_{x2})^2 + \ldots \left(\frac{\delta X}{\delta x_n}\right)(w\sigma_{xn})^2} \qquad (2.2)$$

Useful work done (W) by solar dryer depends on M_w and h_{fg} as in Eq. (2.3)

$$W = f\left(M_W, h_{fg}\right) \qquad (2.3)$$

Uncertainty in work done (W) of the dryer is calculated using Eq. (2.4)

$$[\sigma W]^2 = \left[\frac{\partial W}{\partial M_W} \cdot \sigma M_W\right]^2 + \left[\frac{\partial W}{\partial h_{fg}} \cdot \sigma h_{fg}\right]^2 \qquad (2.4)$$

Rate of heat supplied (Q) to the solar dryer depends on A_d, I and Δt as in Eq. (2.5)

$$Q = f\left(A_d, I, \Delta t\right) \qquad (2.5)$$

Uncertainty in heat supplied (Q) to the dryer is calculated using Eq. (2.6)

$$[\sigma Q]^2 = \left[\frac{\partial Q}{\partial A_d} \cdot \sigma A_d\right]^2 + \left[\frac{\partial Q}{\partial I} \cdot \sigma I\right]^2 + \left[\frac{\partial Q}{\partial \Delta t} \cdot \sigma \Delta t\right]^2 \qquad (2.6)$$

Uncertainties in measuring equipment's are listed in Table 2.2.

2.4 PROPOSED IoT SYSTEM DESIGN

The Internet of things (IoT) is the combination of hardware device embedded with electronics, sensors, software, and connectivity which enables these things to connect exchange and collect and data. Such an embedded unit for monitoring performance of solar dryer is shown in Figure 2.1 This section describes the fundamental components for integrating solar dryer assembly with IoT concept.

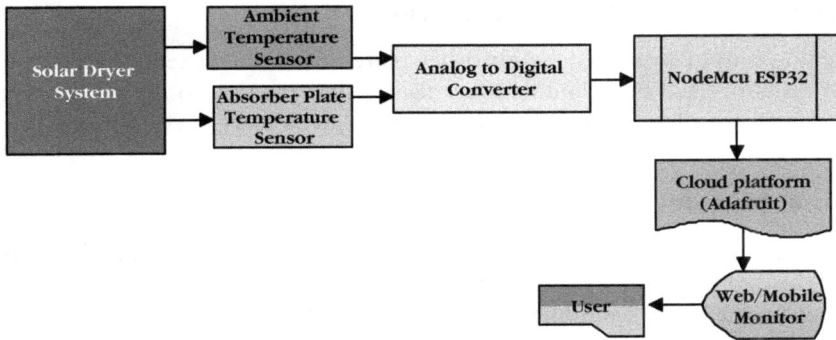

Figure 2.1 IoT-based monitoring and control system for solar dryer system.

2.4.1 Solar dryer setup

Detailed technical specifications and description of solar dryer setup are listed in Section 2.1.

2.4.2 Sensors

A sensor is an electronic device that converts the physical parameter into an output signal which can be measured electrically. In this work two types of sensors are used: (i) temperature sensor for measuring ambient temperature and (ii) temperature sensor for measuring variation in solar dryer absorber plate temperature. The purpose of temperature sensor is to measure temperature variation in temperature from beginning to the end of the experiment. Technical specifications of temperature sensors are listed in Table 2.3.

Table 2.3 Hardware components used in Internet of Things–based heat exchanger system

Sl. no	Name of the components	Specifications	Requirements
1	NodeMcu	ESP8266 CP2102	2
2	Temperature sensor	DS18B20 Digital Temperature Sensor −55°C to + 125°C	2
3	Power Supply	Output 500 mA, 2.5 Watts	1
4	Ethernet Cable	2 ft or 3 ft	1
5	LED Monitor	Dell 18.5 inch	1
6	USB Mouse	2.4 GHz Wireless connection	1
7	USB Keyboard	USB 2.0	1
8	Jumper Wires	150 mm	1
9	Battery	12V, Rechargeable	1

2.4.3 Analog to digital converter

Analog output produced by absorber plate and ambient temperature sensors is converted into digital values using ADC (analog to digital converters). The digital output produced by ADC is given to NodeMcu processor.

2.4.4 NodeMcu processor

NodeMcu is an open-source IoT platform. It includes microcode that runs on the ESP8266 Wi-Fi SoC from Espressif Systems, and hardware that relies on the ESP-12 module. The term "NodeMcu" by default refers to the microcode instead of the event kits. The firmware uses the Lua scripting language. It is supported the eLua project, and designed on the Espressif Non-OS SDK for ESP8266. It uses several open-source projects like lua-cjson and SPIFFS.

2.4.5 Cloud storage

Employing a network of remote servers hosted on the web to store, manage, and method knowledge, instead of an area server or a private PC, is termed cloud computing. The main advantages of such cloud computing are cost savings, reliable, unlimited data storage, limited control, being highly secure, etc.

2.4.5.1 Adafruit

In order to develop Internet of Things applications in the cloud, a free web service was developed by Adafruit. The work flow in Adafruit includes the following:

1. **Creating channels and collecting data**: For data collection and analytics, Adafruit acts as the platform for IOT. Channel is the fundamental element of Adafruit activity. It contains fields of data, fields of location, and field of status. Adafruit channel creation is followed by writing data to the channel, processing of data, and viewing data with Python software code and react to the data with alerts and tweets.
2. **Analyzing and visualizing the data**: Adafruit allows us to analyze and visualize our data. The analysis and visualization phase includes a template code that assists us with basic operations on *historic or live data. To analyze data Adafruit serves as a bridge which is capable of connecting sensors with data analytics software.*
3. **Acting on the data:** It includes setting threshold limits on data to send a feedback according to the required situation.

2.5 INVESTIGATING THE IoT-BASED DATA TRANSFER MODEL ACCURACY AND ERROR WITH EXPERIMENTAL DATA SETS

Sridharan et al. [14] in their comparative study measured deviation of predicted results obtained using XGBoost algorithm from experimental values as per Eq. (2.7):

$$\text{Error}\left(e_r\right) = \frac{\text{Predicted value} - \text{Measured value}}{\text{Measured value}} \tag{2.7}$$

Individual percentage accuracy (A_{in}) is then calculated using Eq. (2.8):

$$A_{in} = e_r * 100\% \tag{2.8}$$

Overall model accuracy is then calculated using average of individual accuracy given by Eq. (2.9):

$$A_o = \frac{A_{i1} + \dots A_{in}}{n} \tag{2.9}$$

2.6 RESULTS AND DISCUSSION

2.6.1 Solar dryer part

In this study, the transient performance variation of the solar dryer was observed using the carrot as a product to be dried. Experiment is conducted at Saranathan college of Engineering, Tiruchy, Tamilnadu, India. The variation of output parameters with respect to the time duration in hours is graphically represented in Figures 2.2–2.5. Same values are numerically presented in Tables 2.4 and 2.5.

From Figure 2.4 and Table 2.4, it is observed that the atmospheric ambient temperature is maximum during 03:00 P.M. and minimum when irradiance is minimum during 10.00 hours. Variation of atmospheric ambient temperature is between a minimum of 26°C and the maximum of 30.9°C.

Specifically from Figure 2.3 and Table 2.5, it is clear that the drying chamber temperature is maximum when irradiance is maximum (during 12.00 P.M.) and minimum when irradiance is minimum (during 04.00 P.M.). Variation of drying chamber temperature is directly proportional to hourly varied solar irradiance.

2.6.2 Internet of Things aspect

In the solar dryer assembly, when the absorber plate temperature and ambient temperature value vary, instantaneously temperature sensor sends data

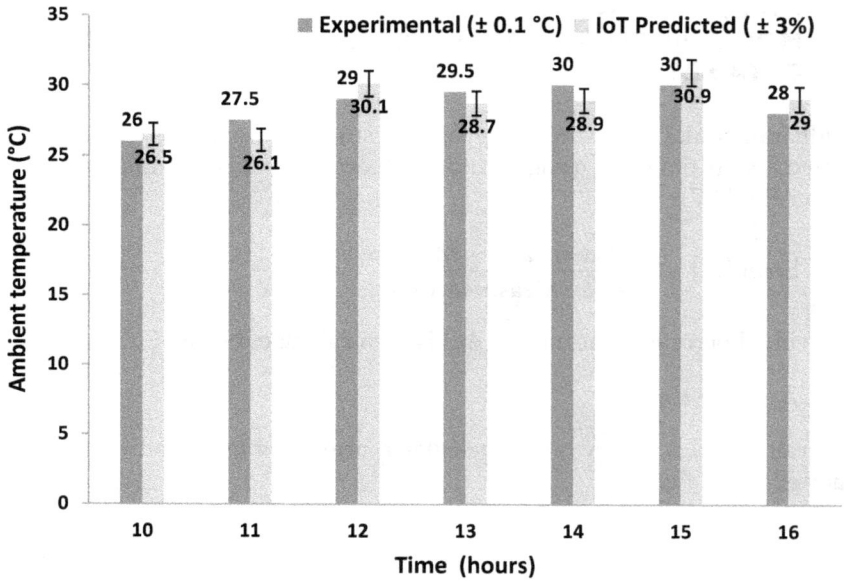

Figure 2.2 Variation in experimental ambient temperature and IoT-based ambient temperature values.

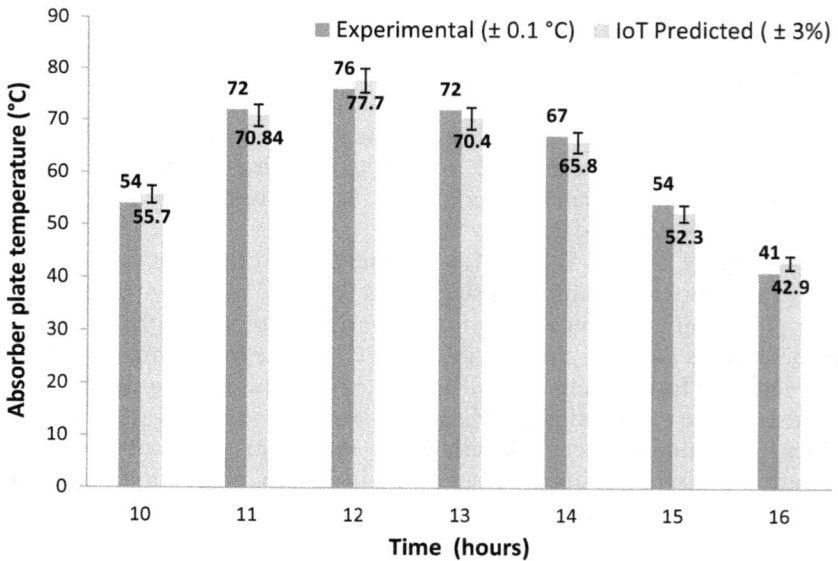

Figure 2.3 Variation in absorber plate temperature and IoT-based in absorber plate temperature values.

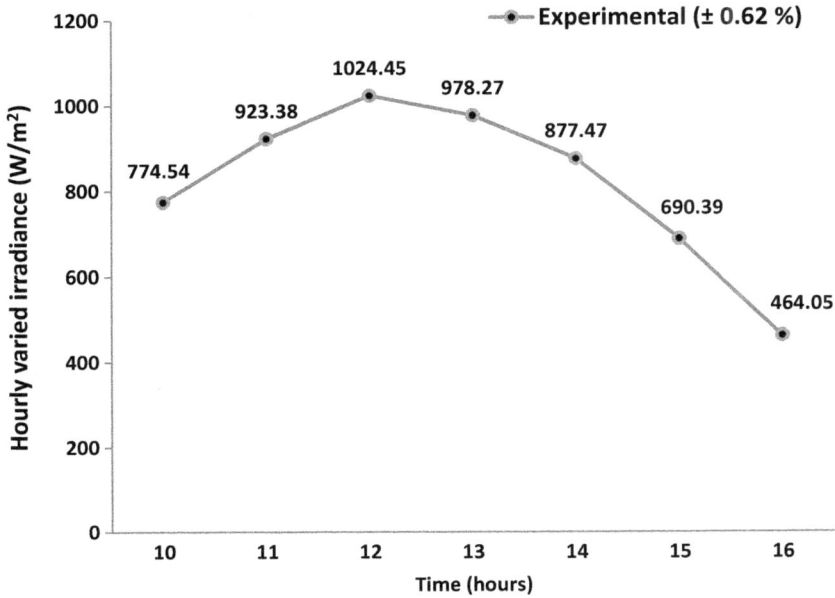

Figure 2.4 Hourly varied averaged solar irradiance.

Table 2.4 Hourly averaged measured ambient temperatures in comparison with IoT-based value

Sl.no	Time	Irradiance	Experimental ambient temperature	IoT ambient temperature	Accuracy
	(Hours)	(W/m²)	(°C)	(°C)	(%)
1	10	774.54	26	26.5	98.12
2	11	923.38	27.5	26.1	94.91
3	12	1024.45	29	30.1	96.35
4	13	978.27	29.5	28.7	97.29
5	14	877.47	30	28.9	96.34
6	15	690.39	30	30.9	97.09
7	16	464.05	28	29	96.56
Average					96.66

Table 2.5 Hourly averaged measured absorber temperatures in comparison with IoT-based value

Sl.no	Time	Irradiance	RTD plate temperature	IoT plate temperature	Accuracy
	(Hours)	(W/m²)	(°C)	(°C)	(%)
1	10	774.54	54	55.7	96.95
2	11	923.38	72	70.84	98.37
3	12	1024.45	76	77.7	97.82
4	13	978.27	72	70.4	97.73
5	14	877.47	67	65.8	98.18
6	15	690.39	54	52.3	96.75
7	16	464.05	41	42.9	95.58
Average					97.34

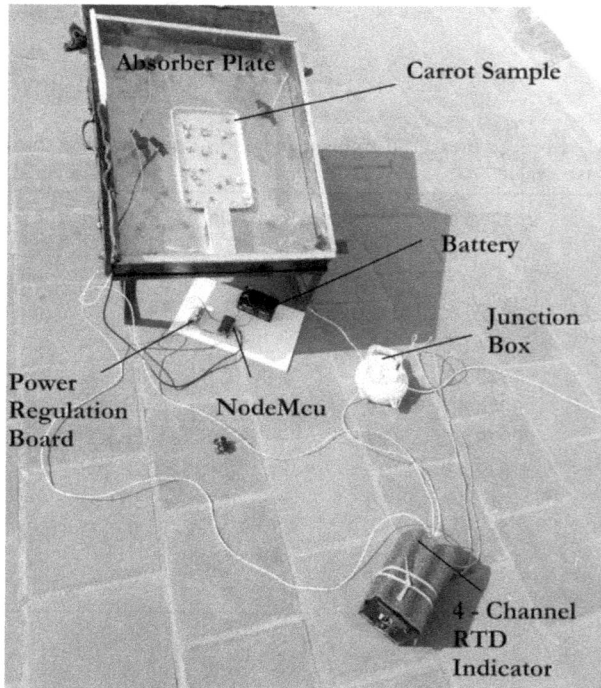

Figure 2.5 Real-time experimental solar dryer system integrated with IoT kit

to the cloud data base (Web-centric server). The response time taken by proposed IoT system with 4G internet speed is 10 seconds. The proposed IoT-based monitoring system for solar dryer assembly is capable of monitoring the system with reduced abnormalities and malfunctions. From Tables 2.4 and 2.5, it is clear that the proposed IoT system is capable of monitoring temperature variation in solar dryer more accurately with an accuracy of 96.66% for ambient temperature and 97.34% for absorber plate temperature.

2.7 CONCLUSIONS

In this chapter, a novel IoT-based temperature monitoring system for the solar dryer is proposed. Such a system was successfully installed at the solar thermal power laboratory at Saranathan College of Engineering, Trichy, India.

The following conclusions are observed from the comparative experiments between the manually (offline) observed and IoT-based (online) process monitoring study:

   i. Solar irradiance increases during the first half of the day and decreases during the second half of the experimental day.

   ii. Similarly, the absorber plate temperature irradiance increases during the first half of the day and decreases during the second half of the experimental day.

   iii. Surrounding ambient temperature is high during noon hours than morning hours.

   iv. The performance of proposed IoT-based process monitoring system completely depends on the data transfer rate of the network provider.

   v. Overall prediction accuracy of proposed IoT-based system in monitoring the variations in the ambient temperature is 96.66%.

   vi. Overall prediction accuracy of proposed IoT-based system in monitoring the variations in the absorber plate temperature is 97.34%.

NOMENCLATURE

A_d area of the drying chamber in m^2
M_w mass of moisture removed in kg
h_{fg} latent heat in kJ/kg
I irradiance in w/m^2
Δt time duration in hours
Q heat supplied in Watts
A_{in} individual accuracy in %
A_O overall model accuracy in %
e predicted results error in %

ABBREVIATION

IoT   Internet of Things
n   number of data sets

REFERENCES

[1] M. Sridharan, R. Devi, C. S. Dharshini, M. Bhavadarani, IoT based performance monitoring and control in counter flow double pipe heat exchanger.

[2] T. P. Fowdur, Y. Beeharry, V. Hurbungs, Performance analysis and implementation of an adaptive real-time weather forecasting system, *Internet of Things* 3–4 (2018) 12–33.

[3] Sepideh Radmannia, Milad Naderzad, IoT-based electrosynthesis ecosystem, *Internet of Things* 3–4 (2018) 46–51.

[4] Dragos Mocrii, Yuxiang Chen, Petr Musilek, IoT-based smart homes: A review of system architecture, software, communications, privacy and security, *Internet of Things* 1–2 (2018) 81–98.

[5] Gabriel Machado Lunardi, Fadi Al Machot, IoT-based human action prediction and support, *Internet of Things* 3–4 (2018) 52–68.

[6] Ming Yang, Smart metal forming with digital process and IoT, *International Journal of Lightweight Materials and Manufacture*, 1 (2018) 207–214. https://doi.org/10.1016/j.ijlmm.2018.10.001

[7] Channa Mallikarjuna Mattihalli, Edemialem Gedefaye, Fasil Endalamaw, Adugna Necho, Plant leaf diseases detection and auto-medicine, *Internet of Things* 1–2 (2018) 67–73.

[8] Morteza Hadipour, Javad Farrokhi Derakhshandeh, Automatic washing system of LED street lighting via Internet of Things, *Internet of Things* 1–2 (2018) 74–80.

[9] M. Chandrasekar, T. Senthilkumar, B. Kumaragurubaran, J. Peter Fernandes, Experimental investigation on a solar dryer integrated with condenser unit of split air conditioner (A/C) for enhancing drying rate, *Renewable Energy* 122, July 2018, 375–381.

[10] B. M. A. Amer, M. A. Hossain, K. Gottschalk, Design and performance evaluation of a new hybrid solar dryer for banana, *Energy Conversion and Management* 51 (2010) 813–820.

[11] Wei Wang, Ming Li, Reda Hassanien Emam Hassanien, Yunfeng Wang, Luwei Yang, Thermal performance of indirect forced convection solar dryer and kinetics analysis of mango, *Applied Thermal Engineering* 134, April 2018, 310–321.

[12] M. Virbhadra, Arun T. Swamia, Anil T. R. Auteeb, Experimental analysis of solar fish dryer using phase change material, *Journal of Energy Storage* 20 (2018) 310–315.

[13] M. Sridharan, Performance augmentation study on a solar flat plate water collector system with modified absorber flow design and its performance prediction using the XGBoost algorithm: A machine learning approach, *Iran J. Sci. Technol. Trans. Mech. Eng.* (2023). https://doi.org/10.1007/s40997-023-00648-8

[14] M. Sridharan, Application of generalized regression neural network in predicting the performance of natural convection solar dryer, *ASME – J. Sol. Energy Engg. Wind Energy Conv.*, 2020. https://doi.org/10.1115/1.4045384

Chapter 3

Natural convection study in cylindrical annulus through OpenFOAM

Seetu Rana
Government College Hisar, Hisar, India

Bidesh Sengupta
Nanyang Technological University, Singapore

Satyvir Singh
RWTH Aachen University, Aachen, Germany

3.1 INTRODUCTION

In fluid dynamics, natural convection heat transfer within an annulus has attracted attention for many years. Heat transfer through natural convection within confined spaces holds significant importance across various thermal engineering applications, including heat exchangers, solar collectors, and heating or cooling systems. Thus, natural convection constitutes a foundational issue within the realm of fluid mechanics.

In a horizontal cylindrical annulus, natural convection heat transfer has long been studied mathematically and experimentally. Natural convection in the annular space was investigated by Kuehn and Goldstein (1976) both theoretically and practically. Further, Kuehn and Goldstein (1978) conducted experimental research on natural convection in the eccentric and concentric regions and calculated the overall heat transfer coefficient. The outcomes of their study indicated that altering the position of the inner cylinder significantly influences heat transfer, leading to both reductions and enhancements. Additionally, relocating the cylinder to the lower part of the circular space resulted in an approximately 10% increase in the heat transfer rate.

At very low temperatures, McLeod and Bishop (1989) studied the turbulent natural convection in the helium gas concentric annulus. The results of Yoo's (1998) study on mixed convection in a concentric annulus with constant boundary conditions are detailed using flow patterns and isotherms. The findings revealed that the average Nusselt number for all Rayleigh numbers drops as the Reynolds number rises. In the study of natural convection inside the annulus in nanofluid, Abu-Nada et al. (2008) discovered that heat transport is greatly accelerated by high thermal conductivity nanoparticles.

DOI: 10.1201/9781003465171-3

Shu et al. (2001) quantitatively investigated natural convective heat transmission in a horizontal eccentric annulus between a square outer enclosure and a heated circular inner cylinder. Yuan et al. (2015) presented the numerical investigation of free convection in horizontal concentric annuli of varying with inner shape where the inner and outer surfaces are kept at a constant temperature. In a study by Angeli et al. (2010), the progression of flow patterns and temperature distributions in a horizontal annulus was explained as the Rayleigh number increased. At extremely low Rayleigh numbers, the behavior is characterized as pseudo-diffusive, with heat transfer being predominantly governed by conduction. Tsui and Tremblay (1984) investigated the transient natural convection between two horizontal isothermal cylinders.

Although there have been some studies conducted in the past, the available literature on Newtonian fluids in a horizontal concentric cylindrical annulus is limited. Therefore, this study provides the OpenFOAM simulations to investigate the computational analysis of laminar natural convection in a horizontal concentric cylindrical annulus, focusing on the influence of physical configurations and the impact of Rayleigh numbers on the characteristics of momentum and heat transfer. Concentric cylindrical annulus with an inner heating surface is utilized in polymer processing industries to create weld lines using natural convection heat transfer. These geometric configurations find practical implementation in various applications within the polymer processing industry. The presented findings in this study consist of the stream function, isotherms, vorticity, and enstrophy contours near the concentric cylindrical annulus, as well as the local values of the Nusselt numbers. Lastly, this research concludes with some final remarks and suggestions for future studies in this research area.

3.2 PROBLEM SETUP AND MATHEMATICAL FORMULATION

3.2.1 Problem setup

Assume that the natural convective 2D steady flow of an incompressible Newtonian fluid is embedded with the concentric cylindrical annulus. A schematic diagram of the cross section of a horizontal concentric cylindrical annulus is illustrated in Figure 3.1. The radii of the inner and outer cylinders are R_i and R_o, respectively, while the spacing between the two concentric cylinders is L. The uniform temperatures of the inner and outer cylinders are T_i and T_o, $(T_i > T_o)$, respectively. Thus, the cylindrical annulus is filed with fluids. For the numerical simulations, we set $L/R_i=1.6$. Additionally, the gravity force g is set up in the negative y-direction. The presence of temperature gradient leads to thermal expansion of the fluid in the domain, and as a result, the fluid near the heated cylinder will be lighter compared to the fluid far away from the cylinders. Owing to the density gradient, an upward flow

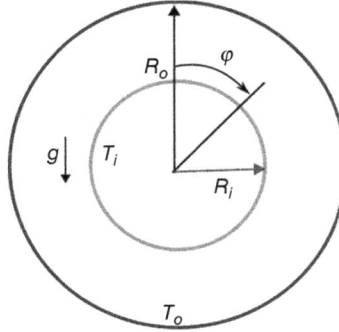

Figure 3.1 Schematic diagram of the problem setup.

or buoyancy will be set up adjacent to the cylinder which, in turn, results in natural convection of the fluid in the domain.

3.2.2 Mathematical formulations

The flow model assumes a non-Newtonian power law fluid with constant properties, except for the density in the body force term of the momentum equation. The Boussinesq approximation is applied to establish a connection between alterations in density and temperature, thereby facilitating the coupling of the temperature field with the flow field. The conservation laws governing natural convection flow, which involve the conservation of mass, momentum, and energy, can be expressed as follows:

$$\frac{\partial u}{\partial x} + \frac{\partial v}{\partial y} = 0, \tag{3.1}$$

$$u\frac{\partial u}{\partial x} + v\frac{\partial u}{\partial y} = -\frac{1}{\rho}\frac{\partial p}{\partial x} + \vartheta\left(\frac{\partial^2 u}{\partial x^2} + \frac{\partial^2 u}{\partial y^2}\right), \tag{3.2}$$

$$u\frac{\partial v}{\partial x} + v\frac{\partial v}{\partial y} = -\frac{1}{\rho}\frac{\partial p}{\partial y} + \vartheta\left(\frac{\partial^2 v}{\partial x^2} + \frac{\partial^2 v}{\partial y^2}\right) + g\beta(T - T_c), \tag{3.3}$$

$$u\frac{\partial T}{\partial x} + v\frac{\partial T}{\partial y} = \alpha\left(\frac{\partial^2 T}{\partial x^2} + \frac{\partial^2 T}{\partial y^2}\right), \tag{3.4}$$

The corresponding boundary conditions are given below

$$u = 0, v = 0, T = T_c \text{ at } r = R_o$$

$$u = 0, v = 0, T = T_h \text{ or } T = T_c + (T_h - T_c)\cos(\varphi) \text{ at } r = R_i \tag{3.5}$$

In the above equations, (u, v) represent the velocity components in the horizontal and vertical directions, ρ is the density, T is the temperature, p is the pressure, g is the gravitational acceleration, and φ is defined as the angle from the top vertical position in the clockwise direction. While β and α are the coefficient of the thermometric expansion and the thermal diffusivity of the fluid, respectively.

Using the following change of variables

$$X = \frac{x}{L}, Y = \frac{y}{L}, U = \frac{uL}{\alpha}, V = \frac{vL}{\alpha}, \theta = \frac{T - T_c}{T_h - T_c}, P = \frac{pL^2}{\rho\alpha^2}, \text{Pr} = \frac{v}{\alpha}, \quad (3.6)$$

$$\text{Ra} = \frac{g\beta(T_h - T_c)L^3 \text{Pr}}{v^2}$$

the abovementioned conservation laws become

$$\frac{\partial U}{\partial X} + \frac{\partial V}{\partial Y} = 0, \tag{3.7}$$

$$U\frac{\partial U}{\partial X} + V\frac{\partial U}{\partial Y} = -\frac{\partial p}{\partial X} + \text{Pr}\left(\frac{\partial^2 U}{\partial X^2} + \frac{\partial^2 U}{\partial Y^2}\right), \tag{3.8}$$

$$U\frac{\partial V}{\partial X} + V\frac{\partial V}{\partial Y} = -\frac{\partial p}{\partial Y} + \text{Pr}\left(\frac{\partial^2 V}{\partial X^2} + \frac{\partial^2 V}{\partial Y^2}\right) + \text{Ra}\,\text{Pr}\,\theta, \tag{3.9}$$

$$U\frac{\partial \theta}{\partial X} + V\frac{\partial \theta}{\partial Y} = \alpha\left(\frac{\partial^2 \theta}{\partial X^2} + \frac{\partial^2 \theta}{\partial Y^2}\right), \tag{3.10}$$

with the boundary conditions

$$U = 0, V = 0, \theta = 0 \text{ at } r = R_o$$

$$U = 0, \theta = 0, \theta = 1 \text{ or } \theta = \cos(\varphi) \text{ at } r = R_i \tag{3.11}$$

3.2.3 Important quantities in natural convection heat transfers

3.2.3.1 Stream function

The fluid motion is displayed using the stream function (ψ) which is obtained from velocity components U and V. The relationships between stream function, ψ, and velocity components for two dimensional flows are:

$$U = \frac{\partial \psi}{\partial Y} \text{ and } V = \frac{\partial \psi}{\partial X}, \tag{3.12}$$

which yield a single equation

$$\frac{\partial^2 \psi}{\partial X^2} + \frac{\partial^2 \psi}{\partial Y^2} = \frac{\partial U}{\partial Y} - \frac{\partial V}{\partial X} = 0. \tag{3.13}$$

Using the above definition of the stream function, the positive sign of ψ denotes anti-clockwave circulation and the negative sign of ψ represents the clock-wise circulation.

3.2.3.2 Vorticity

The vorticity plays a vital role in understanding the heat transfers during the natural convection process that describes the local spinning motion of a continuum near some. It can be defined as the curl of the flow velocity vector.

$$\omega = \nabla \times U = \frac{\partial U}{\partial Y} - \frac{\partial V}{\partial X}. \tag{3.14}$$

3.2.3.3 Enstrophy

The physical phenomena of vorticity generation during the natural convection process interaction can be explained by monitoring the evolution of the enstrophy. The evolution of the enstrophy can be defined as the square of the vorticity in the flow field,

$$\Omega = \frac{1}{2} \omega^2. \tag{3.15}$$

3.2.3.4 Local Nusselt numbers

The heat transfer coefficient along the solid walls is determined in terms of the local Nusselt number (Nu) along the inner and outer cylinders:

$$\text{Nu}_i = -\left[\frac{\partial \theta}{\partial n}\right]_{n=n_i} \text{ and } \text{Nu}_o = -\left[\frac{\partial \theta}{\partial n}\right]_{n=n_o}, \tag{3.16}$$

where n denotes the normal direction on a plane.

3.3 NUMERICAL SOLVER AND VALIDATION

The numerical simulations are performed using the 'incompressible Flow-Solver', which is part of the open source C++ libraries of OpenFOAM solver.

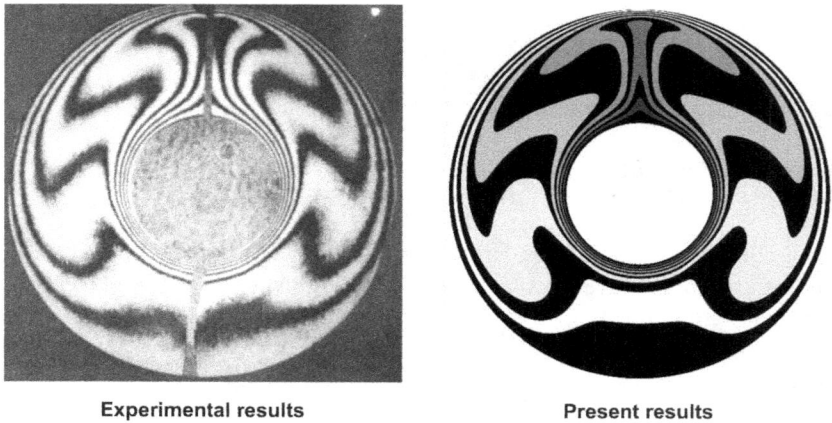

Experimental results   **Present results**

Figure 3.2 Validation study: comparison of interferograms from the experiment (left) and isotherm contours (right) from present numerical simulation for Pr = 0.706 and Ra = 4.7 × 104.

This solver is designed to solve a generic set of partial differential equations using a finite volume computational method. The present numerical solver is validated in the experimental study by Kuehn and Goldstein (1976). In this experimental study, the natural convection flows between the concentric cylindrical annulus were presented with Pr = 0.706, and Ra = 4.7 × 10^4. In the annulus, the hot fluid tends to move up while the cold fluid tends to move down because of the body force and the change of the fluid density in a temperature field. As shown in Figure 3.2, the temperature of the fluid near the hot inner boundary is higher than that near the cold outer boundary, thus the hot fluid moves up along the inner boundary and forms a plume at the top center part of the annulus, while the cold fluid moves down along the outer boundary to the bottom. It can be seen in Figure 3.2 that the OpenFOAM simulation agrees well with the experiment.

3.4 RESULTS AND DISCUSSION

The natural convection heat transfers in a horizontal concentric annulus in the range of $10^2 \leq Ra \leq 10^6$ at Pr = 10 are investigated through a series of simulations. The numerical results of the isotherms, stream functions, vorticity, enstrophy, and local Nusselt numbers at different Rayleigh numbers are illustrated in Figures 3.3 to 3.7.

Figure 3.3 shows the contours of isotherms in the horizontal concentric annulus at different Ra numbers along with the same Pr = 10. At the small Ra number (Ra = 10^2), the conduction dominates the heat transfer in the annulus and the natural convection is very weak, as shown in Figure 3.3(a).

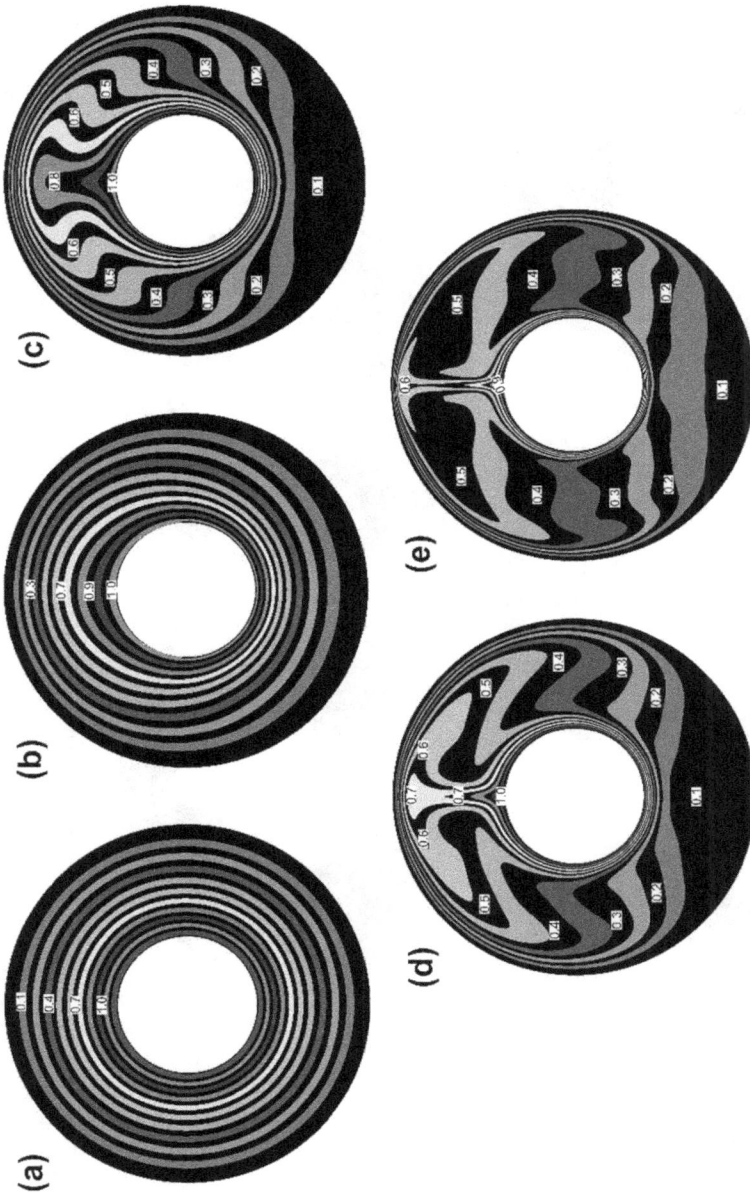

Figure 3.3 Contours of isotherm for Pr = 10 with different Rayleigh numbers (a) Ra = 10^2, (b) Ra = 10^3, (c) Ra = 10^4, (d) Ra = 10^5, and (e) Ra = 10^6.

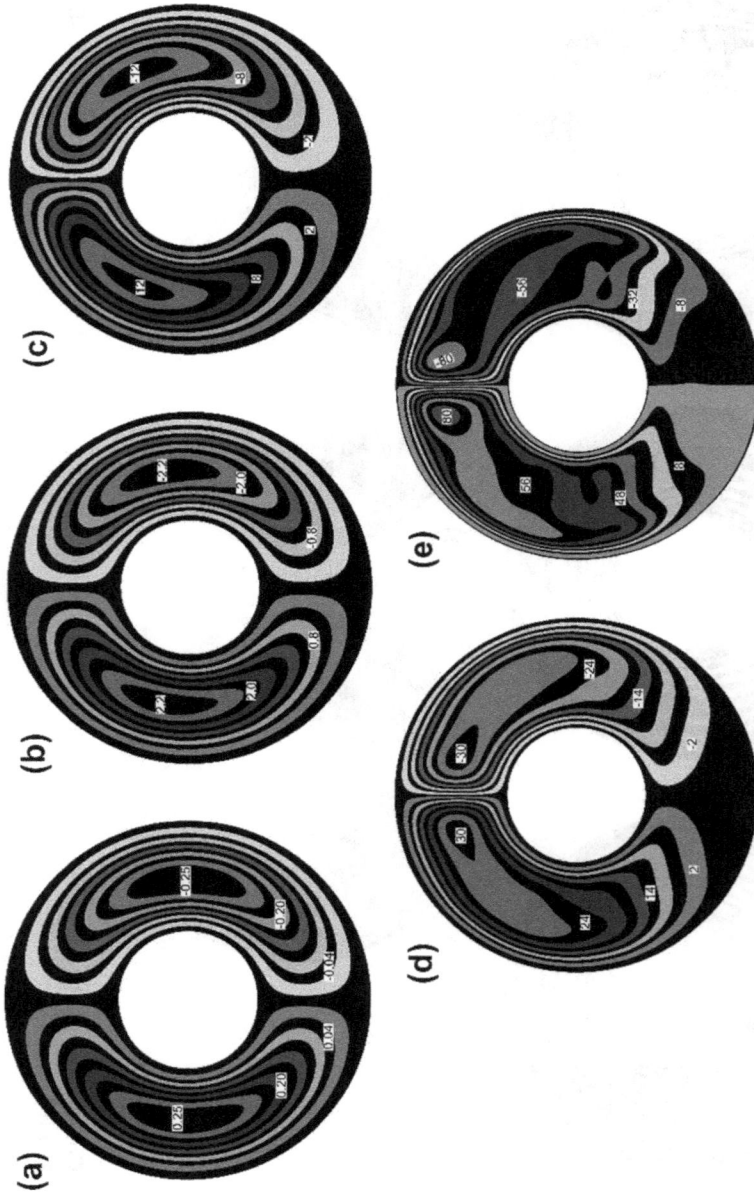

Figure 3.4 Contours of stream function for $Pr = 10$ with different Rayleigh numbers (a) $Ra = 10^2$, (b) $Ra = 10^3$, (c) $Ra = 10^4$, (d) $Ra = 10^5$, and (e) $Ra = 10^6$.

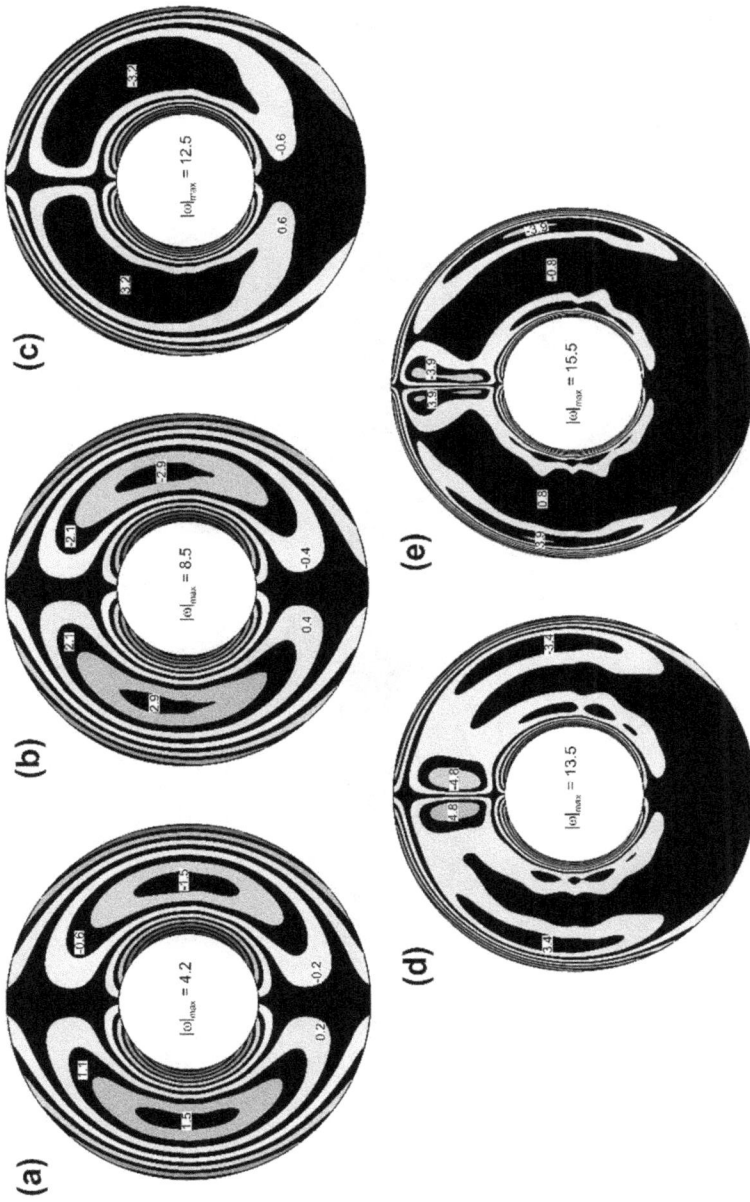

Figure 3.5 Contours of vorticity for Pr = 10 with different Rayleigh numbers (a) Ra = 10^2, (b) Ra = 10^3, (c) Ra = 10^4, (d) Ra = 10^5, and (e) Ra = 10^6.

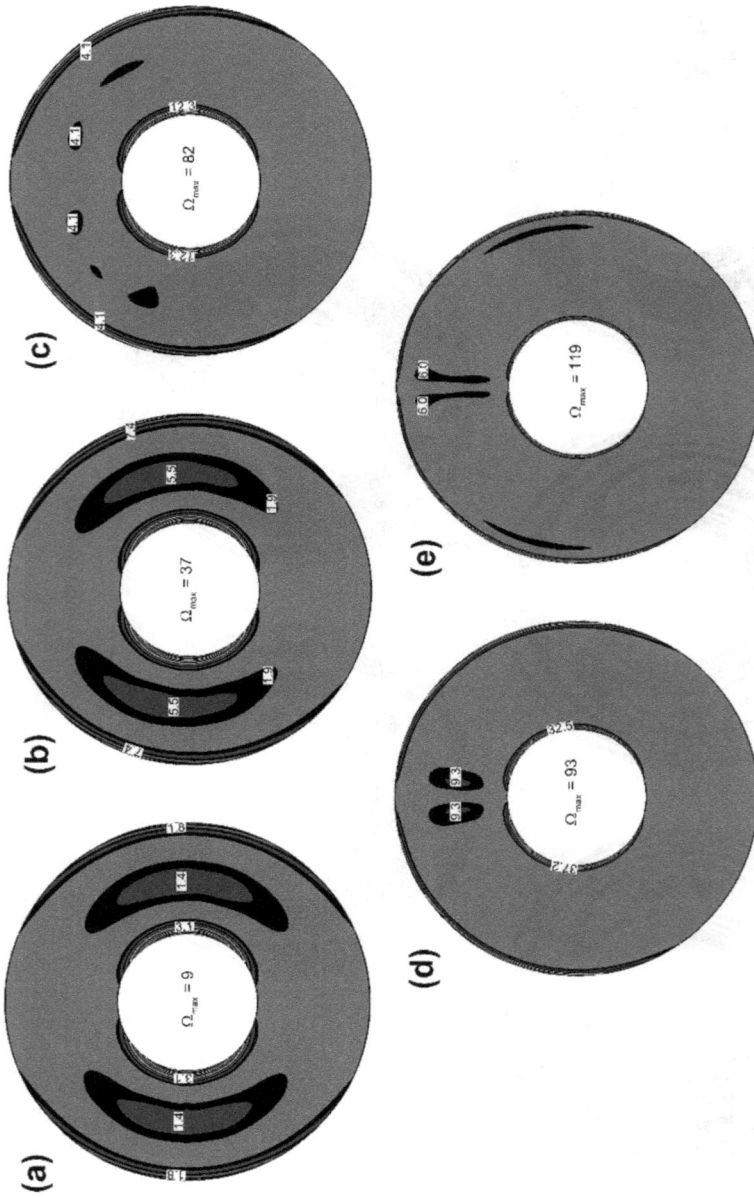

Figure 3.6 Contours of enstrophy for Pr = 10 with different Rayleigh numbers (a) Ra = 10^2, (b) Ra = 10^3, (c) Ra = 10^4, (d) Ra = 10^5, and (e) Ra = 10^6.

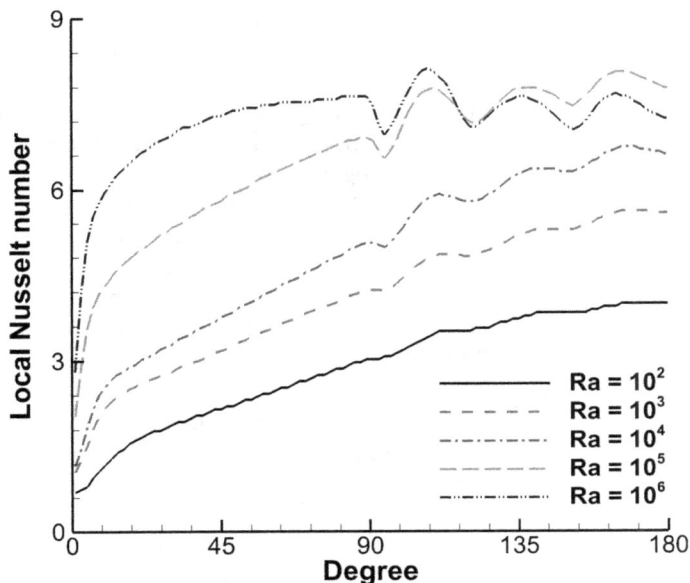

Figure 3.7 Variation of the local Nusselt number along the inner surface of the cylindrical annulus at different Rayleigh numbers.

Thus, isotherms are almost concentric circles. As the Ra number increases, i.e. Ra = 10^3, the natural convection in the annulus becomes stronger, which causes the isotherms move up and alter their shapes, as displayed in Figure 3.3(b). When the Ra number is increased to 10^4, the hot fluid rises and forms plumes in the annulus, as illustrated in Figure 3.3(c). The plume begins from the hot inner cylinder to the cold outer cylinder, and there is a cold fluid layer between the plume and the outer cylinder. Interestingly, both the width of the plume and the thickness of the cold layer decrease with the increase of the Ra number from 10^4 to 10^6, as shown in Figure 3.3(d)–(e). The formed plume of cold fluid which begins from the top of the cold outer cylinder to the hot inner cylinder is formed at Ra = 10^5. This phenomenon is similar to the so-called Rayleigh-Benard convection.

Figure 3.4 shows the contours of stream function in the horizontal concentric annulus at different Ra numbers along with the same Pr = 10. At low Ra number (Ra = 10^2), the natural convection is found to be very weak and the heat transfer is primarily due to conduction. Therefore, two symmetric vertices caused by natural convection can be observed in Figure 3.4(a). As expected, due to the cold outer walls, fluids rise up from middle portion of the inner wall and flow down along the outer walls forming two symmetric rolls with clockwise and anti-clockwise rotations inside the co-centric annulus. Also, the magnitude of stream function is very low, i.e. $|\psi|_{max} = 0.25$. As the Ra number increases, the location of the configuration cell moves up.

In addition, the magnitudes of stream functions enhance. The magnitude values for Ra = 10^3, 10^4, 10^5, and 10^6 are examined as $|\psi|_{max}$ = 2.2, 12, 30, and 80, respectively. Interestingly, when Ra reaches 10^6, the flow begins to show unsteady characteristics which translate to asymmetric and distorted streamlines, as can be seen in Figure 3.4(e). Also, the circulation core is further confined at the top while forming a secondary core.

Figure 3.5 displays the contours of vorticity in the horizontal concentric annulus at different Ra numbers along with the same Pr = 10. Vorticity is a useful tool to understand how the ideal potential flow solutions can be perturbed to model real flows. It is found that when the Ra number increases, the strength of the vortex also increases. For example, the maximum values of vorticity magnitude for Ra = 10^2, 10^3, 10^4, 10^5, and 10^6 are observed $|\omega|_{max}$ = 4.2, 8.5, 12.5, 13.5, and 15.5, respectively.

To better understand the role of Ra numbers in the natural convection process, enstrophy is one of the main fundamental fields in heat transfers flows. Figure 3.6 illustrates the contours of enstrophy in the horizontal concentric annulus at different Ra numbers along with the same Pr = 10. There are significant changes in enstrophy at various Ra numbers during the natural convection. With compared to modest Ra numbers, the flow fields of this quantity are significantly enhanced at high Ra numbers. For example, the maximum value of enstrophy for Ra = 10^2, 10^3, 10^4, 10^5 and 10^6 are observed Ω_{max} = 9, 37, 82, 93, and 119, respectively.

The development of convective flow and heat transfer is now presented by the local Nusselt number along the inner cylinder in this study. Figures 3.7 show the variation of the local Nusselt number along the inner surface of the cylindrical annulus at different Rayleigh numbers with the same Prandtl number Pa = 10. It can be seen from the figure that different Rayleigh numbers have a great effect on the heat transfer. Interestingly, until the fluid particles heated through the inner hot cylinder reach the outer cylinder, the temperature gradient does not develop on the outer cylinder, and therefore the local Nusselt number is near to zero at the initial stage of the flow. After this initial stage, a sudden increase in the local Nusselt number is seen this trend continues at a modest level until the steady state flow condition is reached. In high Ra numbers, the local Nusselt number exhibits an oscillatory behavior, following a primary overshoot, due to the flow bifurcation as the convective flow is unstable.

3.5 CONCLUSION

This study aims to investigate numerically the natural convection Newtonian fluid in a horizontal concentric cylindrical annulus with constant temperature in a steady-state, two-dimensional scenario. The hot and cold temperatures are assigned to the inner and outer surfaces, respectively. The OpenFOAM software is utilized for numerical simulations, where the conservation laws of mass, momentum, and energy are discretized using a finite volume approach to solve the resulting system of equations. The numerical method was validated by comparing numerical results with experimental results from literature, and good agreement was obtained. The investigation focuses on examining the influence of different Rayleigh numbers ($10^2 \leq$ Ra $\leq 10^6$) at Prandtl number Pr = 10, on the heat and fluid flows. The impacts of Ra numbers on the isotherm, stream functions, vorticity, and enstrophy are emphasized. It is found that the flow in the annulus is stable at low Ra number but unstable at high Ra number. Finally, the local Nusselt number is calculated and examined for heat transfer. The results suggest that as the Ra number increases, the local Nusselt number increases, which signifies an enhancement in the heat transfer rate.

REFERENCES

Kuehn, T.H., & Goldstein, R.J. (1976). An experimental and theoretical study of natural convection in the annulus between horizontal concentric cylinders. *Journal of Fluid Mechanics*, 74(4), 695–719.

Kuehn, T.H., & Goldstein, R.J. (1978). An experimental study of natural convection heat transfer in concentric and eccentric horizontal cylindrical annuli. *Communications Journal of Heat Transfer*, 100(4), 635–640.

McLeod, A.E., & Bishop, E.H. (1989). Turbulent natural convection of gases in horizontal cylindrical annuli at cryogenic temperatures. *International Journal of Heat and Mass Transfer*, 32, 1967–1978.

Yoo, J.S. (1998). Mixed convection of air between two horizontal concentric cylinders with a cooled rotating outer cylinder. *International Journal of Heat and Mass Transfer*, 41, 293–302.

Abu-Nada, E., Masoud, Z., & Hijazi, A. (2008). Natural convection heat transfer enhancement in horizontal concentric annuli using nanofluids. *International Communications of Heat and Mass Transfer*, 35, 657–665.

Shu, C., Xue, H., & Zhu, Y.D. (2001). Numerical study of natural convection in an eccentric annulus between a square outer cylinder and a circular inner cylinder using DQ method. *International Journal of Heat and Mass Transfer*, 44, 3321–3333.

Yuan, X., Tavakkoli, F., & Vafai, K. (2015). Analysis of natural convection in horizontal concentric annuli of varying inner shape. *Numerical Heat Transfer, Part A: Applications*, 68(11), 1155–1174.

Angeli, D., Barozzi, G.S., Collins, M.W., & Kamiyo, O.M. (2010). A critical review of Buoyancy-induced flow transitions in Horizontal Annuli. *International Journal of Thermal Sciences*, 49, 2231–2241.

Tsui, Y.T., & Tremblay, B. (1984). On transient natural convection heat transfer in the annulus between concentric – Horizontal cylinders with isothermal surfaces. *International Journal of Heat Mass Transfer*, 27, 103–111.

Chapter 4

MHD forced convection in Casson hybrid nanofluids with Soret and Dufour effects

K. Gnanaprasanna, Abhishek Kumar Singh,
Pankaj Shukla, and Awani Bhushan
School of Advanced Science, Vellore Institute of Technology,
Chennai, India

4.1 INTRODUCTION

Fluids are substances that exist in nature, and they are broadly classified as liquids and gases. Fluids continuously deform their state in the presence of stress and strain rates. Liquids have a large intermolecular force of attraction, while gases have less intermolecular forces. Fluid mechanics is the study of science dealing with fluids in stationary and moving state. Statics is the study of science involving fluids at rest, and kinematics involves the study of fluid flows in the absence of external forces. Fluid dynamics is the branch of science that deals with the study of motion of fluid particles together with the influences of pressure and gravity forces [1–2].

The Greek physicist and mathematician Archimedes was the first person to enunciate the buoyancy law of forces. Sir Isaac Newton analyzed the flow of fluids in various medium and discovered laws of gravity, motion, and many contributions in calculus. Leonhard Euler, a famous Swiss mathematician, pioneered in finding various analytical methods to solve number theory and methods to solve problems in numerical analysis. Later Bernoulli-Euler developed the basic equations incorporating various fluid properties like density and pressure in fluid dynamics.

During the twentieth century, Choi and his research group developed the nanofluid technology. The phenomenon of suspending nanoparticles with the base fluid improves the thermal conductivity of the system. In particular, the MHD hybrid nanofluid is widely spread in nuclear reactors, automotive cooling, industrial sectors, etc.

The effects of variable viscosity for a high Reynolds number occur in a thin layer called boundary layer where the velocity transition takes place. In 1904, Ludwig Prandtl manifested the concept of the velocity changes that occur asymptotically from zero velocity or no slip boundary condition to free stream in a thin layer called boundary layer, and in this region, the viscous effects are neglected. The region in which the frictional effects are negligible and velocity remains constant is termed an inviscid region.

DOI: 10.1201/9781003465171-4

The significant difference between the enhancement in viscosity models leads to heat transfer of hybrid nanofluid across the enclosed flow fields. Many young scientists of the present era are keenly interested in the study of non-Newtonian fluids due to their viscous nature. Certain non-Newtonian fluids, including Casson fluid, Williamson fluid, second grade fluid, etc., possess unique shear stress and strain rates. Sakiadis considered the boundary layer flows of a moving surface. Choi and Eastman experimentally proved that certain metallic or non-metallic particles of size less than 100 nm enhanced the thermal conductivity of the whole system. Later Buongiorno experimented seven elements, namely Brownian diffusion, inertia, diffusiophoresis, thermophoresis, fluid drainage, Magnus effect, and gravity, and proved that thermophoresis and Brownian motion created major impacts in enhancing the characteristics of heat transfer and thermal conductivity [3–5].

The rapid advancement of energy resources in industries paved the way for introducing hybrid nanofluid theory. The combined effects of heat flux generated due to thermal diffusion and diffusion thermal characteristics of hybrid nanofluids have a significant influence on heat and mass transfer and have applications in designing thermal systems, cooling electrical components, electrical conductors, gas turbines, solar collectors, exothermic chemical reaction, etc [6–7]. The primary objective of this numerical study is to analyze the presence of heat flux and mass flux generated due to Soret and Dufour effects with no-slip boundary condition on MHD forced convective Casson hybrid nanofluid flows over a diverging channel. The governing partial differential equations pertaining to the flow, temperature, and concentration are highly nonlinear in nature. The governing boundary layered equations are non-dimensionalized by applying certain non-similar transformations and linearized using the quasilinearization technique. The system of linearized PDEs is solved using the implicit finite difference scheme together with Varga's algorithm.

4.2 MATHEMATICAL FORMULATION OF THE FLOW GEOMETRY

Consider a two-dimensional steady laminar forced convective electrically conducting Casson hybrid nanofluid flows over diverging channel with velocity $u_e = u_\infty(1 - \epsilon e^\xi)$. The fluid flow is considered to be in the horizontal x-axis, and y-axis is observed to be vertically upward perpendicular to it. The constant magnetic field B_0 is set to be applied normal to the flow field. Variable viscosity and thermal conductivity along with Joule heating and radiative heat flux together with the Buongiorno's model are to be utilized in the present work. Table 4.1 portrays the thermophysical values of base fluid ethylene glycol along with copper and alumina hybrid

Table 4.1 The thermophysical properties of ethylene glycol and Cu, Al$_2$O$_3$ Casson hybrid nanofluid properties

Properties	Ethylene glycol	Cu	Al$_2$O$_3$
Cp (J kg^{-1} K^{-1})	2400	385	765
ρ (kg m^{-3})	1110	8933	3970
k (W m^{-1} K^{-1})	0.26	400	40
σ (s m^{-1})	1.07×10^{-4}	59.6×10^6	35×10^6

nanoparticles. The isotropic rheological non-Newtonian Casson hybrid model is expressed as [8–10]:

$$
\tau_{ij} = \begin{cases} 2e_{ij}\left(\mu_a + \dfrac{P_y}{\sqrt{2\pi}}\right), \pi > \pi_c \\[2ex] 2e_{ij}\left(\mu_a + \dfrac{P_y}{\sqrt{2\pi_c}}\right), \pi < \pi_c \end{cases}
$$

where $\pi = e_{ij}e_{ij}$ and e_{ij} is the $[i, j]^{th}$ component of the deformation rate defined as

$$
e_{ij} = \frac{1}{2}\left(\frac{\partial v_i}{\partial x_j} + \frac{\partial v_j}{\partial x_i}\right)
$$

$P_y = \dfrac{\mu_a\sqrt{2\pi}}{\beta}$, yield stress, π_c is the critical value of π and μ_a plastic dynamic viscosity of non-*Newtonian fluid,* the boundary layer equations governing the flow model consist of the continuity, momentum, energy, and concentration equations, along with the dimensional physical quantities included in the boundary conditions are as follows [4, 5]:

The continuity equation is:

$$
\frac{\partial u}{\partial x} + \frac{\partial v}{\partial y} = 0 \tag{4.1}
$$

The momentum equation is:

$$
\left(u\frac{\partial u}{\partial x} + v\frac{\partial u}{\partial y}\right) = \underbrace{\left(u_e\frac{du_e}{dx}\right)}_{\text{Term I}} + \underbrace{\frac{1}{\rho_{hnf}}\left(1 + \frac{1}{\beta}\right)\frac{\partial}{\partial y}\left(\mu_{hnf}\frac{\partial u}{\partial y}\right)}_{\text{Term II}} - \underbrace{\frac{\sigma_{hnf}}{\rho_{hnf}}B_0^2(u - u_e)}_{\text{Term III}}
$$

$$
\tag{4.2}
$$

where Term I denotes the pressure term, Term II denotes the variable viscous term which includes the Casson parameter, and Term III denotes the external magnetic field.

The energy equation is:

$$\left(u\frac{\partial T}{\partial x} + v\frac{\partial T}{\partial y} \right) = \underbrace{\frac{1}{(\rho c_p)_{hnf}}\frac{\partial}{\partial y}\left(k_{hnf}\frac{\partial T}{\partial y} \right)}_{\text{Term I}} + \underbrace{\frac{Q_0}{(\rho c_p)_{hnf}}(T - T_\infty)}_{\text{Term II}} - \underbrace{\frac{1}{(\rho c_p)_{hnf}}\left(\frac{\partial q_r}{\partial y} \right)}_{\text{Term III}}$$

$$+ \underbrace{\frac{\sigma_{hnf}}{(\rho c_p)_{hnf}} B_0^2 (u - u_e)^2 + \left(1 + \frac{1}{\beta} \right)\frac{\mu_{hnf}}{(\rho c_p)_{hnf}}\left(\frac{\partial u}{\partial y} \right)^2}_{\text{Term IV} \qquad\qquad \text{Term V}}$$

$$\underbrace{\frac{(\rho c_p)_s}{(\rho c_p)_{hnf}}\left[D_B\frac{\partial C}{\partial y}\frac{\partial T}{\partial y} + \frac{D_T}{T_\infty}\left(\frac{\partial T}{\partial y} \right)^2 \right]}_{\text{Term VI}} + \underbrace{\frac{D_m K_T \partial^2 C}{C_s C_p \partial y^2}}_{\text{Term VII}}$$

$$(4.3)$$

where Term I denotes the variable thermal conductivity, Term II denotes the chemical reaction parameter, Term III denotes the radiative heat flux, Term IV denotes the Joule heating effect, Term V denotes the viscous dissipation parameter, Term VI denotes the Brownian motion and thermophoresis effects, and Term VII denotes influences of Dufour effects.

The species concentration equation is:

$$\left(u\frac{\partial C}{\partial x} + v\frac{\partial C}{\partial y} \right) = \underbrace{D_B\frac{\partial^2 C}{\partial y^2} + \frac{D_T}{T_\infty}\left(\frac{\partial^2 T}{\partial y^2} \right)}_{\text{Term I}} + \underbrace{\frac{D_m K_T}{T_m}\left(\frac{\partial^2 T}{\partial y^2} \right) - R(C - C_\infty)}_{\text{Term II}} \quad (4.4)$$

where Term I denotes the effects of Brownian motion and thermophoresis, and Term II denotes the Soret effects of the species concentration with boundary conditions:

$$u = 0; v = v_w(x); T = T_w(x); \text{at } y = 0 \tag{4.5}$$

$$u \to u_\infty; T \to T_\infty; \text{ at } y \to y_\infty \tag{4.6}$$

where u,v denote the velocities in x,y directions respectively. $u_e \dfrac{du_e}{dx}$ denotes the pressure term, β denotes the Casson parameter, μ_{hnf} denotes the variable viscosity, B_0 the external magnetic field term in the normal direction, k_{hnf} represents the variable thermal conductivity, Q_0 denotes the chemical reaction parameter, q_r denotes the radiative heat flux parameter, the fourth and the fifth terms on the RHS of the energy equation denote the Joule heating and viscous dissipation effects of the non-Newtonian model. D_B and D_T are respectively the thermophoretic and Brownian motion with no slip boundary condition on velocity gradients. The conventional Casson base fluid and hybrid nanofluid for various physical parameters are valid only for the dilute suspensions of copper and aluminum oxide nanoparticles whose base fluid is ethylene glycol are as follows: Refer [4, 5]

Viscosity: $(D_1) = \dfrac{\mu_{hnf}}{\mu_{bf}}$

$$\frac{\mu_{hnf}}{\mu_{bf}} = \frac{\mu_\infty}{(1 + a_1\theta)(1-\phi)^{2.5}}$$

Density: $(D_2) = \dfrac{\rho_{hnf}}{\rho_{bf}}$

$$\frac{\rho_{hnf}}{\rho_{bf}} = (1-\phi) + \frac{\rho_{cu}\phi_{Cu}}{\rho_{bf}}$$

Electrical conductivity: $(K_1) = \dfrac{\sigma_{hnf}}{\sigma_{bf}}$

$$\frac{\sigma_{hnf}}{\sigma_{bf}} = 1 + \left[\frac{3\phi(\psi_1 - \phi\sigma_{bf})}{\psi_2 - (\psi_1 - \phi\sigma_{bf})\phi}\right]$$

where

$$\psi_1 = (\sigma\phi)_{Cu} + (\sigma\phi)_{Al_2O_3}; \quad \psi_2 = \psi_1 + 2\phi\sigma_{bf}$$
$$\psi_3 = (\phi k)_{Cu} + (\phi k)_{Al_2O_3}; \quad \psi_4 = 2\phi\sigma_{bf}$$

Specific heat capacity: $(D_4) = \dfrac{(\rho C_p)_{hnf}}{(\rho C_p)_{bf}}$

$$\frac{(\rho C_p)_{hnf}}{(\rho C_p)_{bf}} = (1-\phi) + \frac{(\rho C_p)_{Cu} \phi_{Cu}}{(\rho C_p)_f} + \frac{(\rho C_p)_{Al_2O_3} \phi_{Al_2O_3}}{(\rho C_p)_f}$$

Thermal conductivity: $(D_5) = \dfrac{k_{hnf}}{k_{bf}}$

$$\frac{k_{hnf}}{k_{bf}} = \frac{2k_{bf} + \dfrac{\psi_3}{\phi} - 2\left(\phi k_{bf} - \psi_3\right)}{2k_{bf} + \phi k_{bf} + \left(\dfrac{\psi_3}{\phi} - \psi_3\right)}$$

where $\psi_3 = \dfrac{\phi_{Cu} k_{Cu} + \phi_{Al_2O_3} k_{Al_2O_3}}{\phi_C}$; $\phi = \phi_{Cu} + \phi_{Al}$ using non-similar transformations, the basic governing equations are as follows:

$$\xi = \frac{x}{L}; \quad \eta = \left(\frac{u_e v}{x}\right)^{1/2} y; \quad \psi(x,y) = (u_e x v)^{1/2} f; \quad u = \frac{\partial \psi}{\partial y}$$

$$v = -\frac{\partial \psi}{\partial x}; \quad f_\eta = F; \quad \theta(\xi, \eta) = \frac{(T - T_\infty)}{(T_w - T_\infty)};$$

$$u = u_e f_\eta; \quad \epsilon = \frac{U_\infty(x)}{U_w(x) + U_\infty(x)}; \quad Re = \frac{u_\infty L}{v_\infty}$$

$$\beta = \mu_a \frac{\sqrt{2\pi}}{P_y}; \quad v = (-v u_e)^{1/2} x^{-1/2} \left[\frac{(s+1)}{2} f + \frac{(s+1)}{2} \xi f_\xi + \frac{(s-1)}{2} \eta f_\eta\right];$$

$$v_\infty = \frac{\mu_{bf}}{\rho_{bf}}; Pr = \frac{\mu_\infty}{Cp k_\infty}; Ec = \frac{u_e^2(T_w - T_\infty)}{(C_p)_{bf}}; St = \frac{B_0^2 L \sigma_{bf}}{\rho_{bf} u_\infty} \quad R_1 = \frac{16\sigma^* T^3}{3K^* K_\infty};$$

$$N_b = \frac{\tau D_B \Delta C}{v_\infty}; \tau = \frac{(\rho c_p)_{hnf}}{(\rho c_p)_{bf}}; N_t = \tau \frac{D_T \Delta T}{v_\infty T_\infty}; Sc = \frac{v_\infty}{D_B}$$

$$D_f = \frac{D_m K_T}{C_s C_p v_\infty}\left(\frac{\Delta C}{\Delta T}\right) \quad S_r = \frac{D_T K_T \Delta C}{C_s C_p v_\infty \Delta T}$$

$$k_{bf} = k_\infty (1 + b_1 \theta)$$

$$\frac{k_{hnf}}{k_{bf}} = \frac{x_1 \theta + x_2}{x_3 \theta + x_4}$$

$$x_1 = 2\phi k_\infty b_1 - 2\phi^2 k_\infty b_1; \quad x_2 = 2\phi k_\infty + \psi_3 - 2\phi^2 k_\infty + 2\phi \psi_3$$

$$x_3 = 2\phi k_\infty b_1 + \phi^2 k_\infty b_1; \quad x_4 = 2\phi k_\infty + \phi^2 k_\infty + \psi_3 + \psi_3 \phi$$

Applying the congenial non-similar transformations to equations (2) to (4), the dimensionless form of the equations is as follows:

$$s\left(F^2 - 1\right) + \xi FF_\xi - \left(\frac{s+1}{2}\right)fF_\eta - \xi f_\xi F_\eta = -\left[\frac{K_1}{D_1}\xi\frac{(F-1)St}{\left(1 - \epsilon e^\xi\right)}\right]$$

$$+\left[\frac{1}{D_1 D_2}\left(1 + \frac{1}{\beta}\right)\frac{F_{\eta\eta}}{(1 + a_1\theta)}\right] - \left[\frac{1}{D_1 D_2}\left(1 + \frac{1}{\beta}\right)\frac{a_1\theta_\eta F_\eta}{(1 + a_1\theta)^2}\right]$$

$$\xi F\theta_\xi - \xi f_\xi\theta_\eta - \left(\frac{s+1}{2}\right)f\theta_\eta = \frac{1}{D_4}\left[\frac{b_1\theta_\eta^2 x_5}{\Pr}\right] + \frac{1}{D_4}\left[\frac{(1 + b_1\theta)\theta_{\eta\eta}x_5}{\Pr}\right] + \frac{R_1\theta_{\eta\eta}}{D_4\Pr}$$

$$+\frac{1}{D_4}\left[\frac{(1 + b_1\theta)\theta_{\eta\eta}^2 x_6}{\Pr}\right] + \frac{1}{D_4}\left[\frac{\theta\xi qRe}{1 - \epsilon e^\xi}\right] + \frac{1}{D_2 D_4}\left[\frac{F_\eta^2 Ec}{1 + a_1\theta}\right]$$

$$+\frac{K_1}{D_1}\left[F^2\xi\left(1 - \epsilon e^\xi\right)StEc\right] - \frac{K_1}{D_1}\left[F\xi\left(1 - \epsilon e^\xi\right)StEc\right]$$

$$+\frac{D_3}{D_1}\left[\xi\left(1 - \epsilon e^\xi\right)StEc\right] + \frac{\left(1 + \frac{1}{\beta}\right)F_\eta^2 Ec}{D_1 D_4(1 + a_1\theta)} + N_b\theta_\eta H_\eta + N_T\theta_\eta^2 + D_f\phi_{\eta\eta}$$

$$S_c\left[\xi f_\eta H_\xi - \left(\frac{s+1}{2}\right)fH_\eta - \xi f_\xi H_\eta\right] = H_{\eta\eta} + \left(\frac{N_t}{N_b}\right)\theta_{\eta\eta} + S_r\theta_{\eta\eta}$$

The dimensionless boundary conditions are:

when $\eta = 0$

$$F(\xi,\eta) = 0, \quad \theta(\xi,\eta) = 1 \text{ and } H(\xi,\eta) = 1$$

when $\eta \to \infty$

$$F(\xi,\eta) = 1, \quad \theta(\xi,\eta) \to 0 \text{ and } H(\xi,\eta) \to 0$$

The mathematical computations for the variable physical quantities of skin friction, Nusselt number, and Sherwood number for the current study are as follows: The skin friction coefficient is:

$$C_{fx} = \frac{\mu_{hnf}\left(\dfrac{\partial u}{\partial y}\right)_{y=0}}{\dfrac{1}{2}\rho_{bf}u_e^2} \qquad C_{fx} = 2F_\eta \, \mathrm{Re}_x^{-1/2}\,\frac{D_1}{D_2}$$

The Nusselt number is given by

$$N_{ux} = -\frac{x q_w}{k_f\left(T_w - T_\infty\right)} \quad \text{where } q_w = k_{hnf}\left(\frac{\partial T}{\partial y}\right)_{y=0} \qquad N_{ux} = -\theta_\eta \, \mathrm{Re}_x^{1/2}\,D_5$$

The Sherwood number is given by:

$$Sh_x = -\frac{x\left(\dfrac{\partial C}{\partial y}\right)_{y=0}}{\left(C_w - C_\infty\right)} \qquad Sh_x = -H_\eta \, \mathrm{Re}_x^{1/2}$$

4.3 SOLUTION METHODOLOGY

In the present study, the author employs the stable implicit finite difference scheme for the dimensionless partial differential equations in combination with the quasilinearization technique. This technique linearizes the coupled nonlinear PDEs. The linearized equations follow a system of equations when the central difference scheme is applied to the ξ direction and the backward difference in η direction. The iterated system of the block tri-diagonal structured matrix is solved by the method of Varga's algorithm. The linearized form of the nonlinear partial differential equations along with the boundary conditions is as follows:

$$L_1^i F_{\eta\eta}^{i+1} + L_2^i F_\eta^{i+1} + L_3^i F_{\xi i}^{i+1} + L_4^i F^{i+1} + L_5^i \theta_\eta^{i+1} + L_6^i F_\eta^{i+1} = L_7^i$$

$$M_1^i \theta_{\eta\eta}^{i+1} + M_2^i \theta_\eta^{i+1} + M_3^i \theta_\xi^{i+1} + M_4^i \theta^{i+1} + M_5^i F_\eta^{i+1} \\ + M_6^i F^{i+1} + M_7^i H_\eta^{i+1} + M_8^i H_{\eta\eta}^i + 1 = M_9^i$$

$$N_1^i H_{\eta\eta}^{i+1} + N_2^i H_\eta^{i+1} + N_3^i H_\xi^{i+1} + N_4^i \theta_{\eta\eta}^{i+1} + N_5^i F^{i+1} = N_6^i$$

where

$$L_1 = \frac{1}{N_1 N_2}\left(1 + \frac{1}{\beta}\right)\frac{1}{1 + a_1\theta}$$

$$L_2 = \frac{1}{N_1 N_2}\left(1 + \frac{1}{\beta}\right)\frac{a_1\theta_\eta}{(1 + a_1\theta)^2} + \xi F + \left(\frac{s+1}{2}\right)f$$

$$L_3 = -\xi F$$

$$L_4 = \xi F_\eta - \xi F_\xi - 2sF - \frac{K_1}{N_1} St\left(\frac{\xi}{1 - \epsilon e^\xi}\right)$$

$$L_5 = \frac{-1}{ss1}\left(\frac{1}{1 + \beta}\right)\frac{a_1 F_\eta}{(1 + a_1\theta)^2}$$

$$L_6 = \frac{-a_1 F_{\eta\eta}}{(1 + a_1\theta)^2}\frac{1}{ss1}\left(\frac{1}{1 + \beta}\right) + \frac{2a_1^2 F_\eta\theta_\eta}{(1 + a_1\theta)^3}\frac{1}{ss1}\left(\frac{1}{1 + \beta}\right)$$

$$L_7 = -sF^2 - \xi FF_\xi - \frac{1}{N_1 N_2}\left(1 + \frac{1}{\beta}\right)\frac{a_1 F_\eta\theta_\eta}{(1 + a_1\theta)^2} - \frac{1}{N_1 N_2}\left(1 + \frac{1}{\beta}\right) - \frac{a_1 F_{\eta\eta}\theta}{(1 + a_1\theta)^2}$$

$$+ \frac{1}{N_1 N_2}\left(1 + \frac{1}{\beta}\right)\frac{2a_1^2 F_\eta\theta\theta_\eta}{(1 + a_1\theta)^3} - s - \frac{K_1}{N_1}\frac{St\xi}{\left(1 - \epsilon e^\xi\right)}$$

$$M_1 = \frac{(1 + b_1\theta)x_5}{\Pr N_4} + \frac{R_1}{\Pr N_4}$$

$$M_2 = \xi f_\xi + \left(\frac{s+1}{2}\right)f + \frac{2b_1\theta_\eta x_5}{N_4 \Pr} + \frac{1 + b_1\theta x_5}{N_4 \Pr} + N_b H_\eta + 2N_t\theta_\eta$$

$$M_3 = -\xi F$$

$$M_4 = \frac{b_1 x_5\theta_{\eta\eta}}{N_4 \Pr} + \frac{b_1\theta_\eta^2 x_6}{N_4 \Pr} + \frac{\xi Q Re}{\left(1 - \epsilon e^\xi\right)N_4} - \frac{a_1 F_\eta^2 Ec}{(1 + a_1\theta)^2 N_2 N_4}$$

$$M_5 = \frac{2F_\eta Ec}{(1 + a_1\theta)N_2 N_4}$$

$$M_6 = \frac{2N_3 F\xi StEc\left(1 - \epsilon e^\xi\right)}{N_1} - \frac{2N_3\xi StEc\left(1 - \epsilon e^\xi\right)}{N_1} - \xi\theta_\xi$$

$$M_7 = N_b\theta_\eta$$

$$M_8 = Df$$

$$M_9 = \frac{-N_3}{N_1}\xi\left(1 - \epsilon e^\xi\right)StEc + \frac{N_3}{N_1}F^2\xi\left(1 - \epsilon e^\xi\right)StEc + \frac{F_\eta^2 Ec}{(1 + a_1\theta)N_2 N_4} + \frac{(1 + b_1\theta)\theta_\eta^2}{N_4 \Pr}$$

$$+ \frac{b_1\theta_\eta^2 x_5}{N_4 \Pr} - \frac{a_1\theta F_\eta^2 Ec}{(1 + a_1\theta)^2 N_2 N_4} + \frac{b_1\theta_\eta^2\theta x_6}{N_4 \Pr} + \frac{b_1 x_5\theta_{\eta\eta}\theta}{N_4 \Pr} - \xi F\theta_\xi + N_b H_\eta\theta_\eta + N_t\theta_\eta^2$$

$$N_1 = 1$$

$$N_2 = S_c\left(\frac{s+1}{2}\right)f + S_c\xi f_\xi$$

$$N_3 = -S_c\xi F$$

$$N_4 = \frac{N_t}{N_b}$$

$$N_5 = -S_c\xi H_\xi$$

$$N_6 = -S_c\xi FH_\xi$$

4.4 MATHEMATICAL COMPUTATION OF FINITE DIFFERENCE SCHEME

The quasi-linearized dimensionless PDEs are converted into a system of linear equations using the implicit finite difference scheme in which the central difference and backward difference operators are applied in ξ,η directions, respectively. The emerged matrix structured in a block tri-diagonal manner. As the step size of any experiment plays a major role, all the important results should be chosen wisely. The author chooses 0.05 and 0.01 as the step size for $\Delta\xi$ and $\Delta\eta$ in the ξ and η directions, respectively. When the relative difference between the two successive values lies below 0.0001, the iterative process is terminated. The numerical scheme employed to achieve the stable criteria of convergence is as follows:

$$\max\left|\left(F_\eta\right)_w^{i+1} - \left(F_\eta\right)_w^{i+1}\right|, \left|\left(G_\eta\right)_w^{i+1} - \left(G_\eta\right)_w^{i+1}\right|, |\left(H_\eta\right)_w^{i+1} - \left(H_\eta\right)_w^{i+1}| \leq 10^{-4}$$

$$\begin{bmatrix} F_{m,n+1} \\ \theta_{m,n+1} \\ H_{m,n+1} \end{bmatrix} \times \begin{bmatrix} a_{11} & a_{12} & a_{13} \\ a_{21} & a_{22} & a_{23} \\ a_{31} & a_{32} & a_{33} \end{bmatrix} + \begin{bmatrix} F_{m,n} \\ \theta_{m,n} \\ H_{m,n} \end{bmatrix} \times \begin{bmatrix} b_{11} & b_{12} & b_{13} \\ b_{21} & b_{22} & b_{23} \\ b_{31} & b_{32} & b_{33} \end{bmatrix} + \begin{bmatrix} F_{m,n-1} \\ \theta_{m,n-1} \\ H_{m,n-1} \end{bmatrix} \times \begin{bmatrix} c_{11} & c_{12} & c_{13} \\ c_{21} & c_{22} & c_{23} \\ c_{31} & c_{32} & c_{33} \end{bmatrix} = \begin{bmatrix} d_1 \\ d_2 \\ d_3 \end{bmatrix}$$

4.5 RESULT AND DISCUSSIONS

A steady laminar forced convective Casson hybrid nanofluid flows exposed to magnetized together with Joule heating effects and Soret and Dufour effects for varying radiative heat flux variations of velocity temperature, skin friction, and heat transfer coefficients are briefly discussed in the present model. Figure 4.1 demonstrates the physical model of the coordinate system. ξ,η are respectively the dimensionless variables along the direction of the fluid flow and in the normal direction. Copper and aluminum oxide are the hybrid nanoparticles suspended with the base Casson fluid ethylene glycol in order to enhance the thermal conductivity of the system. Figure 4.2

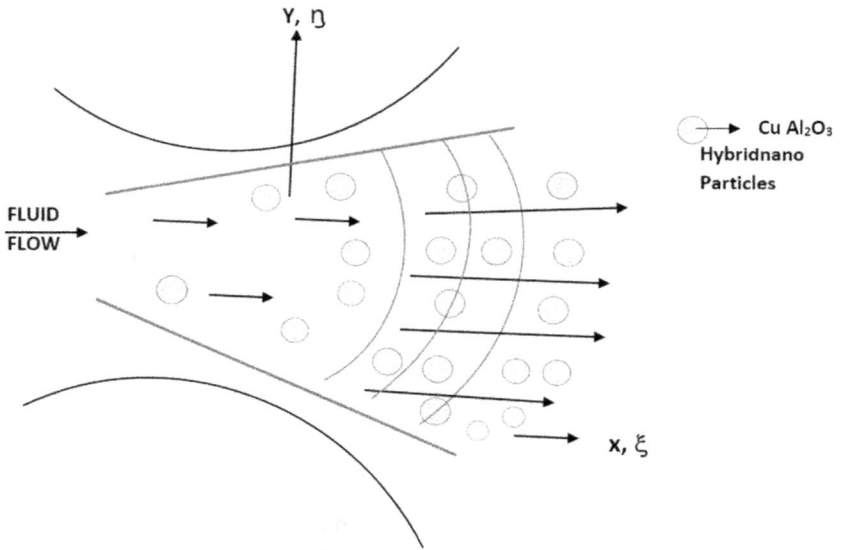

Figure 4.1 Physical geometry of the co-ordinate system.

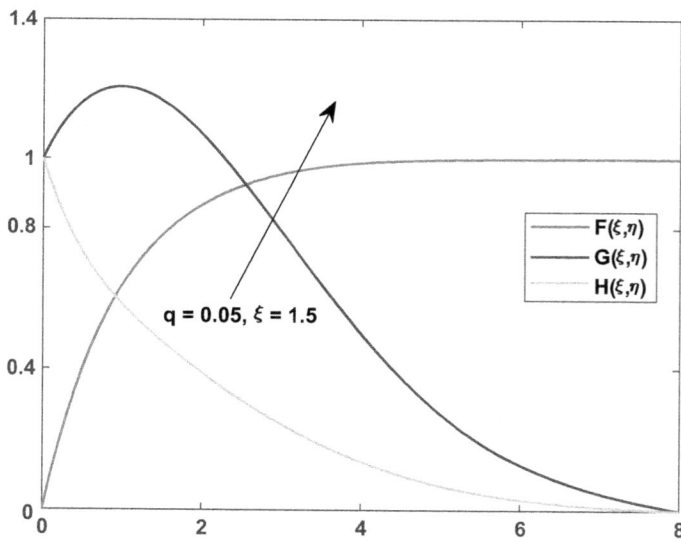

Figure 4.2 Soret and Dufour effects on velocity, temperature, and concentration profiles.

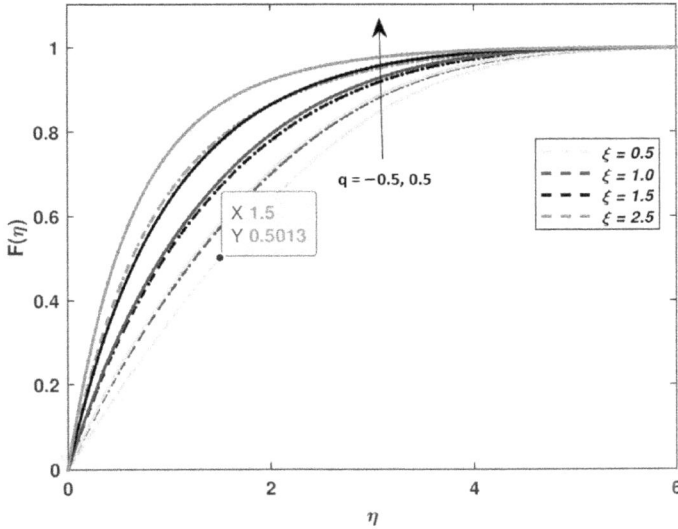

Figure 4.3 Impact of heat flux parameter on velocity profiles.

establishes the combined profiles of velocity, temperature, and concentration for a fixed radiative heat flux parameter. It is to be noted that the thermal profile is enhanced near the wall region. The velocity profile is enhanced slowly and reaches a steady state due to the resistive Lorentz force created by the magnetic parameter.

Figure 4.3 manifests the velocity variations for varying values of ξ in the range $0.5 \leq \xi \leq 2.5$ for positive and negative values of radiative heat flux parameter (q). Figure 4.4 portrays the impacts of temperature variations for $q = -0.5$ and 0.5 along with ξ variations. It is clearly picturized that lower values of heat flux reduce the thermal profile, and for positive values of q a slight uplift is observed. The physical implications suggest that the Soret or thermal diffusion effect physically defines the variations of temperature gradient and causes mass flux and thermal diffusion effect, or the Dufour effect creates an energy flux due to variations of concentration gradients. In particular, Soret and Dufour effects help to enhance the heat flux and mass flux of the hybrid nanoparticles. Figure 4.5 reveals the variations of concentration gradients due to radiative heat flux parameter together with ξ variations.

Figures 4.6–4.10 demonstrate the surface plots of skin friction coefficient and heat transfer coefficient for varying values of radiative heat flux parameter and dimensionless Soret number and Dufour number. Figure 4.6 illustrates the Soret and Dufour effects on heat transfer coefficients. The profile is enhanced for increasing values of the radiative heat flux parameter. The impacts of Soret and Dufour effects on skin friction profiles are observed in

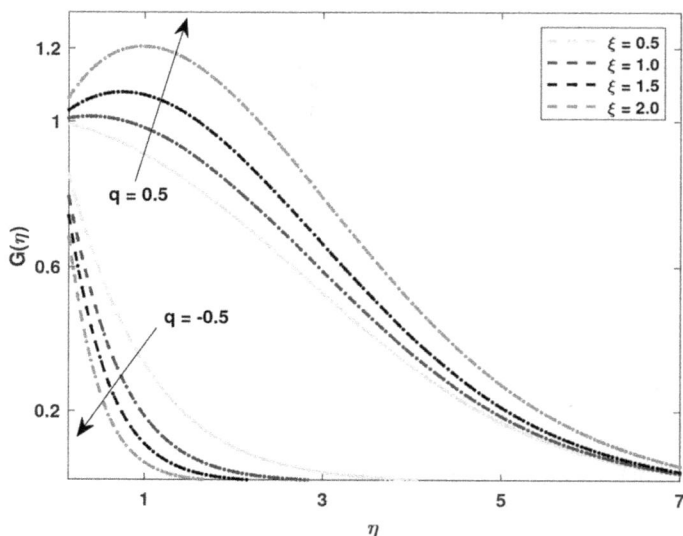

Figure 4.4 Impact of heat flux parameter on temperature profiles.

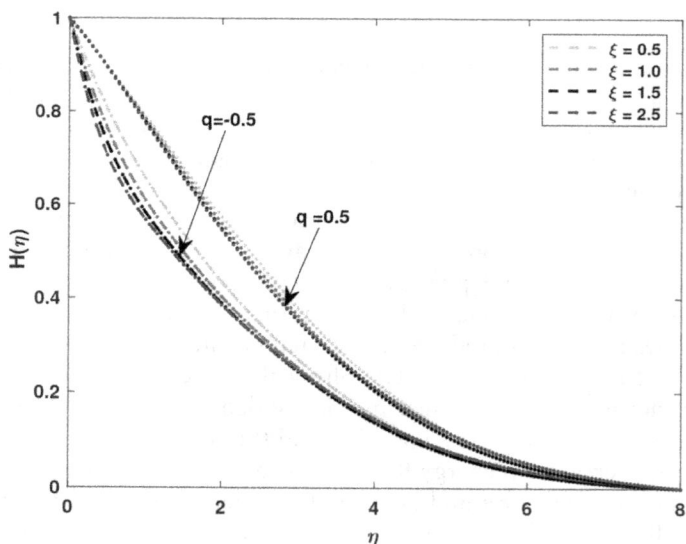

Figure 4.5 Impact of heat flux parameter on concentration profiles.

Figures 4.7 and 4.8. The profile is declined for increasing values of q. The physical interpretation suggests that the flow increases the shear stress rates due to Casson parameter β. Figures 4.9 and 4.10 evidence the heat transfer profiles for varying values of non-dimensional Soret number and radiative heat flux parameter (q). This is because the Soret number increases the temperature gradients at the surface.

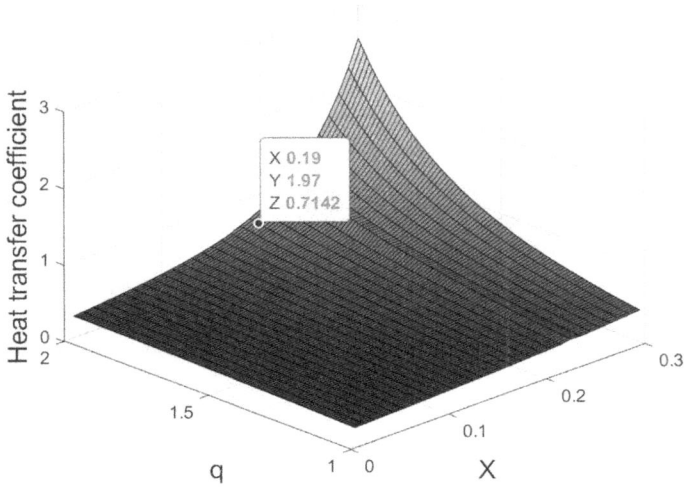

Figure 4.6 Impact of Soret and Dufour effects on mass transfer coefficient profiles.

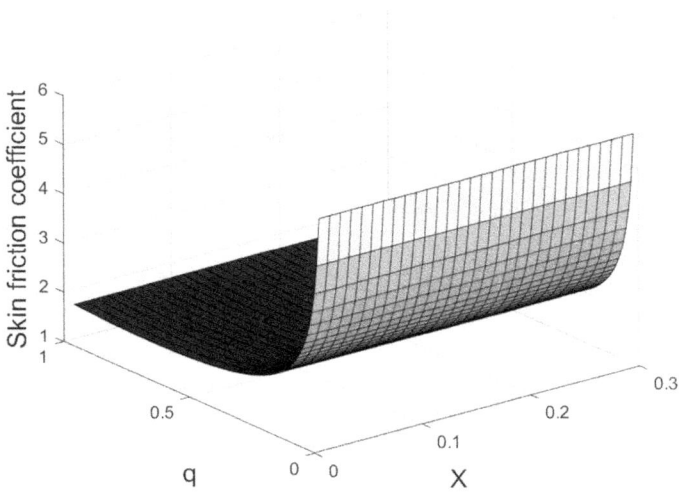

Figure 4.7 Impact of Soret and Dufour effects on skin friction coefficient profiles.

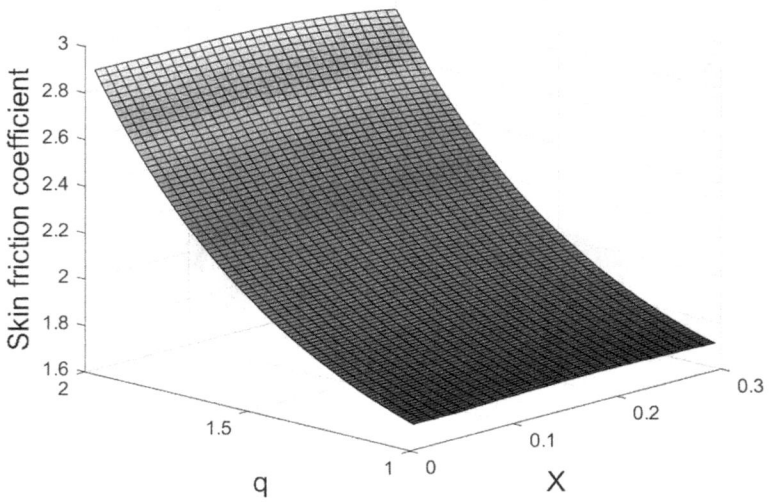

Figure 4.8 Impact of heat flux parameter on skin friction coefficient profiles.

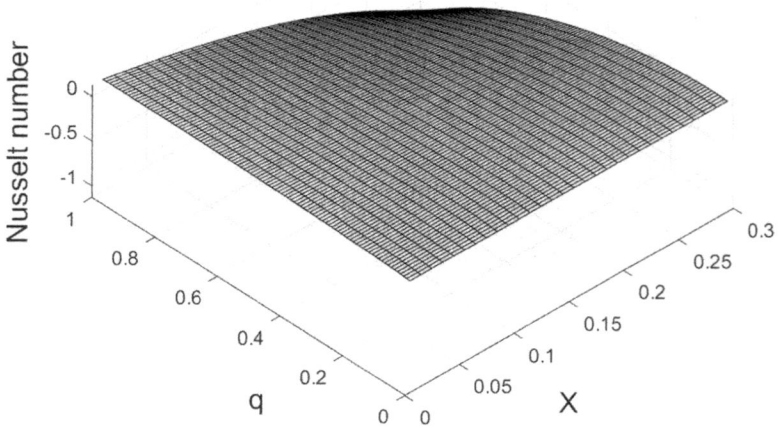

Figure 4.9 Impact of Soret and Dufour effects on heat transfer coefficient profiles.

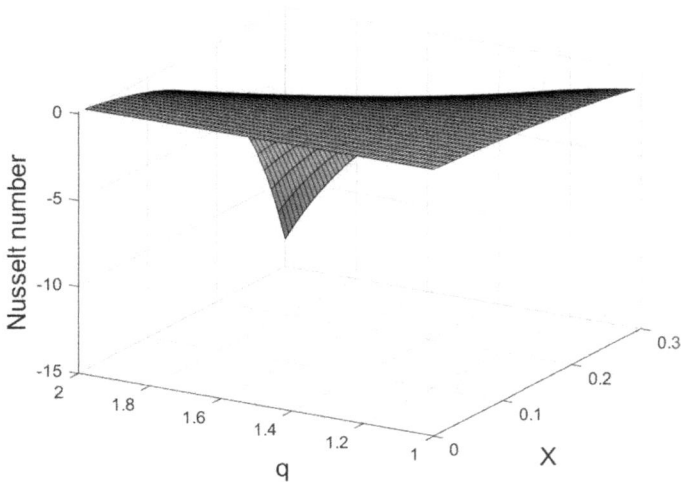

Figure 4.10 Impact of heat flux parameter on heat transfer coefficient profiles.

4.6 CONCLUSIONS

- Thus the numerical study of heat flux and mass flux generated due to Soret and Dufour effects on MHD forced convective Casson hybrid nanofluid flows over diverging channel is discussed in the present model.
- Soret number helps to enhance the thermal boundary layer, and Dufour number helps to increase the concentration boundary layer.
- The physical influences of viscous dissipation parameter (Ec), Brownian motion (Nb), and diffusive parameter (Sc) of $Cu-Al_2O_3$ hybrid nanoparticles suspended with non-Newtonian Casson base fluid for velocity, temperature, concentration, skin friction, Nusselt number, entropy production rates, and Bejan lines are discussed elaborately in this chapter.
- The upsurged values of the velocity profile and the diminished values of thermal profiles are due to the presence of an applied magnetic field which creates a resistive Lorentz force.

APPENDIX

$$a_{11} = L_1 - \frac{L_2 \Delta \eta}{2}$$

$$a_{12} = -\frac{L_5 \Delta \eta}{2}$$

$$a_{13} \quad = \quad 0$$

$$a_{21} \quad = \quad -\frac{M_5 \Delta\eta}{2}$$

$$a_{22} \quad = \quad M_1 - \frac{M_2 \Delta\eta}{2}$$

$$a_{23} \quad = \quad -\frac{M_7 \Delta\eta}{2}$$

$$a_{31} \quad = \quad 0$$

$$a_{32} \quad = \quad N_4$$

$$a_{33} \quad = \quad N_1 - \frac{N_2 \Delta\eta}{2}$$

$$b_{11} \quad = \quad -2L_1 + \frac{L_3 \Delta\eta^2}{\Delta\xi} + L_4 \Delta\eta^2$$

$$b_{12} \quad = \quad L_6 \Delta\eta^2$$

$$b_{13} \quad = \quad 0$$

$$b_{21} \quad = \quad M_6 \Delta\eta^2$$

$$b_{22} \quad = \quad -2M_1 + \frac{M_3 \Delta\eta^2}{\Delta\xi} + M_4 \Delta\eta^2$$

$$b_{23} \quad = \quad 0$$

$$b_{31} \quad = \quad N_5 \Delta\eta^2$$

$$b_{32} \quad = \quad -2N_4$$

$$b_{33} \quad = \quad -2N_1 + \frac{N_3 \Delta\eta^2}{\Delta\xi}$$

$$c_{11} \quad = \quad L_1 + \frac{L_2 \Delta\eta}{2}$$

$$c_{12} \quad = \quad \frac{L_5 \Delta\eta}{2}$$

$$c_{13} \quad = \quad 0$$

$$c_{21} \quad = \quad \frac{M_5 \Delta\eta}{2}$$

$$c_{22} \quad = \quad M_1 + \frac{M_2 \Delta\eta}{2}$$

$$c_{23} \quad = \quad \frac{M_7 \Delta\eta}{2}$$

$$c_{31} \quad = \quad 0$$

$$c_{32} \quad = \quad N_4$$

$$c_{33} = N_1 + \frac{N_2 \Delta \eta}{2}$$

$$d_1 = L_7 \Delta \eta^2 + \frac{L_3 \Delta \eta^2}{\Delta \xi} F_{m-1,n}$$

$$d_2 = M_8 \Delta \eta^2 + \frac{M_3 \Delta \eta^2}{\Delta \xi} \theta_{m-1,n}$$

$$d_3 = N_6 \Delta \eta^2 + \frac{N_3 \Delta \eta^2}{\Delta \xi} H_{m-1,n}$$

NOMENCLATURE

β	Casson fluid parameter
b_1	Thermal conductivity parameter
B_0	External magnetic field in the normal direction (Tesla)
C	Species concentration (mol/m^3)
C_w	Concentration at the wall (mol/m^3)
C_∞	Ambient species concentration (mol/m^3)
C_{fx}	Skin friction coefficient
D_B	Brownian diffusion coefficient
D_T	thermophoretic diffusion coefficient
Ec	Eckert number
F	dimensionless velocity (m/s)
f	dimensionless stream function
g	acceleration due to gravity (m/s^2)
H	dimensionless concentration
k_∞	Ambient thermal conductivity parameter
$K*$	Rosseland mean absorption coefficient
L	Length of the channel (m)
MHD	Magnetohydrodynamics
Nb	Brownian motion parameter
Nt	thermophoresis parameter
N_{ux}	Nusselt number
Pr	Prandtl number
q_r	radiation heat flux
Q_0	Chemical reaction parameter
Re_x	Reynolds number
Rd	Thermal radiation flux
R_1	Richardson's number
St	Stuart number
Sh_x	Sherwood number
Sc	Schmidt number
T	dimensionless temperature of the nanofluid (K)
T_w	uniform channel temperature

T_∞	Ambient species temperature
u	velocity component in the x direction (m/s)
u_e	velocity component at the edge of the boundary layer
U_w	moving channel velocity
U_∞	free stream velocity
U	reference velocity
V	velocity component in the y direction

GREEK SYMBOLS

ψ	dimensionless stream function
θ	dimensionless temperature function
τ	ratio between heat capacity of nanoparticle and base fluid
σ_{hnf}	electrical conductivity of hybrid nanofluid
k_{hnf}	Thermal conductivity of hybrid nanofluid
μ_{hnf}	dynamic viscosity of hybrid nanofluid (Nsm^{-2})
ν	kinematic viscosity (m^{2}s^{-1})
ρ_{hnf}	density of hybrid nanofluid (kgm^{-3})
$(\rho_c)_{bf}$	specific heat capacity of the base fluid (JK^{-1})
$(\rho c)hnf$	specific heat capacity of the hybrid nanofluid (JK^{-1})
$\sigma*$	Stefan-Boltzmann constant
ϵ	the ratio between free stream velocity and the reference velocity
ξ,η	Transformed variables

REFERENCES

[1] Omid Mahian, Ali Kianifar, Clement Klein Streuer, Mohd A. Al-Nimr, Ioan Pop, Ahmet Z. Sahin, Somchi Wongwises et al. A reivew of entropy generation in nanofluid flow. *Int Heat Mass Transf*. 2013;6:5514–5532.

[2] Adewoye Raphel Adebisi et al. An Analysis of Boundary layer flow over a flat plate using Modified Adomain Decomposition method. *Sch J Eng Tech*. 2015;3(4c):520528.

[3] Tapas Barman, S. Roy, Ali J. Chamkha et al. Analysis of entropy production in a bi-convective magnetized and radiative hybrid nanofluid flow using temperature sensitive base fluid (water) properties. *Sci Rep*. 2022;12:11831.

[4] Tapas Barman, S. Roy, Ali J. Chamkha et al. A bi convective magnetized Hybrid nanofluid flow along with thermal radiation in an adverse pressure field using Temperature sensitive base fluid (water) properties. *J Nanofluids*. 2022;11:142–153.

[5] Tapas Barman, S. Roy, Ali J. Chamkha et al. Magnetized Bi convective nanofluid flow and entropy production using temperature sensitive base fluid properties: A unique approach. *J Appl Comput Mech*. 2022;8(4):1163–1175.

[6] Tapas Barman, S. Roy, Ali J. Chamkha et al. *The role of non-erratic slot-mass disposal in a hybrid nanofluid flow due to source/sink and radiation*. Waves Random Complex Media. 2022;

[7] Tapas Barman, S. Roy, Ali J. Chamkha et al. Entropy generation analysis of MHD hybrid nano fluid flow due to radiation with non-erratic slot wise mass transfer over a rotating sphere. *Alexandria Eng J*. 2023; 67:271–286.

[8] Ankita Bist and Rajesh Sharma et al. Non-similar solution of Casson nanofluid flow with variable viscosity and variable thermal conductivity. *Int J Numer Method H*. 2020;30(8):3919–3938.

[9] Roy, S and Saikrishnan, P. Non-uniform slot injection (suction) into steady laminar boundary layer flow over a rotating sphere. *Int. J. Heat Mass Transf*. 2003;46:338996.

[10] P.M. Patil, S. Roy and E. Momoniat. Thermal diffusion and diffusion-thermo effects on mixed convection from an exponentially impermeable stretching surface. *Int. J. Heat Mass Transf*. 2016;100:482–489.

Chapter 5

Boundary layer flow in aerospace applications

Mrityunjai Verma and Varun Pratap Singh

School of Advanced Engineering, UPES, Dehradun, India

5.1 INTRODUCTION

The performance and efficiency of aerospace systems are heavily influenced by boundary layer flow, making it a crucial issue in aerospace engineering. In this chapter, we will delve into boundary layer flow's theoretical underpinnings, mathematical models, and aerospace-specific applications. Aerodynamics and heat transmission in aircraft parts like wings and aerofoils can be predicted with a deeper understanding of boundary layers. Parameters such as Reynolds number and surface roughness are discussed in relation to the different forms of boundary layer flows and their impacts on aerodynamics and heat transfer. In addition, this chapter covers computational techniques and design solutions for improving boundary layer flow, providing useful information to academics and engineers. The overall goal of this chapter is to improve the effectiveness, safety, and performance of future aircraft systems through contributing to the development of aerospace technology. For aerospace engineers looking to optimize aircraft design, fuel efficiency, and emissions reduction by careful management of boundary layer flow across a wide range of operating conditions, this book is indispensable [1].

In order to perform efficient and cost-effective boundary layer flow analysis, this chapter provides readers with computational methods and design ideas. These resources can be used by researchers, engineers, and students in the fields of computational fluid flow and heat transfer to improve aeronautical technology. In conclusion, the information presented in this chapter improves our knowledge of boundary layer flow and provides useful recommendations for implementing it in aircraft. It benefits those working in aerospace research, design, and engineering by providing new avenues for enhancing efficiency, safety, and performance in aircraft systems (Figure 5.1).

DOI: 10.1201/9781003465171-5

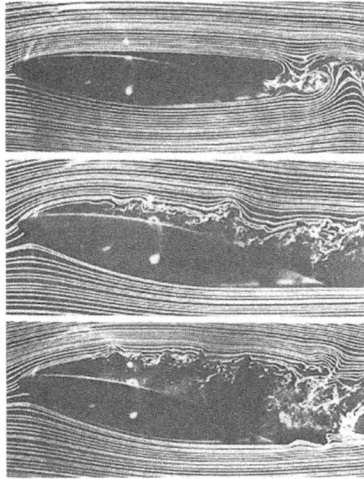

Figure 5.1 An illustration of boundary layer formation and flow.

5.2 FUNDAMENTALS OF BOUNDARY LAYER FLOW

When a fluid, like air, travels over a solid surface, like an airplane wing, a critical phenomenon known as boundary layer flow takes place. Here, we will go into the basics of boundary layer flow and discuss why it is so important in aeronautical settings. The major goal is to comprehend the boundary layer's behaviour, its growth, and the effects it has on the aerodynamics and heat transfer properties of aeronautical components (Figure 5.2).

In the boundary layer, an extremely thin region immediately adjacent to the solid surface, the fluid velocity transitions from zero at the surface to the freestream velocity of the approaching flow. A gradual transition in velocity from the surface to the outer flow is brought about by viscosity forces in the boundary layer. This area is crucial to the aerodynamic performance of aerospace systems. Both laminar and turbulent growth of the boundary layer is possible. The boundary layer starts thin and flows smoothly. However, flow

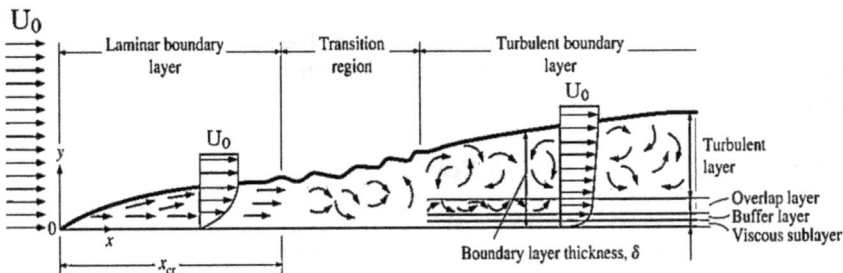

Figure 5.2 A depiction of the various stages of the development of flow in a fluid.

perturbations can cause a turbulent boundary layer as the surface distance increases. The boundary layer's aerodynamic and heat transmission qualities change as laminar flow becomes turbulent, affecting aircraft and other aerospace components. Researchers utilize the thickness of the boundary layer (represented by "δ.") to explain its behaviour. It depicts the distance from the surface where fluid velocity is 99% of freestream velocity. Aircraft drag and heat transfer depend on boundary layer thickness [2].

5.2.1 Concept of boundary layer

The boundary layer concept is fundamental in aerospace applications, representing the thin layer of fluid that forms next to a solid surface when fluid is in motion. This layer adheres to the surface due to the no-slip condition, transitioning from stationary to freestream velocity as we move away from the surface. It effectively separates the surface-adjacent fluid from the unaffected freestream flow and exists in two types: laminar with smooth streamlines and turbulent with irregular fluctuations. Understanding the boundary layer is crucial as it directly influences an aircraft's drag, impacting fuel consumption and efficiency. By giving researchers and engineers a thorough understanding of boundary layer flow principles, such as classification, boundary layer thickness, and laminar versus turbulent flows, this section will enable them to maximize aerodynamic performance in aerospace design [3].

5.2.2 Boundary layer thickness

The significance of boundary layer thickness (δ) in aeronautical applications is discussed in this section. Thickness of the boundary layer affects heat transmission and the aerodynamics of aerospace components. At this distance from the solid surface, the fluid's velocity reaches 99% of the freestream velocity. Aircraft and aerospace vehicle drag depends on boundary layer thickness. Reynolds number (Re) and surface roughness affect boundary layer thickness along the component's surface. At lower Reynolds numbers, the boundary layer remains laminar and thin, reducing drag. At higher Reynolds numbers, the boundary layer is more prone to become turbulent, increasing drag. To optimize aircraft systems, researchers and engineers must analyse and anticipate boundary layer thickness. Simulating and forecasting boundary layer behaviour over complex geometries requires CFD. Accurate boundary layer thickness prediction enables effective design alterations to reduce drag, increase lift, and improve aircraft performance [4] (Figure 5.3).

5.2.3 Velocity profiles and flow characteristics

This section discusses boundary layer velocity profiles and flow. Fluid velocity changes from zero at the surface to freestream velocity as it moves along the surface. Changing velocity distribution is captured by the velocity profile. Laminar boundary layers have a smooth velocity profile with a maximum at the surface and a decreasing velocity towards the outer flow.

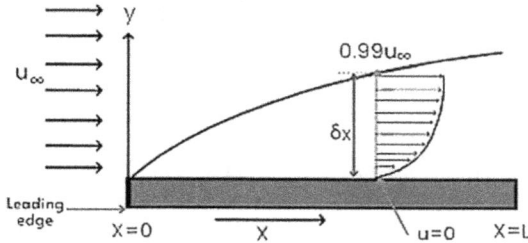

Figure 5.3 An explanatory diagram for the thickness of boundary layer.

In turbulent boundary layers, fluctuations and eddies complicate the velocity profile and flow properties. Turbulent boundary layers feature a steeper velocity gradient towards the surface, increasing mixing, momentum, and heat transfer. Assessing aeronautical component drag, lift, and performance requires understanding velocity profiles and flow parameters. It also helps anticipate skin friction, pressure distribution, and surface heat transfer. This understanding can help engineers optimize aircraft wings, aerofoils, and other aerospace structures for aerodynamic efficiency and stability [5].

5.2.4 Boundary layer separation

Boundary layer separation is important in aerospace. When flow curvature or pressure gradient is sufficiently high, the boundary layer detaches from the surface. A wake zone with low pressure and flow reversal arises after separation. When boundary layers separate, aircraft performance suffers, including higher drag, lift, and stability. Complex flow patterns and high angles of attack make it difficult to control and anticipate [6]. Researchers and engineers use boundary layer control, new airfoil designs, and wing platform optimization to reduce boundary layer separation. Aerospace systems can improve aerodynamic performance, fuel efficiency, and flight safety by regulating boundary layer separation [7] (Figure 5.4).

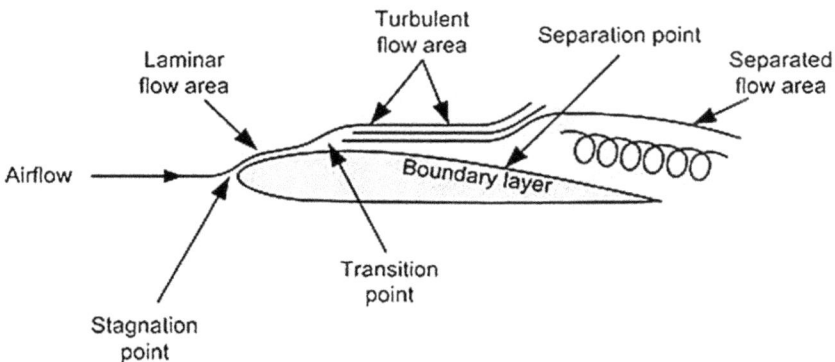

Figure 5.4 Boundary layer separation.

5.3 MATHEMATICAL MODELLING OF BOUNDARY LAYER FLOW

In aerospace engineering, mathematical modelling of boundary layers is crucial for understanding and analysing fluid flow near solid surfaces. The boundary layer is the thin fluid layer that forms next to a solid mass in air or water. The no-slip situation at the solid surface gradually gives way to freestream flow far away in this stratum. Boundary layer mathematical modelling helps engineers anticipate and optimize the aerodynamic performance of aerospace components including wings, fuselages, and turbine blades [8].

Navier-Stokes equations are used to develop boundary layer modelling equations. Due to the complexity of the full Navier-Stokes equations, simplified versions are typically used for computing efficiency and practicality. The boundary layer theory, which states that viscous forces dominate the flow inside the boundary layer but not the inviscid flow outside, is frequently employed. This enables boundary layer equations that explain velocity and pressure profiles. Many mathematical approaches are used to solve boundary layer equations, from analytical for basic geometries to numerical for complicated ones. Due to the development and implementation of accurate numerical solvers, engineers can now simulate and analyse intricate flow patterns with high precision. Computational fluid dynamics (CFD) has also expanded boundary layer modelling, allowing engineers to explore different flow scenarios under different operating conditions [9] (Figure 5.5).

Boundary layer mathematical modelling has a major significance for aircraft engineering. Understanding boundary layer behaviour helps develop and optimize aerodynamic designs for drag reduction and fuel economy. Predicting flow separation and stall, which affect aeroplane stability and performance, requires exact boundary layer properties. Studying boundary layers optimizes turbine blade and nozzle aerodynamics, which helps build sophisticated propulsion systems like turbofans and supersonic engines [10].

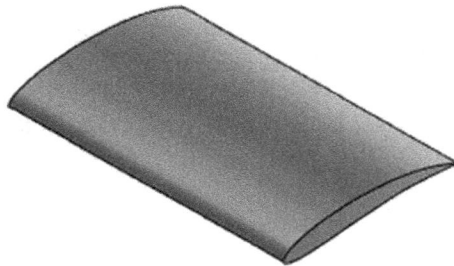

Figure 5.5 NACA 6612 airfoil developed via CAD.

5.3.1 Navier-Stokes equations and simplifications

Navier-Stokes equations are a set of partial differential equations that describe the behaviour of fluid flow. The 19th-century fluid mechanics pioneers George Gabriel Stokes and Claude-Louis Navier are honoured in their names. These formulas are essential for understanding fluid motion and are applied widely in many disciplines, including meteorology, mechanical engineering, and aerospace engineering [11].

The Navier-Stokes equations are commonly represented in the vector form, and they can be defined as follows in three dimensions (x, y, and z):

Continuity equation (Conservation of Mass):

$$\partial\rho / \partial t + \nabla \cdot (\rho v) = 0$$

Momentum equation (Newton's second law):

$$\rho \, \partial v / \partial t + \rho (v \cdot \nabla) v = -\nabla P + \mu \nabla^2 v + \rho g$$

Energy equation (first law of thermodynamics):

$$\partial (\rho E) / \partial t + \nabla \cdot (\rho E v) = \nabla \cdot (k \nabla T) + \mu (v \cdot \nabla)^2 + Q$$

Here, ρ represents the fluid density, t is the time, v is the velocity vector of the fluid, P is the pressure, μ is the dynamic viscosity of the fluid, g is the acceleration due to gravity, E is the specific total energy (sum of internal energy and kinetic energy per unit mass), k is the thermal conductivity, T is the fluid temperature, and Q represents any additional heat sources or sinks acting on the fluid.

The continuity equation guarantees mass conservation by equating the change in mass within a fluid volume to the net mass flow in or out of that volume. Meanwhile, the momentum equation applies Newton's second law to fluid motion, considering fluid particle acceleration due to factors like pressure, viscosity, and external forces such as gravity. These equations are fundamental in fluid dynamics, underpinning our understanding of mass and motion within fluids. The term $\rho(v \cdot \nabla)v$ is the convective acceleration, which represents the acceleration of a fluid parcel as it moves through a flow field. The energy equation is derived from the first law of thermodynamics. It relates the rate of change of total energy of the fluid to various energy transfers, including work done by pressure and viscous forces, heat transfer, and any external sources or sinks of energy [12].

5.3.1.1 Simplifications of the Navier-Stokes equations

The full Navier-Stokes equations, which describe fluid motion, are often exceedingly complex and difficult to solve analytically or computationally for many real-world flow problems [13]. As a result, simplifications and assumptions are frequently employed to derive more manageable forms of these equations. Common simplifications encompass:

- **Incompressible Flow**: When dealing with flows where density variations are negligible (such as most low-speed fluid flows), the continuity equation can be simplified by assuming $\nabla \cdot \mathbf{v} = 0$. This results in incompressible Navier-Stokes equations.
- **Steady-State Flow**: For problems where the flow parameters do not change with time, the time derivatives in the Navier-Stokes equations ($\partial/\partial t$ **terms**) become zero, simplifying the equations.
- **Laminar Flow**: In certain cases, the effects of turbulence can be neglected, simplifying the momentum and energy equations, and leading to the laminar flow solutions.
- **Simplified Boundary Conditions**: Applying appropriate boundary conditions based on the specific problem and geometry can simplify the solution process.

These simplifications make the Navier-Stokes equations easier to solve analytically or numerically, helping engineers and scientists understand fluid behaviour and optimize designs for aerospace engineering. However, these simplifications may not adequately reflect all flow phenomena, especially in high-speed or turbulent flows, which require more advanced turbulence models or CFD simulations. The boundary layer equations are simplified Navier-Stokes equations for fluid flow near a solid surface. This is the boundary layer. By assuming specific flow characteristics within the boundary layer, the boundary layer equations can be used to analyse fluid flow near solid objects like aircraft wings or turbine blades [14].

5.3.2 Boundary layer equations

Based on the previously discussed assumptions, the boundary layer equations are derived by simplifying the Navier-Stokes equations. For a two-dimensional flow (x and y directions), the boundary layer equations are typically expressed in terms of the velocity components u and v in the x and y directions, respectively, and the pressure P [15]. The momentum equations for the boundary layer can be written as follows:

x-Direction (u-velocity):

$$u \partial u / \partial x + v \partial u / \partial y = -1 / \rho \, \partial P / \partial x + v(\partial^2 u / \partial y^2$$

y-Direction (v-velocity):

$$u \partial v / \partial x + v \partial v / \partial y = -1 / \rho \, \partial P / \partial y + v \left(\partial^2 v / \partial y^2 \right)$$

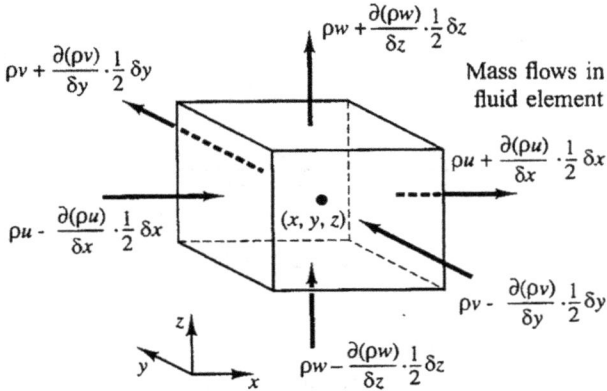

$$\rho w + \frac{\partial(\rho w)}{\delta z} \cdot \frac{1}{2} \delta z$$

$$\rho v + \frac{\partial(\rho v)}{\delta y} \cdot \frac{1}{2} \delta y$$

Mass flows in fluid element

$$\rho u + \frac{\partial(\rho u)}{\delta x} \cdot \frac{1}{2} \delta x$$

$$\rho u - \frac{\partial(\rho u)}{\delta x} \cdot \frac{1}{2} \delta x$$

(x, y, z)

$$\rho v - \frac{\partial(\rho v)}{\delta y} \cdot \frac{1}{2} \delta y$$

$$\rho w - \frac{\partial(\rho w)}{\delta z} \cdot \frac{1}{2} \delta z$$

Figure 5.6 Visualization of Navier-Stokes equations.

Here, u and v are the velocity components in the x and y directions, respectively, P is the pressure, ρ is the fluid density, ν is the kinematic viscosity (μ/ρ, where μ is the dynamic viscosity) (Figure 5.6).

The boundary layer equations elucidate changes in velocity components within the boundary layer due to viscous effects and pressure gradients. Solving these equations offers valuable insights into flow behaviour near solid surfaces, including velocity profiles, boundary layer thickness, and skin friction (shear stress) at the wall. Understanding aerodynamic properties in engineering applications such as drag, lift, and heat transfer requires knowledge of this material. It is important to recognize, nevertheless, that these equations are valid only in the boundary layer region, and they might not adequately explain the flow at distances from the surface. Potential flow theory and boundary layer equations are frequently used in conjunction to address this. For complex or high-speed flows, advanced computational techniques like computational fluid dynamics (CFD) are necessary for precise predictions [16].

5.3.3 Boundary conditions

In aerospace engineering, boundary layers are of paramount importance in shaping the aerodynamic characteristics of aircraft and aerospace vehicles. To accurately predict and optimize parameters such as aerodynamic drag, lift, and heat transfer, which directly impact the efficiency and performance of aerospace systems, a thorough understanding of the governing boundary conditions is indispensable. These boundary conditions pertain to both the solid surface (wall) and the far-field (freestream) and exert significant influence on the development and characteristics of the boundary layer. Key boundary conditions encompass the fundamental "no-slip condition" at the solid surface, which assumes zero fluid velocity relative to the surface.

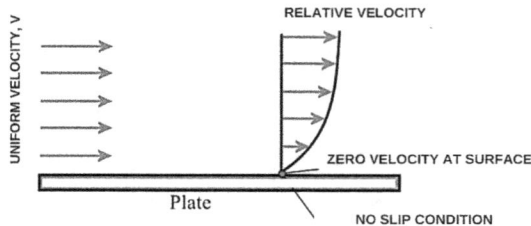

Figure 5.7 Illustrative depiction of no slip condition.

Adiabatic wall conditions often assume a constant temperature at the solid surface, signifying no heat transfer to or from the fluid. Freestream conditions, representing uniform fluid velocity and temperature far from the surface, serve as input for boundary layer equations. Additionally, considerations such as the hydrostatic pressure gradient, relevant for flows over curved or inclined surfaces, and turbulent intensity in the freestream are vital boundary conditions, collectively forming the foundation for accurate modelling and optimization in aerospace engineering applications [17] (Figure 5.7).

Solving the boundary layer equations and obtaining the velocity and temperature profiles within the boundary layer require these boundary conditions. The skin friction (shear stress) and boundary layer thickness (all three) can be predicted with more precision with their help. Another crucial feature of boundary layer flow in aerospace engineering is the way in which the boundary conditions affect the onset of turbulent flow within the boundary layer. Boundary conditions in engineering practice need to be determined using a mix of theory, experiments, and computer simulations. By studying the effects of various boundary conditions on boundary layer growth, aerospace engineers may better optimize aircraft designs, reduce aerodynamic drag, and improve aircraft performance [18].

5.3.4 Numerical methods for solving boundary layer equations

Numerical methods are indispensable in aerospace engineering for solving the complex boundary layer equations encountered in various flow scenarios. Analytical solutions are often impractical due to the equations' complexity. Numerical techniques offer efficient and accurate ways to approximate solutions, enabling engineers to predict boundary layer characteristics and optimize aerospace designs. Common numerical methods include the finite difference method (FDM), which discretizes equations into a grid and iteratively solves them; the finite volume method (FVM), which applies conservation principles within control volumes; the finite element method (FEM), suitable for complex geometries; the shooting method, useful for specified

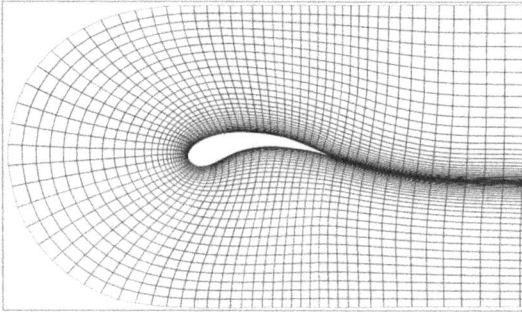

Figure 5.8 Meshed two-dimensional airfoil profile.

far-field conditions; and relaxation methods, like Gauss-Seidel and SOR, for solving discretized equations iteratively. High-order methods, such as spectral methods, enhance accuracy. In summary, numerical methods are essential for tackling complex aerospace flow problems, thus facilitating insight into boundary layer behaviour and optimizing aerospace systems [19] (Figure 5.8).

A meticulous exploration of numerical analysis for solving boundary layer equations in aerospace engineering involves a multifaceted approach. First and foremost, grid generation sets the stage, requiring the creation of a suitable grid or mesh to discretize the computational domain, particularly crucial in the realm of complex aerospace geometries. The choice of a numerical scheme looms large, with methods like finite difference or finite element playing pivotal roles in achieving accuracy, efficiency, and stability, especially in scenarios marked by steep velocity gradients. Precise representation of boundary conditions, including the no-slip and adiabatic wall conditions, is paramount to ensure the realism of the modelling process. Proper initialization and the establishment of stringent convergence criteria lay the foundation for solution stability and consistency. Turbulence modelling becomes essential in many aerospace scenarios, demanding the selection of an apt turbulence model for the accurate prediction of turbulence effects. Validation against experimental data and verification with known analytical solutions become touchstones for method reliability. Adequate computational resources, such as high-performance clusters, become prerequisites for handling resource-intensive aerospace simulations. Sensitivity analysis, probing the impact of various parameters, guides optimization endeavours, while effective interpretation and visualization tools are indispensable for unravelling complex flow patterns and boundary layer behaviour, aiding informed design decisions. Through this comprehensive approach, aerospace engineers gain the tools and insights needed to optimize designs, thereby enhancing overall performance and efficiency, underpinned by a profound understanding of boundary layer phenomena [20].

5.4 AERODYNAMIC CHARACTERISTICS OF BOUNDARY LAYER FLOW

In the realm of aerospace engineering, boundary layer flow emerges as a pivotal influencer, significantly shaping the aerodynamic characteristics of aircraft and other aerospace vehicles. This impact extends to fundamental parameters such as lift, drag, and heat transfer, rendering it an indispensable consideration in the design and optimization of aerospace components. The profound understanding of the aerodynamic characteristics within the boundary layer holds the key to enhancing aircraft performance, fuel efficiency, and overall flight safety. Among the crucial facets are boundary layer thickness, which delineates the transition from no-slip conditions to freestream flow and affects pressure distribution; boundary layer separation, pivotal for optimal lift and stall avoidance; laminar-turbulent transition, impacting skin friction drag; boundary layer drag, with skin friction drag as a dominant component; boundary layer control techniques that delay separation and boost lift; management of heat transfer in high-speed applications; handling of vortex shedding and buffeting; and the mitigation of boundary layer-induced aerodynamic noise. These factors collectively underscore the intricate dance between the boundary layer flow and aerospace performance, guiding engineers in the pursuit of superior aircraft efficiency, manoeuvrability, and safety [21] (Figure 5.9).

Boundary layer thickness, separation, transition, drag, control, heat transfer, vortex shedding, and noise are among the key factors that aerospace

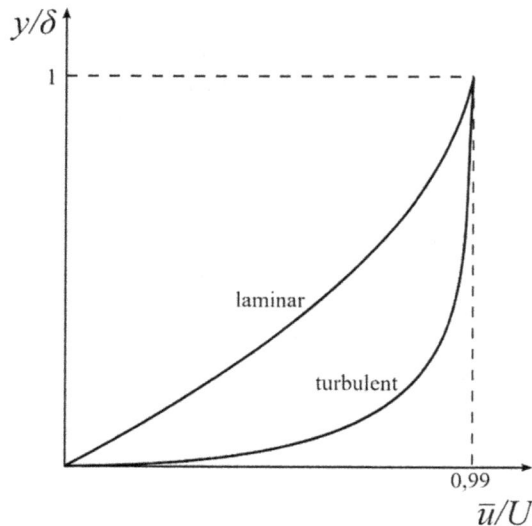

Figure 5.9 Graphical representation of the difference between laminar and turbulent flows.

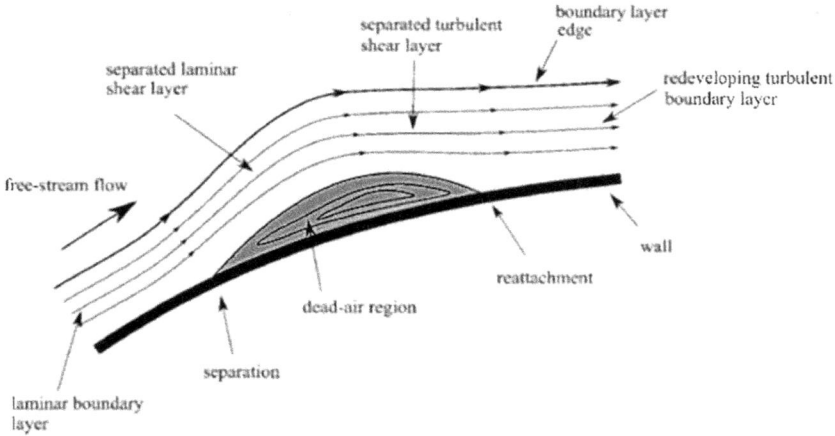

Figure 5.10 Schematic representation of boundary layer thickness.

engineers carefully analyse and optimize to achieve superior aerodynamic performance. By leveraging insights gained from numerical analysis and experimental data, engineers can make informed decisions to design innovative and efficient aerospace components that lead to advancements in aerospace technology and flight capabilities [22] (Figure 5.10).

5.4.1 Boundary layer thickness and growth

Boundary layer thickness is an important aspect of boundary layer flow in aerospace engineering applications. Between the solid surface and the point at which the fluid's velocity reaches a specific percentage of the freestream velocity, it describes the distance. The performance of aerodynamic components is significantly influenced by the thickness of the boundary layer, which is a critical factor in determining the behaviour of the boundary layer. Let us examine the concept of boundary layer growth and thickness in more detail [23].

5.4.1.1 Boundary layer thickness (δ)

The distance from the solid surface at which the fluid velocity reaches 99% of the freestream velocity (U_∞) is the boundary layer thickness (δ) for an incompressible flow over a flat plate. In terms of math, it can be stated as:

$$U(y = \delta) = 0.99^* U_\infty$$

where $U(y = \delta)$ is the velocity of the fluid at the boundary layer thickness (δ) from the solid surface.

For a laminar boundary layer, the velocity profile follows a smooth parabolic distribution, and the boundary layer thickness (δ) can be determined using the following expression:

$$\delta = 5.0 * \left[\frac{(v * x)}{U_\infty} \right]^{\left(\frac{1}{2}\right)}$$

where v is the kinematic viscosity of the fluid and x is the distance along the flat plate from the leading edge.

For a turbulent boundary layer, the velocity profile is flatter and thicker compared to the laminar case. The boundary layer thickness (δ) in a turbulent flow can be estimated using the empirical relation known as the "Prandtl's mixing length theory":

$$\delta = 0.37 * x * \text{Re}_x^{\frac{1}{5}}$$

where Re_x is the Reynolds number based on the distance x from the leading edge, defined as:

$$\text{Re}_x = (U_\infty * x) / v$$

Growth of the Boundary Layer: As fluid moves across the body's surface, the thickness of the boundary layer rises. The boundary layer is growing as a result of this rise in thickness. In general, a number of variables, such as flow conditions, Reynolds number, surface roughness, and boundary layer control strategies, affect how quickly the boundary layer grows. The thickness of the boundary layer grows with the square root of the distance from the leading edge (x) for laminar boundary layers, which grow relatively slowly. For laminar boundary layers, the previously described formula accurately predicts the growth. According to Prandtl's mixing length theory, the formation of turbulent boundary layers is quicker than that of laminar flows and has a power law connection with the distance from the leading edge (x).

In aerospace engineering, controlling boundary layer growth is crucial for designing streamlined and efficient aerodynamic shapes. Engineers seek to prolong laminar flow before its transition to turbulence to minimize skin friction drag. Techniques like laminar flow control and boundary layer suction are used to maintain a laminar boundary layer over a longer surface distance, reducing boundary layer thickness and improving aerodynamic performance. A grasp of boundary layer thickness and its growth is fundamental for optimizing aerospace designs, particularly in low-drag configurations like laminar flow wings and aerofoils.

5.4.2 Skin friction and drag in boundary layer flow

In boundary layer flow, skin friction and drag are pivotal aerodynamic forces directly influencing aerospace vehicle performance. Skin friction drag, acting parallel to the surface due to boundary layer viscosity, results from velocity gradients generating tangential shear stresses within the boundary layer. It increases with larger surface areas and is more prominent in turbulent boundary layers. Minimizing skin friction drag is essential, achieved through techniques like laminar flow control, promoting laminar flow to reduce fuel consumption.

Drag encompasses various components, including skin friction and pressure drag, impacting overall aerodynamic resistance. Pressure drag is influenced by the pressure distribution over the surface, often exacerbated by boundary layer separation. Optimizing boundary layer behaviour and minimizing drag is crucial for improved efficiency. Engineers employ streamlined shapes, surface treatments, and advanced flow control to manage boundary layers effectively, reducing drag and enhancing aerospace vehicle performance, fuel efficiency, and sustainability [24].

5.4.3 Boundary layer control techniques

By regulating the growth and thickness of the boundary layer, boundary layer management techniques are utilized by aerospace engineers to maximize the aerodynamic performance and efficiency of aircraft and aerospace vehicles. These methods are intended to improve lift generation, decrease skin friction drag, and postpone boundary layer separation. It is essential to comprehend and apply boundary layer control techniques in order to achieve safe and stable flight operations. Now let's look at a few well-known boundary layer control methods in aeronautical engineering.

The goal of laminar flow control is to spread the flow over a greater area of the aircraft's surface. Compared to turbulent flows, laminar boundary layers have less skin friction drag, which results in higher range and better fuel economy. To accomplish LFC, a variety of techniques are used, including sophisticated surface treatments, naturally laminar flow aerofoils, and specifically shaped airfoil forms. To keep laminar flow going, surfaces need to be sleek and smooth. Another method used by engineers to remove energy from the boundary layer and slow down the transition to turbulence is suction via microscopic surface slots. To manage the boundary layer, flow control devices are aerodynamic or mechanical components that are carefully positioned on the surface of the aircraft. Micro-vanes, boundary layer gates, and vortex generators are a few types of flow control devices [25] (Figure 5.11).

Vortex generators energize the boundary layer, delaying separation and improving control authority, particularly at high angles of attack. Boundary layer fences act as barriers to prevent cross flow and ensure smoother airflow over the surface. Micro-vanes are small, triangular structures designed

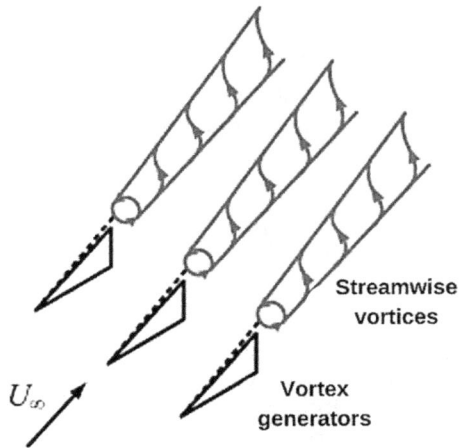

Figure 5.11 A schematic representation of the working of vortex generators.

Figure 5.12 Riblets for turbulent fluid flow control.

to control the boundary layer by redirecting the flow. Riblets are small, stream-wise grooves etched into the aircraft's surface. These rib-like structures induce a micro-turbulence in the boundary layer, reducing skin friction drag. Riblets are particularly effective in turbulent flow conditions and have been used in high-performance aircraft to improve aerodynamic efficiency [26] (Figure 5.12).

Active flow control uses actuators and sensors to control boundary layer behaviour in real time. These approaches are flexible and responsive to flight conditions. Active flow control includes boundary layer suction, which extracts air through microscopic surface holes to maintain a stable boundary layer. Other methods energize the boundary layer and delay separation with synthetic jets or plasma actuators. Aerospace engineering requires boundary layer control to optimize growth and thickness. Laminar flow control, flow control devices, riblets, and active flow control delay boundary layer separation, reduce skin friction drag, and improve aeroplane performance. Aerospace engineers develop fuel-efficient, high-performance aircraft that are safe and stable in varied flying conditions by carefully using these techniques. Research and development in the boundary layer control shape aircraft technology and sustainability [27].

5.5 BOUNDARY LAYER FLOW OVER AEROSPACE COMPONENTS

In aerospace engineering, understanding boundary layer flow over various components is crucial for optimizing the aerodynamic performance of aircraft and aerospace vehicles. The boundary layer behaviour over key components, such as aerofoils, wings, and control surfaces, significantly impacts lift, drag, and control effectiveness.

5.5.1 Boundary layer flow over aerofoils

Aerofoils, also known as airfoils, are specialized shapes designed to produce lift when air flows over them. The boundary layer flow over aerofoils is of paramount importance, as it directly affects the lift and drag forces acting on the aircraft. Boundary layer flow over aerofoils can be both laminar and turbulent, depending on the freestream conditions and the aerofoil's shape (Figure 5.13).

At low angles of attack and low speeds, the boundary layer over the upper surface of the aerofoil is typically laminar. Turbulent flow may replace the boundary layer when the flow approaches the trailing edge. Because turbulent boundary layers permit larger lift coefficients and are more resistant to flow separation, they are preferred for aerofoil designs. When designing aerofoils, boundary layer separation is a crucial consideration. Reduced lift

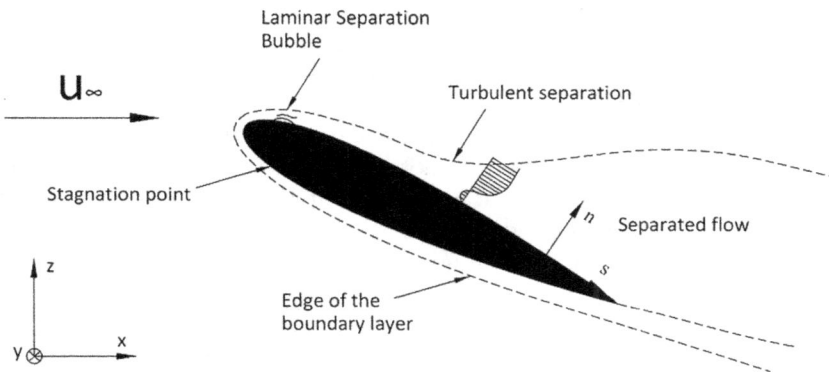

Figure 5.13 Depiction of flow over an airfoil at a high angle of attack.

and flow separation result from the boundary layer separating from the aerofoil's upper surface, which also produces a low-pressure area. Engineers employ strategies like wing shaping and boundary layer management to prevent separation by preserving connected flow throughout a wider range of angles of attack [28].

5.5.2 Boundary layer flow over wings

The wing serves as the principal aerodynamic component responsible for generating lift in an airplane. The flow of the boundary layer over wings is of utmost importance in the determination of the lift, drag, and overall aerodynamic performance of an aircraft. Similar to aerofoils, the boundary layer that forms above wings can exhibit either laminar or turbulent characteristics, contingent upon the prevailing flying circumstances. The study of boundary layer flow over wings holds great importance in the field of aerospace engineering, since it exerts a substantial impact on the aerodynamic performance and efficiency of aircraft. The phenomenon of air flowing across the surface of a wing results in the formation of a narrow region referred to as the boundary layer. The behaviour of the boundary layer has a direct impact on the aerodynamic parameters of the wing, including lift, drag, and overall performance [29].

During low-speed flying circumstances and at low angles of attack, the boundary layer that forms over the top surface of the wing is generally laminar in nature, exhibiting a smooth and streamlined flow. Nevertheless, when the fluid moves closer to the trailing edge, there is a possibility for the boundary layer to undergo a transition into turbulent flow. Turbulent boundary layers exhibit enhanced resistance to flow separation, rendering them advantageous in aerofoil configurations due to their capacity to provide larger lift coefficients. The phenomenon of boundary layer separation is of utmost importance in the field of wing design. When the boundary layer detaches from the upper surface of the wing, it generates a zone of decreased pressure, resulting in flow separation and diminished lift. Engineers utilize a variety of techniques to minimize separation and sustain connected flow throughout a wider spectrum of angles of attack. In order to enhance the occurrence of laminar flow and mitigate flow separation, engineers employ several wing design elements, including laminar flow airfoils, wing sweep, and winglets. Furthermore, in order to sustain laminar flow during take-off and landing or at high angles of attack, active flow management methods such as boundary layer suction and vortex generators are utilized. The optimization of boundary layer flow over wings is a crucial aspect in strengthening an aircraft's lift capabilities, lowering drag, and improving its overall aerodynamic efficiency. Engineers consistently endeavour to regulate the behaviour of boundary layers through the utilization of sophisticated aerodynamic shaping techniques and active flow control technologies [30].

This pursuit aims to enhance the fuel efficiency, manoeuvrability, and overall performance of wings employed in aeronautical engineering applications. Wings are commonly engineered with aerodynamic profiles that prioritize smooth and streamlined geometries, hence promoting laminar flow to postpone flow separation and minimize skin friction drag. Nevertheless, the attainment of laminar flow throughout the entirety of the wing's surface might pose difficulties as a result of defects and surface roughness. Engineers employ many wing design elements, including laminar flow airfoils, wing sweep, and winglets, to manipulate the behaviour of the boundary layer and enhance the aerodynamic efficiency. In order to sustain laminar flow and postpone flow separation, active flow management methods such as boundary layer suction and vortex generators are utilized, particularly in high-lift systems during the phases of take-off and landing [31].

5.5.3 Boundary layer flow over control surfaces

Control surfaces, including flaps, ailerons, elevators, and rudders, are dynamic components located on the wings and tail of an aircraft. These devices are employed to regulate the orientation and motion of the aircraft. The impact of the boundary layer flow on control surfaces is of great importance as it plays a substantial role in determining their efficacy and controllability. The study of boundary layer flow over control surfaces holds significant importance in the field of aeronautical engineering, as it plays a central role in determining the controllability and manoeuvrability of aircraft. Control surfaces, including flaps, ailerons, elevators, and rudders, are dynamic components situated on the wings and tail of an aircraft, for the purpose of regulating the aircraft's orientation and manoeuvrability. The efficacy and control authority of these surfaces are strongly influenced by the flow of the boundary layer [32].

During aerial manoeuvres, the control surfaces undergo fluctuations in angles of attack, resulting in alterations in the behaviour of the boundary layer. During high-deflection manoeuvres, the flow through control surfaces might exhibit local flow separation within the boundary layer. Flow separation can potentially have negative consequences on the efficiency of control mechanisms and can lead to the occurrence of control surface stall, hence restricting the aircraft's ability to manoeuvre and maintain stability. Aerospace engineers utilize a range of aerodynamic design elements and flow control techniques to efficiently regulate the boundary layer flow across control surfaces. Control surfaces are meticulously crafted with precisely contoured forms and streamlined edges in order to limit the occurrence of flow separation and enhance the effectiveness of control authority. In order to optimize the performance of control surfaces, engineers employ several techniques such as including slotted flaps and implementing control surface fairings. These features play a crucial role in managing the boundary layer flow and mitigating flow separation, particularly in key flying manoeuvres like take-off and landing. Furthermore, the utilization of active flow control techniques is implemented in order to optimize the behaviour of the boundary layer. In order to provide smooth flow and prevent flow separation, the use of techniques such as boundary layer suction is employed. This technique involves the extraction of air through minuscule apertures or slits located on the control surface, even when the aircraft is subjected to high angles of attack [33].

A comprehensive comprehension and effective management of boundary layer flow over control surfaces are important in order to guarantee accurate and dependable airplane control. Aerospace engineers enhance the control authority of aircraft by implementing sophisticated aerodynamic design elements and flow control methods. This results in improved manoeuvrability, responsiveness, and safety across a range of flying situations and mission objectives. During high-deflection manoeuvres, the flow through control surfaces in a boundary layer might potentially encounter instances of local flow separation. Flow separation has the potential to diminish the efficacy of control mechanisms and trigger the stalling of control surfaces, imposing restrictions on the aircraft's ability to perform manoeuvres. In order to enhance the performance of control surfaces, engineers utilize various aerodynamic design elements such as slotted flaps, control surface fairings, and boundary layer control devices. The aforementioned characteristics contribute to the effective management of the boundary layer flow, postponement of separation, and augmentation of control authority, particularly in situations involving high angles of attack and key flight stages [34].

5.6 FACTORS AFFECTING BOUNDARY LAYER FLOW

In aerospace engineering, several critical factors influence boundary layer flow, significantly impacting aircraft and aerospace vehicle aerodynamic performance and efficiency. These factors include freestream conditions, airfoil or wing shape, Reynolds number, surface roughness, angle of attack, viscosity, surface contamination, and the use of flow control devices. Freestream velocity and temperature, wing shape, and Reynolds number directly affect boundary layer behaviour, while surface roughness, angle of attack, and viscosity play crucial roles. Additionally, surface contamination and flow control devices can manipulate the boundary layer flow. Engineers use advanced simulations and testing to optimize these factors, ensuring efficient and safe aerospace vehicle design and operation, especially in varying flight conditions and mission requirements [35].

5.6.1 Reynolds number and its influence

In relation to boundary layer flow across aircraft surfaces, the Reynolds number is a key dimensionless parameter in aerospace engineering.

The Reynolds number, named after its inventor Osborne Reynolds and originally used to quantify the relative importance of inertial and viscous forces in fluid flow, was first used in the 1880s. When trying to foretell how the change from laminar to turbulent flow within the boundary layer will affect aerodynamic performance, knowing how the Reynolds number works is crucial.

The Reynolds number (Re) is calculated as the ratio of inertial forces to viscous forces and can be expressed as:

$$R_e = \frac{(\rho * V * L)}{\mu}$$

where ρ is the fluid density, V is the characteristic velocity (e.g., freestream velocity), L is the characteristic length (e.g., chord length of an airfoil), and μ is the dynamic viscosity of the fluid.

The Reynolds number greatly affects the boundary layer flow. The laminar-to-turbulent transition is predicted using it. The transition is crucial because turbulent boundary layers mix more and dissipate more energy, increasing skin friction drag yet resisting flow separation better than laminar boundary layers [36] (Figure 5.14).

For Reynolds numbers below 5,000, boundary layer flow is laminar. Smooth, streamlined laminar flow reduces skin friction drag. As Reynolds number grows, flow is more susceptible to disturbances and perturbations, and the boundary layer is more likely to become turbulent. Transitional flow may occur in the boundary layer at moderate Reynolds numbers, such as 5,000 to 10,000. This transitional regime has elements of laminar and turbulent flows. The flow becomes increasingly turbulent as the Reynolds

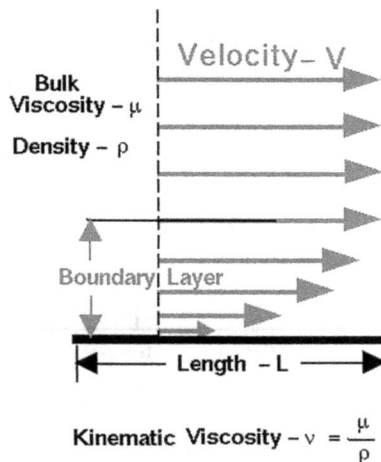

Figure 5.14 The impact of Reynolds number on boundary layer.

number rises above 10,000, especially under practical airplane operation conditions. Turbulent boundary layers are thicker and have more skin friction drag than laminar flows. Turbulent flows resist flow separation better, resulting in larger lift coefficients and control authority [37].

5.6.1.1 Influence on aerodynamic design

The Reynolds number is crucial to aerodynamic design. When constructing aircraft wings and airfoils, engineers must consider its effect. For example, wings are designed to sustain laminar flow as long as possible to minimize skin friction drag. Airfoil and wing geometry that delay turbulent flow may be chosen. The Reynolds number sensitivity of an airfoil or wing shape is also important. Some airfoils perform differently at varying Reynolds numbers, requiring careful design optimization for diverse flight situations. Reynolds number predicts boundary layer behaviour and aerodynamic performance. As Reynolds number grows, laminar to turbulent flow increases, affecting skin friction drag, flow separation, and lift. Aerospace engineers optimize aircraft designs by balancing laminar and turbulent flow to produce efficient and stable aerodynamic performance over a range of flight circumstances and Reynolds numbers [38] (Figure 5.15).

5.6.2 Surface roughness effects

Surface roughness is a critical factor in aerospace engineering, significantly affecting the behaviour of boundary layer flow over aircraft surfaces. Maintaining smooth and streamlined surfaces is paramount for optimizing

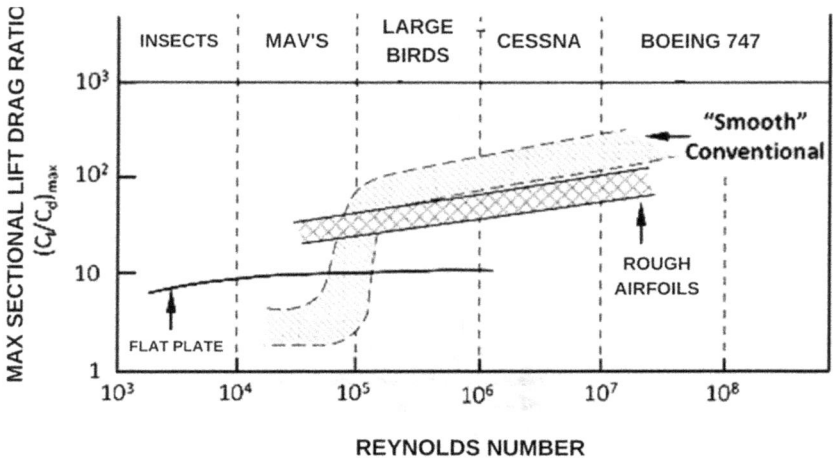

Figure 5.15 A representation of the action of Reynolds number on objects and aerodynamic bodies of different shapes, sizes and profiles.

aerodynamic performance and fuel efficiency. Even minor imperfections, like rivet heads or paint irregularities, can disrupt the boundary layer's flow, leading to increased skin friction drag and alterations in flow patterns. Surface roughness induces premature boundary layer transition from laminar to turbulent, resulting in a thicker, more energy-dissipative turbulent boundary layer and higher drag. Engineers employ advanced surface treatments and flow control devices to mitigate these adverse effects, promoting laminar flow and reducing skin friction drag for improved aircraft efficiency and performance. Understanding and managing surface roughness are crucial for safe, efficient, and sustainable aerospace operations [39].

5.6.3 Temperature effects

In boundary layer flow, temperature significantly influences boundary layer growth and thickness over aerospace surfaces. Temperature gradients within the boundary layer affect fluid behaviour, thermal characteristics, and aerodynamic performance. Higher temperature gradients lead to increased thermal expansion, altering density and viscosity and influencing velocity profiles and shear stresses. The thickness of boundary layers and the level of mixing are influenced by temperature, with lower temperatures leading to greater thickness and less mixing, and higher temperatures resulting in decreased thickness and increased mixing. The determination of the thickness of the thermal boundary layer in aerospace applications, particularly those involving high speeds and temperatures, is primarily influenced by the fluid characteristics and surface thermal conditions. Differences in temperature gradients can lead to an increased thickness of the thermal boundary layer in comparison to the velocity boundary layer, hence affecting the total thickness. In order to ensure safe and efficient operations, engineers must take into account these effects during the design of aircraft surfaces, with the aim of optimizing heat dissipation and effectively managing heat transmission. Temperature fluctuations have a significant influence on the process of heat transfer across surfaces, hence exerting an impact on cooling mechanisms, engine functionality, and the thermal strains experienced by various components of aircraft. In order to maintain optimal aircraft performance, many thermal management strategies are utilized, including the implementation of coatings, barriers, and heat exchangers. These techniques are applied with the objective of controlling the expansion of the boundary layer and facilitating efficient operation of the aircraft. A comprehensive comprehension of the influence of temperature on the behaviour of the boundary layer is necessary in order to optimize aerodynamic efficiency and ensure safety in aerospace engineering, particularly under diverse flying situations and temperature spectrums [40].

5.6.4 Compressibility effects

The influence of compressibility on the flow of boundary layers across aircraft surfaces is of utmost importance, especially under conditions of high velocities and altitudes when significant variations in air density occur. Comprehending these impacts is crucial for the optimization of aerodynamic designs and the assurance of safe flight operations. In the context of compressible flow, the density of the fluid exhibits variations in response to changes in velocity and pressure, whereby the Mach number (Ma) assumes a significant role as a fundamental quantity. In the realm of subsonic velocities, characterized by Mach numbers less than 1, the behaviour of fluid flow is akin to that of an incompressible medium. Conversely, when velocities exceed the speed of sound, resulting in Mach numbers greater than 1, the influence of compressibility becomes more pronounced. Shock waves are generated when an object moves at speeds above the speed of sound, and these shock waves have a significant impact on the boundary layer and the resulting aerodynamic forces. Aircraft shapes are engineered with the objective of minimizing the occurrence of shock-induced boundary layer separation. The compressibility of a substance also has an influence on the rates of heat transport, which in turn can result in the development of thermal stresses. The management of heat transport is facilitated by thermal protection systems. Transonic flight, with both subsonic and supersonic regions, faces challenges like drag divergence near Mach 1. Engineers optimize airfoil shapes and control surfaces to manage transonic effects effectively. Compressibility is a critical consideration for high-speed and high-altitude aerospace operations, affecting aerodynamic performance. Engineers employ computational tools and testing to manage these effects, ensuring safe and efficient flight operations in diverse regimes, contributing to the advancement of aerospace technology [41].

5.7 DESIGN STRATEGIES FOR BOUNDARY LAYER FLOW CONTROL

Design strategies for boundary layer flow control are of paramount importance in various engineering applications, as they play a crucial role in enhancing aerodynamic performance and mitigating drag. These strategies encompass a range of techniques aimed at achieving effective boundary layer flow control. Advanced airfoil shapes and meticulously crafted surface contours are employed to mitigate adverse pressure gradients and maintain the attachment of the boundary layer, thus delaying separation. Precise management of surface roughness through innovations like riblets and dimples serves to minimize skin friction and encourage laminar flow. Modifications to streamline shapes, achieved through techniques, such as fairings, fillets, and streamlined body designs, effectively reduce drag and postpone flow separation. Manipulating the pressure gradient along the airfoil or surface aids in preserving attached flow and further delaying separation. Additionally, the application of hydrophobic coatings plays a pivotal role in preventing water accumulation and ice formation, particularly in aviation applications, ensuring optimal boundary layer control and safety in adverse weather conditions [42].

5.7.1 Passive flow control techniques

Passive flow control techniques constitute a fundamental category of methods meticulously designed to exert influence over the boundary layer's behaviour without necessitating external energy sources. These techniques hold critical significance in diverse engineering applications where enhancing aerodynamic performance and mitigating drag are paramount objectives. The repertoire of passive methods encompasses several key strategies, including the strategic placement of vortex generators on surfaces to induce controlled vortices, thereby energizing and delaying boundary layer separation. Additionally, passive techniques involve tripping the laminar boundary layer into turbulence through the deployment of roughness elements or strips, thereby promoting an earlier transition to turbulent flow. Surface texture control, achieved through microstructures like riblets and dimples, actively reduces skin friction, fostering the propagation of laminar flow and thereby enhancing aerodynamic efficiency. The strategic placement of bumps or surface protrusions serves to modify pressure distributions, delaying boundary layer separation, and facilitating optimized flow control. Finally, the introduction of porous materials into surface structures allows for controlled mass transfer through the surface, affording a means to influence and manage boundary layer behaviour. Given their ability to achieve substantial improvements in aerodynamic performance while operating independently of external energy inputs, passive flow control techniques remain a cornerstone in the realm of boundary layer flow control [43].

5.7.2 Active flow control techniques

Active flow control techniques represent a dynamic and sophisticated approach to influencing the behaviour of the boundary layer by harnessing external energy sources, often requiring the integration of sensors and actuators for precision. These techniques encompass several prominent strategies. Blowing and suction, for instance, entail the deliberate injection or extraction of fluid through strategically positioned apertures on surfaces, effectively reshaping the velocity profile of the boundary layer. The deployment of dielectric barrier discharge plasma actuators generates localized forces within the boundary layer, actively modifying its characteristics and enhancing control over flow separation. Piezoelectric actuators induce controlled vibrations, finely tuning the boundary layer's properties.

The application of magnetic fields interacts with conducting fluids in boundary layers, especially in aerospace contexts, offering a novel means of flow control through magnetic forces. Fluidic oscillators, by generating periodic fluidic jets, energize the boundary layer, postponing flow separation and extending favourable flow characteristics. These active flow control techniques, characterized by their dynamic nature and energy dependency, hold significant promise for optimizing aerodynamic performance and mitigating flow separation challenges in complex engineering systems that demand precise control [44].

5.7.3 Optimizing boundary layer flow control systems

Efficiency in boundary layer flow control systems hinges on a rigorous optimization process, guided by a multifaceted set of considerations. Central to this endeavour is sensitivity analysis, which discerns the system's responsiveness to diverse control parameters, spotlighting the most influential factors. The deployment of computational fluid dynamics (CFD) simulations enables predictive modelling of the system's performance, offering invaluable insights into the impact of various control strategies. Real-world validation is achieved through wind tunnel tests and experiments, solidifying the effectiveness and reliability of selected techniques. The integration of feedback control, facilitated by sensors and actuators, ensures real-time adaptability to evolving conditions, upholding optimal performance. A cost-benefit analysis further refines decisions by weighing factors like fuel savings and performance enhancements. Concomitantly, environmental considerations, appraised through assessments of reduced fuel consumption and emissions, align with the growing emphasis on sustainability and regulatory compliance. Ultimately, the optimization of boundary layer flow control systems emerges as an iterative process, synthesizing design, analysis, and experimentation to attain the coveted enhancements in performance, harmonizing with the pursuit of enhanced aerodynamic efficiency and diminished environmental impact [45].

5.8 FUTURE DIRECTIONS AND RESEARCH IN BOUNDARY LAYER FLOW

Future directions and research in boundary layer flow control promise significant advancements across several key domains. In the pursuit of more comprehensive solutions, multidisciplinary approaches will integrate knowledge from fluid dynamics, materials science, and control theory. Sustainable transport will remain a focal point, with a concerted effort to develop boundary layer control strategies that align with global sustainability objectives by reducing energy consumption and emissions. Nature's ingenious adaptations will inspire bio-inspired solutions, while adaptive flow control systems capable of real-time adjustments will come to the forefront. Nanomaterials and nanofluids are anticipated to play a pivotal role in achieving superior boundary layer control. These developments underscore the exciting future of boundary layer flow research, where innovation is driven by emerging technologies, advanced computational methods, and the collaborative efforts of researchers across diverse fields. While challenges such as turbulent flows' complexity and environmental considerations persist, the opportunities for energy efficiency, emissions reduction, and improved engineering system performance are both significant and inspiring [46].

The future of boundary layer flow research holds great potential, driven by emerging technologies, computational advancements, and interdisciplinary collaboration. Addressing challenges and seizing opportunities will shape the development of innovative flow control solutions with broad applications across industries.

REFERENCES

[1] A. K. Kundu, M. A. Price, and D. Riordan, *Theory and Practice of Aircraft Performance*. John Wiley & Sons, 2016.

[2] H. Schlichting, K. Gersten, H. Schlichting, and K. Gersten, "Fundamentals of boundary-layer theory," *Boundary-layer Theory*, pp. 29–49, 2000.

[3] H. Schlichting and K. Gersten, *Boundary-layer Theory*. Springer, 2016.

[4] G. A. Marxman, C. E. Wooldridge, and R. J. Muzzy, "Fundamentals of hybrid boundary-layer combustion," in *Progress in Astronautics and Rocketry*, vol. 15, Elsevier, 1964, pp. 485–522.

[5] M. Lewis, S. P. Neill, P. Robins, M. R. Hashemi, and S. Ward, "Characteristics of the velocity profile at tidal-stream energy sites," *Renew. Energy*, vol. 114, pp. 258–272, 2017.

[6] V. P. Singh and S. Jain, "Economic analysis of a large scale solar updraft tower power plant," *Sustain. Energy Technol. Assessments*, vol. 58, p. 103325, 2023, https://doi.org/10.1016/j.seta.2023.103325

[7] F. T. Smith, "Steady and unsteady boundary-layer separation," *Annu. Rev. Fluid Mech.*, vol. 18, no. 1, pp. 197–220, 1986.

[8] N. Dutt, A. Jageshwar, H. Ashwani, K. Mukesh, K. Awasthi, and V. Pratap, "Thermo - hydraulic performance of solar air heater having discrete D - shaped ribs as artificial roughness," *Environ. Sci. Pollut. Res.*, 2023, https://doi.org/10.1007/s11356-023-28247-9

[9] O. A. Oleinik and V. N. Samokhin, *Mathematical Models in Boundary Layer Theory*, vol. 15. CRC Press, 1999.

[10] V. P. Singh and G. Dwivedi, "Technical analysis of a large – Scale solar updraft tower power," *Energies*, vol. 16, no. 1, pp. 1–29, 2023, https://doi.org/10.3390/en16010494

[11] U. S. Vevek, H. Namazi, R. Haghighi, and V. V. Kulish, "Analysis & validation of exact solutions to Navier-Stokes equation in connection with quantum fluid dynamics.," *Math. Eng. Sci. Aerosp.*, vol. 7, no. 2, 2016.

[12] B. Müller, "Low-Mach-number asymptotics of the Navier-Stokes equations," *Floating, Flowing, Fly. Pieter J. Zandbergen's Life as Innov. Inspir. Instigator Numer. Fluid Dyn.*, pp. 97–109, 1998.

[13] K. Chauhan and V. P. Singh, "Proceedings prospect of biomass to bioenergy in India: An overview," *Mater. Today Proc.*, no. xxxx, 2023, https://doi.org/10.1016/j.matpr.2023.01.419

[14] R. W. Zimmerman and I.-W. Yeo, "Fluid flow in rock fractures: From the Navier-Stokes equations to the cubic law," *Geophys. Monogr. Geophys. Union*, vol. 122, pp. 213–224, 2000.

[15] N. Jiang and N. Masmoudi, "Boundary layers and incompressible Navier-Stokes-Fourier limit of the Boltzmann equation in bounded domain I," *Commun. Pure Appl. Math.*, vol. 70, no. 1, pp. 90–171, 2017.

[16] M. Raissi, A. Yazdani, and G. E. Karniadakis, "Hidden fluid mechanics: Learning velocity and pressure fields from flow visualizations," *Science (80-.).*, vol. 367, no. 6481, pp. 1026–1030, 2020.

[17] V. P. Singh, C. S. Meena, A. Kumar, and N. Dutt, "Double pass solar air heater: A review," *Int. J. Energy Resour. Appl.*, vol. 1, no. 2, pp. 22–43, 2022, https://doi.org/10.56896/IJERA.2022.1.2.009

[18] J. G. Heywood, R. Rannacher, and S. Turek, "Artificial boundaries and flux and pressure conditions for the incompressible Navier–Stokes equations," *Int. J. Numer. methods fluids*, vol. 22, no. 5, pp. 325–352, 1996.

[19] M. R. Malik, "Numerical methods for hypersonic boundary layer stability," *J. Comput. Phys.*, vol. 86, no. 2, pp. 376–413, 1990.

[20] V. P. Singh et al., "Nanomanufacturing and design of high-performance piezoelectric nanogenerator for energy harvesting," in *Nanomanufacturing and Nanomaterials Design: Principles and Applications*, 2022, pp. 241–272.

[21] E. J. Plate, "Aerodynamic Characteristics of Atmospheric Boundary Layers," Argonne National Lab., Ill. Karlsruhe Univ.(West Germany), 1971.

[22] Y. G. Ashwani Kumar, Arun Kumar Singh Gangwar, Avinash Kumar, Chandan Swaroop Meena, Varun Pratap Singh, Nitesh Dutt, Arbind Prasad, "Biomedical study of femur bone fracture and healing," in *Adv. Mater. Biomed. Appl.*, 2022, pp. 283–298.

[23] M. Matsubara and P. H. Alfredsson, "Disturbance growth in boundary layers subjected to free-stream turbulence," *J. Fluid Mech.*, vol. 430, pp. 149–168, 2001.

[24] M. R. Abbassi, W. J. Baars, N. Hutchins, and I. Marusic, "Skin-friction drag reduction in a high-Reynolds-number turbulent boundary layer via real-time control of large-scale structures," *Int. J. Heat Fluid Flow*, vol. 67, pp. 30–41, 2017.

[25] T. C. J. Cousteix and J. Cebeci, *Modeling and Computation of Boundary-layer Flows*, Berlin, Germany: Springer, 2005.

[26] A. Datta, A. Kumar, A. Kumar, A. Kumar, and V. P. Singh, "Advanced materials in biological implants and surgical tools," in *Advanced Materials for Biomedical Applications*, 1st ed., N. D. Ashwani Kumar, Yatika Gori, Avinash Kumar, Chandan Swaroop Meena, Ed. 2022, pp. 283–298.

[27] R. K. Shubham Srivastava, Deepti Verma, Shreya Thusoo, Ashwani Kumar, Varun Pratap Singh, "Nanomanufacturing for Energy Conversion and Storage Devices," in *Nanomanufacturing and Nanomaterials Design: Principles and Applications*, 1st ed., S. Singh, S. K. Behura, A. Kumar, K. Verma, Ed. 2022, pp. 165–174.

[28] L. D. Kral, "Recent experience with different turbulence models applied to the calculation of flow over aircraft components," *Prog. Aerosp. Sci.*, vol. 34, no. 7–8, pp. 481–541, 1998.

[29] N. Dutt, A. Binjola, A. J. Hedau, A. Kumar, V. P. Singh, and C. S. Meena, "Comparison of CFD results of smooth air duct with experimental and available equations in literature," *Int. J. Energy Resour. Appl.*, vol. 1, no. 1, pp. 40–47, 2022, https://doi.org/10.56896/IJERA.2022.1.1.006

[30] R. Vinuesa, S. M. Hosseini, A. Hanifi, D. S. Henningson, and P. Schlatter, "Pressure-gradient turbulent boundary layers developing around a wing section," *Flow. Turbul. Combust.*, vol. 99, pp. 613–641, 2017.

[31] V. P. Singh, A. Karn, G. Dwivedi, T. Alam, and A. Kumar, "Experimental assessment of variation in open area ratio on thermohydraulic performance of parallel flow solar air heater," *Arab. J. Sci. Eng.*, vol. 41, no. 12, pp. 1–17, 2022, https://doi.org/10.1007/s13369-022-07525-7

[32] M. Saini, A. Sharma, V. P. Singh, S. Jain, and G. Dwivedi, "Solar thermal receivers — A review," *Lect. Notes Mech. Eng.*, vol. II, p. 1_25, 2022, https://doi.org/10.1007/978-981-16-8341-1

[33] A. Pandey, K. M. Casper, R. Spillers, M. Soehnel, and S. Spitzer, "Hypersonic shock wave-boundary-layer interaction on the control surface of a slender cone," in *AIAA SciTech 2020 Forum*, 2020, p. 815.

[34] V. P. Singh, S. Jain, and J. M. L. Gupta, "Analysis of the effect of variation in open area ratio in perforated multi-V rib roughened single pass solar air heater- Part A," *Energy Sources, Part A Recover. Util. Environ. Eff.*, vol. 44, no. Jan, pp. 1–21, 2022, doi: 10.1080/15567036.2022.2029976

[35] A. L. Braslow, "A review of factors affecting boundary-layer transition," 1966.

[36] V. P. Singh, S. Jain, and J. M. L. Gupta, "Performance assessment of double-pass parallel flow solar air heater with perforated multi-V ribs roughness — Part B," *Exp. Heat Transf.*, vol. 00, no. 00, pp. 1–18, 2022, https://doi.org/10.1080/08916152.2021.2019147

[37] M. M. Metzger and J. C. Klewicki, "A comparative study of near-wall turbulence in high and low Reynolds number boundary layers," *Phys. Fluids*, vol. 13, no. 3, pp. 692–701, 2001.

[38] J. Niu, X. Liang, and D. Zhou, "Experimental study on the effect of Reynolds number on aerodynamic performance of high-speed train with and without yaw angle," *J. Wind Eng. Ind. Aerodyn.*, vol. 157, pp. 36–46, 2016.

[39] V. P. Singh et al., "Recent developments and advancements in solar air heaters: A detailed review," *Sustain.*, vol. 14, no. 19, pp. 1–57, Sep. 2022, https://doi.org/10.3390/su141912149

[40] V. P. Singh, S. Jain, and A. Kumar, "Establishment of correlations for the Thermo-Hydraulic parameters due to perforation in a multi-V rib roughened single pass solar air heater," *Exp. Heat Transf.*, vol. 35, no. 5, pp. 1–20, 2022, https://doi.org/10.1080/08916152.2022.2064940

[41] V. P. Singh, S. Jain, A. Karn, A. Kumar, and G. Dwivedi, "Mathematical modeling of efficiency evaluation of double pass parallel flow solar air heater," *Sustainability*, vol. 14, no. 17, pp. 1–22, 2022, https://doi.org/10.3390/su141710535

[42] R. Messing and M. J. Kloker, "Investigation of suction for laminar flow control of three-dimensional boundary layers," *J. Fluid Mech.*, vol. 658, pp. 117–147, 2010.

[43] V. P. Singh, S. Jain, and J. M. L. Gupta, "Analysis of the effect of perforation in multi-v rib artificial roughened single pass solar air heater: - Part A," *Exp. Heat Transf.*, pp. 1–20, Oct. 2021, https://doi.org/10.1080/08916152.2021.1988761

[44] V. P. Singh et al., "Heat transfer and friction factor correlations development for double pass solar air heater artificially roughened with perforated multi-V ribs," *Case Stud. Therm. Eng.*, vol. 39, no. September, p. 102461, 2022, https://doi.org/10.1016/j.csite.2022.102461

[45] M. D. Gunzburger, *Perspectives in Flow Control and Optimization*. SIAM, 2002.

[46] J. Svorcan, J. M. Wang, and K. P. Griffin, "Current state and future trends in boundary layer control on lifting surfaces," *Adv. Mech. Eng.*, vol. 14, no. 7, p. 16878132221112160, 2022.

Chapter 6

Sensitivity analysis for aerospace engineering applications

L. Prince Raj, P. M. Mohamed Abubacker Siddique, and G. S. Charana

Indian Institute of Engineering Science and Technology, Howrah, India

6.1 INTRODUCTION

Complex systems and processes involved in engineering analysis need a deep comprehension of how various variables and factors interact (Oakley & O'Hagan, 2004; Prince Raj et al., 2020; Sudret, 2008). Sensitivity analysis emerges as a useful method that allows engineers to evaluate the effects of changing input parameters on system outputs. Sensitivity analysis assists in optimizing design, assessing performance, reducing risks, and influencing decision-making processes by methodically examining the correlations between variables. Sensitivity analysis emerges as a potent technique for comprehending and optimizing aircraft behaviour in flight in the aerospace industry, where aerodynamic performance is essential. Engineers and researchers can evaluate the effects of design decisions, improve performance, and ensure safe and effective flying by methodically assessing the impact of input factors on aerodynamic properties (Oakley & O'Hagan, 2004; Mohamed Abubacker Siddique & Raj, 2023).

Multielement airfoils are often used in large jet aeroplanes for better take-off and landing performance and to achieve necessary lift requirements at low velocities. In general, multielement airfoil is made up of two or more airfoil sections that are put next to one another. This configuration is particularly helpful in high-performance and specialized aircraft because it provides better lift, less drag, and increased manoeuvrability. The basic idea of the multielement airfoil design is to control the airflow across the wing at various stages of flight to maximize lift and reduce drag. Due to high design, manufacturing, and maintenance costs, many commercial aircraft use three-element multielement airfoils.

In general, the three-element airfoil systems are composed of main, flap, and slat elements, which are placed suitably for better performance. The research on the three-element airfoils started during the 1970s, and the optimizations are still ongoing for better performance and flow physics investigations. The interaction of flows, boundary layer, and the element gaps complicates the three-element airfoil flow physics (Mohamed Abubacker

DOI: 10.1201/9781003465171-6

Siddique & Raj, 2023; Prince Raj et al., 2019; Smith, 1975). Wild (2013) experimentally investigated the stall behaviour of high-lift airfoil at various Mach and Reynolds numbers. Balaji et al. (2006) investigated the parametric analysis of flap and slat riggings on the three-element airfoil system. Anderson et al. (1995) investigated the multielement airfoil flow using the Navier-Stokes solver and validated the solver by experimental results. A numerical simulation of the flow over a three-element airfoil using the zonal detached eddy simulation (ZDES) investigation for aircraft applications was investigated by Deck and Laraufie (2013). Further, Zhang and Zerihan (2003) investigated the aerodynamics of a two-element wing under the ground effect to investigate the ground proximity and flap settings. Pagani et al. (2017) investigated experimentally the slat configuration on the aeroacoustics of a three-element airfoil (MD30P30N). Further, other research was conducted on multielement airfoils for several physical aspects (Chin et al., 1993; Nakayama et al., 1990; Pomeroy et al., 2012; Rogers, 1994; Rumsey et al., 1998; Spaid & Lynch, 1996) and optimizations (Grendysa & Olszański, 2021; Kim et al., 2002; Li, 2013; Nakayama et al., 2006; Valarezo et al., 1991).

This chapter discusses the sensitivity analysis methodology for the aerospace application, the three-element airfoil, in detail. The 30P30N airfoil is selected for the sensitivity analysis of the geometrical parameters (slat and flap positions). An unstructured-based finite volume method solver is utilized to simulate the 30P30N airfoil at various flap and slat positions. Comparing the computational results with the results of the experiments serves to validate the computational model. The Sobol sequence sampling generator generates the position of slat and flap, and the results from simulations are used to generate the metamodel by radial basis function, and finally, the sensitivity indices are calculated using Sobol's method.

6.2 COMPUTATIONAL MODELLING

A well-known steady-state two-dimensional Reynolds averaged Navier-Stokes equations are utilized to compute the multielement airfoil. It is possible to use different RANS models such as Spalart Allmaras, k-ϵ, k-ω SST for the current simulations. However, from literature studies, it is found that k-ω SST can be suitable for this simulation owing to the capability and accuracy. The turbulent kinetic energy and specific dissipation rate can be given as:

$$\frac{\partial(\rho k)}{\partial t} + \frac{\partial(\rho \bar{u}_i k)}{\partial x_i} = \frac{\partial}{\partial x_j}\left[\left(\mu + \frac{\mu_t}{\sigma_k}\right)\frac{\partial k}{\partial x_j}\right] + \bar{P}_k - \rho\varepsilon \tag{6.1}$$

$$\frac{\partial(\rho\omega)}{\partial t} + \frac{\partial(\rho\bar{u}_i\omega)}{\partial x_i} = \frac{\partial}{\partial x_j}\left[\left(\mu + \frac{\mu_t}{\sigma_\omega}\right)\frac{\partial\omega}{\partial x_j}\right] + \bar{P}_\omega + \frac{\rho\beta^*}{\omega}\varepsilon - \rho\beta\omega^2 \tag{6.2}$$

More details about the current computational model can be found in the literature. In this current work, a well-known commercial software, ANSYS Fluent, is used for the current simulations.

6.3 GLOBAL SENSITIVITY ANALYSIS

Global sensitivity analysis (GSA), one of the main methods in sensitivity studies, investigates the impact of the input parameters on the model output (Saltelli & Homma, 1996; Sobol, 2011; Soulat et al., 2016). Several global sensitivity analysis techniques are available to address various issues (Chen et al., 2005; Werbos, 2005). The major groups of GSA methods include analysis of variance (ANOVA), decomposition of the response variance, and the derivative-based approach. The variance of the outcome is broken down by the ANOVA approach as the total contributions from the input parameters. Then, Sobol's sensitivity indices can be utilized to compute the individual contribution of the input parameter (Saltelli & Homma, 1996). The derivative-based strategy concentrates on the overall effect of the factors. However, due to the high cost of computing, its use is restricted to computationally challenging problems (Sudret, 2008). Therefore, the GSA methods can be used for the engineering applications with low to medium computations (Pehlivanoglu & Yagiz, 2012; Saltelli, 2002). The Sobol sequence sampling-based techniques can considerably enhance the performance of the GSA methodology at low computation cost.

After the generation of the metamodel, the sensitivity indices can be determined by Sobol's or Monte Carlo techniques. The radial basis function (RBF) (Mullur & Messac, 2005), the Gaussian process (Oakley & O'Hagan, 2004), and Kriging (Martin & Simpson, 2005) are a few of the metamodels that may be used with the surrogate-based approach. A metamodel can be built using the polynomial chaos expansion strategy to lower the computational expense of GSA approaches (Sudret, 2008). Orthogonal polynomial interpolation surrogate model construction can also be used to compute the sensitivity indices from a metamodel. The disadvantage of such models is that they poorly handle non-linear, high-dimensional issues. The RBF method is utilized for building the metamodel, and the Sobol's method is implemented to calculate the sensitivity indices.

6.3.1 Sampling methods for the sensitivity analysis

Any sensitivity study typically takes three steps: selecting the sample points, building the metamodel, and determining the sensitivity indices. The distribution of design points is greatly influenced by sampling, a procedure for exploring the design space. In building the metamodel, the sample size and distribution are crucial. For sampling the points, various techniques are available, including Latin hypercube sampling (LHS), pseudo-random, stratified, Sobol sequence, and Hammersley. The Sobol sequence (Burhenne et al., 2013) is employed in the current investigation primarily because of its low discrepancy sampling capacity, as described in the preceding section.

The variables x_1, x_2, and x_3 are displayed against one another for the sample design points in Figure 6.1. The Sobol sequence approach, as can be seen, uniformly distributes the sample points over each design area. Compared to other sampling strategies, this method is recognized to have a reduced discrepancy. Additionally, this method's sampling can achieve genuine mean convergence significantly more quickly than previous approaches. The current study considers sampling points with four parameters the flap and slat overhang and gap.

6.3.2 RBF-based metamodelling

Metamodelling can approximate a system using a few of the system's produced samples. The RBF technique was developed to investigate the land surface contours from the geographic data (Hardy, 1971). The RBF approach successfully captured the arbitrary contours produced by the system's deterministic and stochastic response functions. The expression for RBF is given as:

$$Y = F(x) = \sum_{n=1}^{N} w_n \varphi(\|x - x_n\|) \tag{6.3}$$

Here the symbols φ, x and x_n represent the basis function vector, design variables, and the design variable vector at the nth sampling point, respectively. The unknown is w_n, and the Euclidean norm is represented as $\|x - x_n\|$. Now, w can be computed from the following expression as,

$$w = A^{-1}F \tag{6.4}$$

where

$$w = \begin{bmatrix} w_1 \\ w_2 \\ \vdots \\ w_N \end{bmatrix}, \quad A = \begin{bmatrix} \varphi_{11} & \varphi_{12} & \cdots & \varphi_{1,N} \\ \varphi_{21} & \varphi_{22} & \cdots & \varphi_{2N} \\ \vdots & \vdots & \ddots & \vdots \\ \varphi_{N1} & \varphi_{N2} & \cdots & \varphi_{NN} \end{bmatrix}, \quad F = \begin{bmatrix} f_1 \\ f_2 \\ \vdots \\ f_N \end{bmatrix} \tag{6.5}$$

6.3.3 Sensitivity indices by Sobol's method

The global sensitivity analysis on any input parameters can be started from the following decomposition,

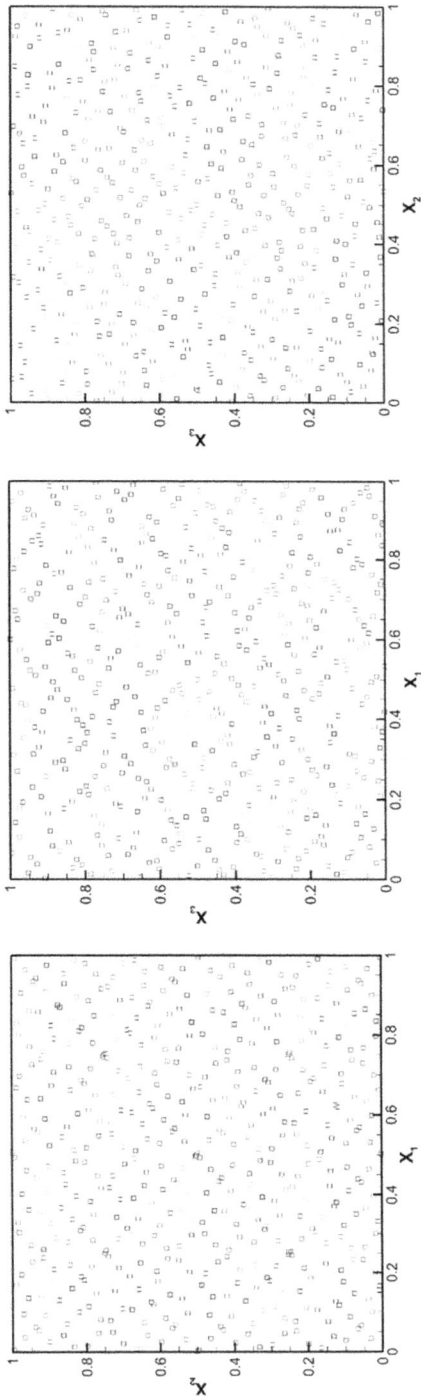

Figure 6.1 Sampling of three parameters plotted against each other based on Sobol sequence-based.

$$F(x) = F_0 + \sum_{N=1}^{n} F_N(x_N) + \sum_{1 \le N < M \le n}^{n} F_{NM}(x_N, x_M) + \cdots + F_{12...n}(x). \qquad (6.6)$$

Here F_0 is a constant, F_N is a function of x_N, and F_{NM} is a function of x_N and x_M. If the following orthogonality condition is enforced,

$$\int_0^1 F_{N_1 N_2 ... N_s}(x_{N_1}, x_{N_2}, ..., x_{N_s}) dx_k = 0, \text{ for } k = N_1, N_2, ..., N_s, \qquad (6.7)$$

The decomposed terms can be constructed,

$$F_0 = \int F(x) dx,$$

$$F_N(x_N) = \int F(x) \prod_{K \ne N} dx_K - F_0,$$

$$F_{NM}(x_N, x_M) = \int F(x) \prod_{K \ne M, N} dx_K - F_0 - F_N(x_N) - F_M(x_M).$$
$$\qquad (6.8)$$

the variance (D) of $F(x)$ can be written as,

$$D = \int F(x)^2 dx - F_0^2. \qquad (6.9)$$

By using the orthogonality property of the decomposition,

$$D = \sum_{i=1}^{k} D_i + \sum_{i<j} D_{ij} + ... + D_{1,2,...,k}. \qquad (6.10)$$

In equation (6.10), the partial variance corresponding to the subset of parameters, $D_{i_1...i_s}$, which is defined as $\int F_{i_1...i_2}^2 (x_{i_1}, ..., x_{i_s}) dx_{i_1}, ..., x_{i_s}$, is the variance of $F_{i_1...i_s}(x_{i_1}, ..., x_{i_s})$.

The sensitivity indices can be computed from the following expression as,

$$S_{i_1...i_s} = \frac{D_{i_1...i_s}}{D}. \qquad (6.11)$$

The main or the first-order effect can be given as:

$$S_i = \frac{D_i}{D}. \qquad (6.12)$$

Further, the interaction or second-order effect can be given as:

$$S_{ij} = \frac{D_{ij}}{D}. \qquad (6.13)$$

Finally, the total sensitivity indices can be given as:

$$S_{Ti} = S_i + \underset{i \neq j}{S_{ij}} + \cdots + S_{1\ldots i\ldots s}. \tag{6.14}$$

The main or the first-order effect shows the influence of a single parameter to the output variables. On the other hand, the total or second-order effects show the influence of parameter interactions to the output variables. In general, the total sensitivity indices are the sum of first-order, second-order, and higher-order effects. It is worth mentioning that higher the values of sensitivity indices indicated, the higher the impact of parameters.

6.4 MESH REFINEMENT ANALYSIS AND VALIDATION

A three-element airfoil 30P30N was used to investigate the sensitivity analysis experimented by NASA (Kim et al., 2002; Valarezo, 1992; Valarezo et al., 1992). Figure 6.2 shows the arrangement of elements in the three-element airfoil; the parameters are similar to the literature. To establish accurate results and keep the calculation cost low, a grid independence analysis was carried out. The grid independence study used three distinct meshes with 1.01, 1.03, and 1.08 growth rates. Figure 6.3(a) displays the pressure coefficients for three different meshes used for the grid refinement analysis. The difference between the results generated by mesh-B and mesh-C grids was insignificant, allowing mesh-B to be used for further calculations. Figure 6.3 (right) depicts the mesh (mesh-B) used for the current simulations.

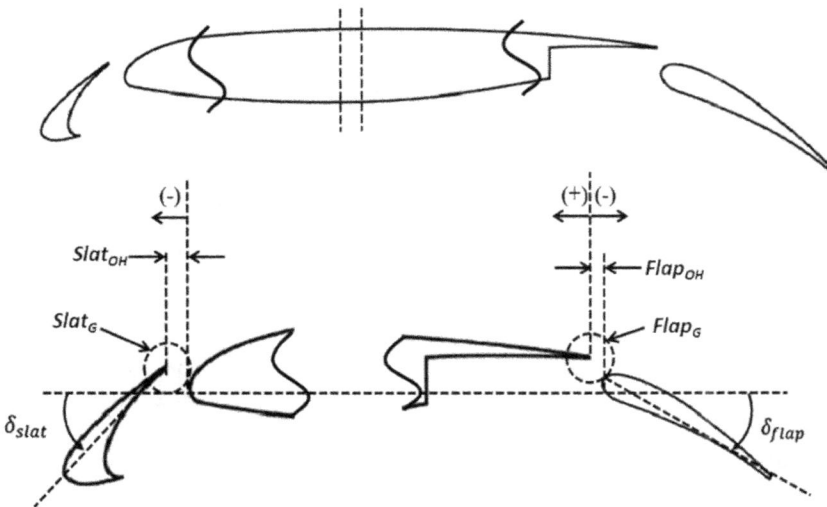

Figure 6.2 30P30N airfoil geometry and elements arrangements.

a)

b)

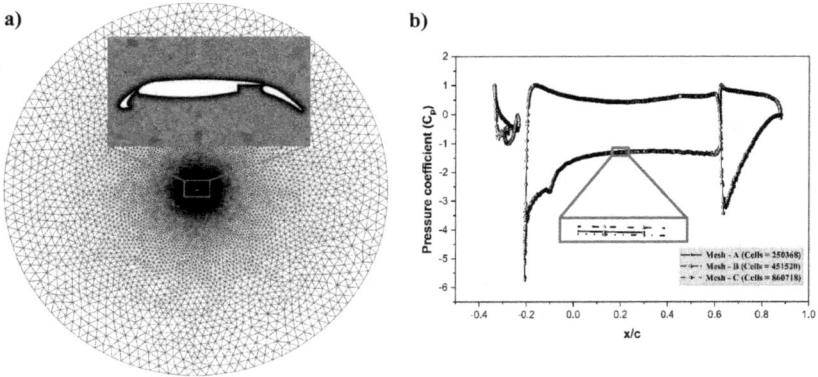

Figure 6.3 Mesh distribution (a) and the pressure distribution (b) around the 30P30N airfoil at a velocity of 68.6 m/s and an angle of attack 8°.

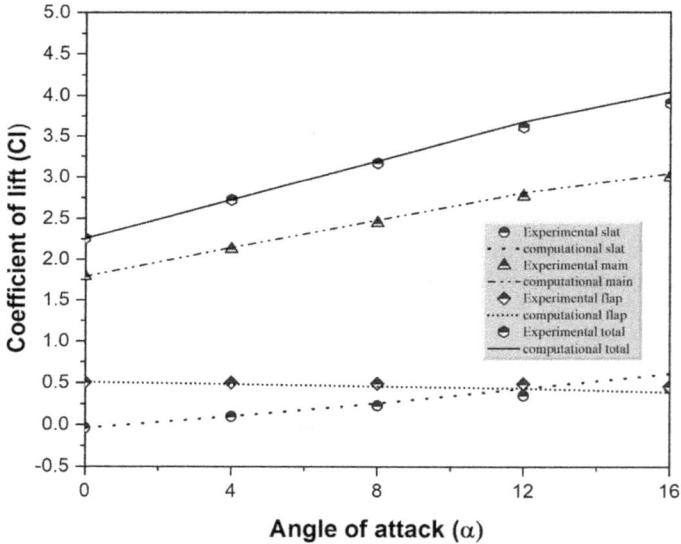

Figure 6.4 Lift coefficient comparison at velocity 68.6 m/s for MDA 30P/30N airfoil.

Figure 6.3(b) shows the pressure distribution along the 30P30N airfoil at an angle of attack of 16° and a velocity of 68.6 m/s. The computational results show good agreement with the experimental results, which shows the suitability of the current computational model for the current simulations.

Further, Figure 6.4 compares the computed lift coefficient with the experimental results, and Figure 6.5 shows the velocity profile at two locations of the multielement airfoil. Figure 6.6 shows the velocity magnitude and

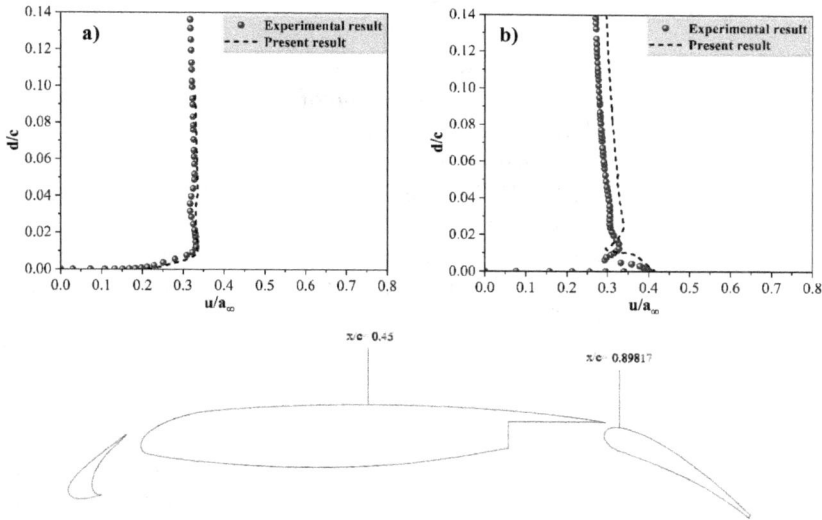

Figure 6.5 Velocity profiles at specific locations around the 30P30N airfoil (a) at x/c = 0.45; (b) at x/c = 0.89817 at a velocity of 68.6 m/s and α = 8°.

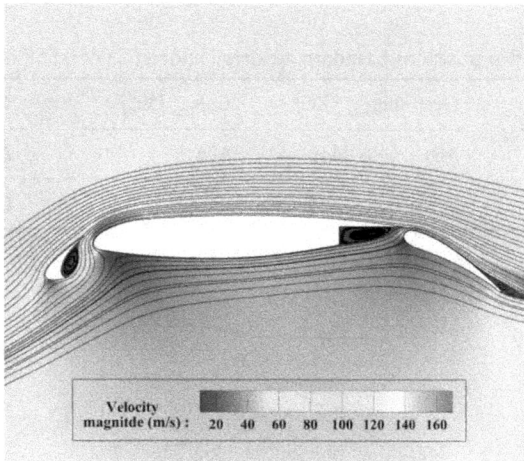

Figure 6.6 Velocity magnitude at a flow velocity of 68.6 m/s and α = 16° around MDA 30P/30N airfoil.

streamlines around the multielement airfoil at 16° angle of attack and velocity 68.6 m/s. It is observed that the current computational solver well captures the flow interactions. Overall, the computational model predicts the experimental results accurately for the given flow conditions. Hence, the present computational model is used for further simulations.

6.5 RESULTS AND DISCUSSION

The sensitivity analysis starts with the sampling point generation of cases based on the maximum and minimum range of variables. Hence, the range for the gap and overhang of the slat and flap is selected from previous research (Chin et al., 1993; Kim et al., 2002; Lin & Dominik, 1997; Spaid & Lynch, 1996; van Dam, 2002). Table 6.1 shows the range of the parameters utilized in this study. Based on the given range, the flap and slat are positioned, and the simulations are conducted for the given flight conditions.

The Sobol sequence method is utilized to generate the sampling points of the slat and flap positions. To generate the flap and slat positions, an envelope is made where the sampling points are plotted (enclosed) inside the envelope, as shown in Figure 6.7. The geometrical positions of the three-element airfoil are based on the sampling points, and the mesh is generated. The simulation is conducted using the developed computational model, and the results are then utilized to determine the sensitivity indices. It is worth mentioning that the sensitivity indices are computed based on variance-based methods.

The flow physics changed depending on the flap's position and slat on a three-element airfoil system. Figure 6.8 shows the velocity contour of the slat while it is placed at different locations in the envelope. Figure 6.8(a)

Table 6.1 Slat and flap positions based on an airfoil chord (C) of 0.559 m

Gap_{slat} [%C]		$OverHang_{slat}$ [%C]		Gap_{flap} [%C]		$OverHang_{flap}$ [%C]	
Min	Max	Min	Max	Min	Max	Min	Max
−3.7	−1.5	−3.5	−1.5	−0.6	0.5	0.8	2.2

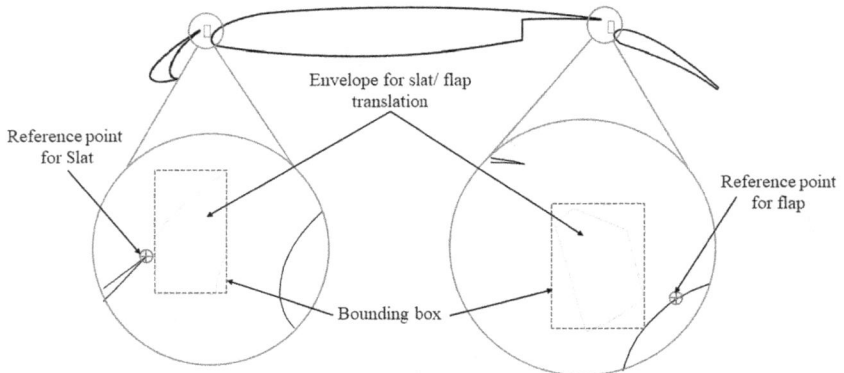

Figure 6.7 Position envelope of the elements in the three-element airfoil (Mohamed Abubacker Siddique & Raj, 2023).

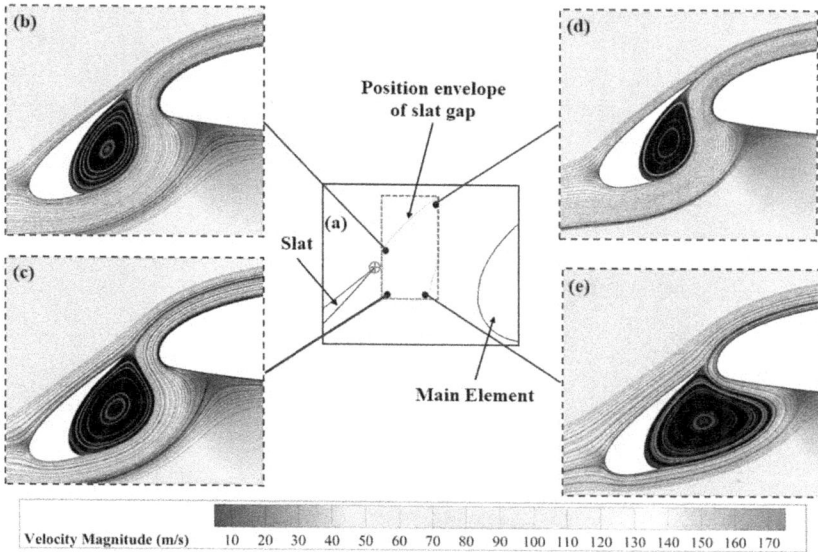

Figure 6.8 (a)–(e) The extreme sampling points slat locations on the envelope and flow field (velocity magnitude) around the slat at a velocity of 68.6 m/s and an angle of attack 8°.

shows the position envelope of the slat along with the slat and main-element airfoil positions. Figure 6.8(b) shows the velocity contour and streamlines in the vicinity of the slat when it is placed at maximum x and y directions. It is noticed that the recirculation area in the slat cove is changing significantly concerning the position of the slat. Figure 6.8(b) shows the flow filed around the slat when the slat is positioned at a high overhang and gap. It is observed that the slat wake zone is less when compared with the flow for the slat position at the smaller gap and overhang, as shown in Figure 6.8(c).

Further, the decrease in the gap and overhang decreased the slat wake zone for the higher gap case (seen in Figure 6.8(d)), while the slat wake zone increased significantly for the lower gap and overhang, as shown in Figure 6.8(e). It was found that the flow through the slot was slowed down by the low slat overhang positions, which pushed the stagnation point downstream. It is possible that the negative lift generated by the slat can be reduced by the high wake region and reduced flow speed. When the mainelement is close to the slat, the slat receives a cushioning effect when it is close to it. Further, it is found that the slat wake zone is reduced with the increase of overhang distance and the flow through the slot is also accelerated. Moreover, it may be possible to generate high drag on the slat element owing to the high recirculation velocity (lower static pressure) on the pressure side of the slat when the slat overhang is high.

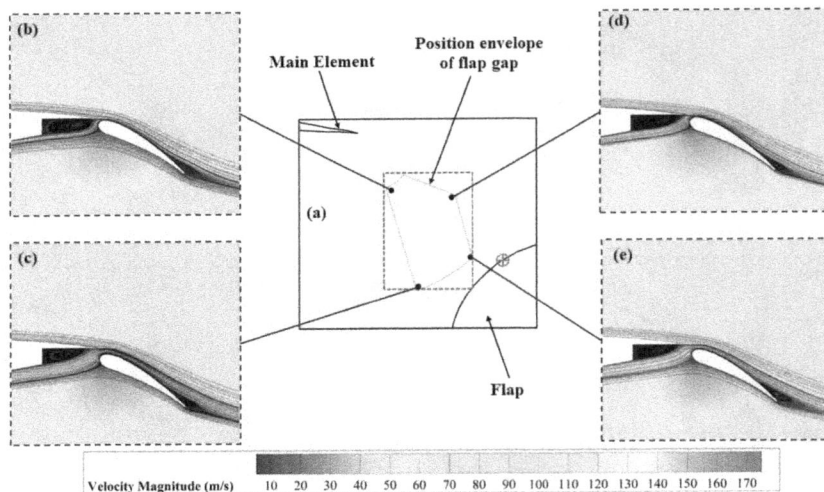

Figure 6.9 (a)–(e) The extreme sampling points flap locations on the envelope and flow filed (velocity magnitude) around the slat at a velocity of 68.6 m/s and an angle of attack 8°.

Figure 6.9 shows the velocity contour in the vicinity of the flap while it is placed at different locations in the envelope. Figure 6.9(a) shows the position envelope of the flap along with the flap and main-element airfoil positions. It is found that the wake zone in the main-element cove is significantly high at minimum flap overhang and gap, as shown in Figure 6.9(b). The flow may get restricted when the flap element is placed close to the main-element, which may reduce the lift. Figure 6.9(c) shows that the wake area at the main-element cove region is considerably reduced at the lower flap gap.

Further, Figure 6.9(d) shows the velocity contour in the vicinity of the flap at the higher overhang and lower gap. The flow is accelerated at a high gap and overhang that may increase lift on the flap and main-element. It is worth mentioning that the flow accelerated in the flap slot and main-element increases the suction on the main-element, which may increase the lift on the main-element. The velocity contour in the vicinity of the flap at a higher gap and overhang shows that the wake region in the main-element cove is reduced considerably, and the circulation on the flap element is increased, as shown in Figure 6.9(e). The parameters range given in Table 6.1 is used as input parameters for the sensitivity analysis. RBF model generated the metamodelling of the system based on the Sobol sequence-generated sampling points. The accuracy of the metamodelling is estimated by R square value as follows:

$$R^2 = 1 - \frac{\sum_{i=1}^{n}(y_i - \hat{y}_i)^2}{\sum_{i=1}^{n}(y_i - \bar{y}_i)^2} = 1 - \frac{\text{Mean Square Error}}{\text{Variance}}, \quad (6.15)$$

where $\hat{y}_i$ is the predicted value for the observed value and $\bar{y}_i$ is the mean variance of the observed values. It is worth mentioning that a more accurate metamodel has a higher R square value. The R square value of the metamodel produced by the present RBF metamodelling, which can be used to calculate the sensitivity indices, is 0.96.

In general, the individual input parameter effects are referred to as main effects, and the interaction effects of input parameters are called total effects. For understanding purposes, the slat's lift, drag, and moment coefficients are given as S_Cl, S_Cd, and S_Cm, respectively. Further, the lift, drag, and moment coefficients of the flap are given as F_Cl, F_Cd, and F_Cm, respectively. The lift, drag, and moment coefficients of the main-element are given as M_Cl, M_Cd, and M_Cm, respectively. Finally, the total lift, drag, and moment coefficients are given as T_Cl, T_Cd, and T_Cm, respectively.

The sensitivity index of the $Slat_{x\_trans}$ on the aerodynamic coefficients is shown in Figure 6.10(a), and the effects of $Slat_{y\_trans}$ on aerodynamic coefficients are shown in Figure 6.10(b). Figure 6.10(a) shows that the movement of the slat in the horizontal direction ($Slat_{x\_trans}$) predominantly influences the aerodynamics of all elements in a three-element airfoil system. Moreover, the $Slat_{x\_trans}$ influenced the aerodynamics of the slat element significantly, and some effects are found on the main-element aerodynamics. It is worth mentioning that the $Slat_{x\_trans}$ influenced almost all the elements; particularly the drag on the main-element (M_Cd) influenced significantly while the effect on total lift (T_Cl) is minimal. As seen in Figure 6.10(b), the parameter $Slat_{y\_trans}$ significantly influences the total drag of the three-element airfoil. When comparing with the $Slat_{x\_trans}$, the

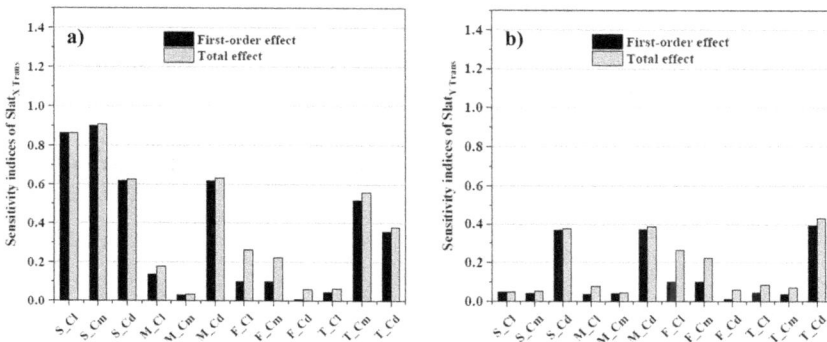

Figure 6.10 Sensitivity indices of (a) $Slat_{x\_trans}$ and (b) $Slat_{y\_trans}$ on the aerodynamic coefficients of the 30P30N airfoil.

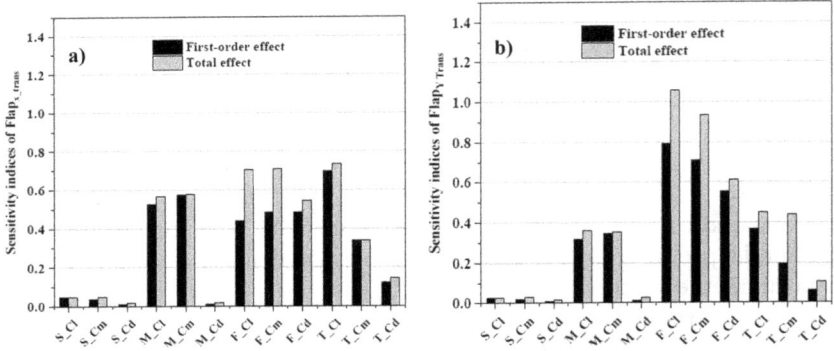

Figure 6.11 Sensitivity indices of (a) Flap$_{x\_trans}$ and (b) Flap$_{y\_trans}$ on the aerodynamic coefficients of the 30P30N airfoil.

influence of Slat$_{y\_trans}$ is minimum, and it is noticed that the effect on F_Cd is negligible. However, the total effects on F_Cl and F_Cm are considerably increased compared to the main effects that indicate the parameters' interaction is high on F_Cl and F_Cm. Drag components on the slat (S_Cd) and main-element (M_Cd) were observed. It is worth mentioning that the influence of Slat$_{y\_trans}$ on M_Cd may be due to the confluent layer interaction with the main-element boundary layer.

Figure 6.11(a) shows the sensitivity indices of Flap$_{x\_trans}$ on the lift, drag, and moment coefficients of the 30P30N airfoil. As expected, it is seen that the Flap$_{x\_trans}$ is not influencing the aerodynamic coefficients of the slat element, while the influence on the M_Cd and M_Cd is significant. Moreover, the main effect of Flap$_{x\_trans}$ on the T_Cl is significantly high when compared with the main effects on the flap's aerodynamic coefficients (F_Cl, F_Cd, and F_Cm). However, the total effects on F_Cl and F_Cm increased considerably, which shows the interaction effects of other parameters on F_Cl and F_Cm. The sensitivity indices of Flap$_{y\_trans}$ on the lift, drag, and moment coefficients of the three-element airfoil are shown in Figure 6.11(b). It is observed that the parameter Flap$_{y\_trans}$ is not affecting the aerodynamics of the slat elements while the main-element lift and moment coefficient got influenced. As expected, the aerodynamics coefficients of the flap (F_Cl, F_Cd, and F_Cm) are significantly influenced by the Flap$_{y\_trans}$. In particular, the total effects are very high, indicating the interaction of parameters on the aerodynamics of the flap element. Further, the T_Cl, T_Cd, and T_Cm are also influenced by the Flap$_{y\_trans}$, while the influence on T_Cd is comparatively less.

6.6 CONCLUDING REMARKS

This chapter introduces the sensitivity analysis on engineering problems. Further, as an example of an engineering problem, the geometric parameters effects on aerodynamics of a multielement airfoil is selected to investigate the sensitivity. A well-validated Navier-Stokes solver and Sobol's sequence sampling generator are utilized for this investigation to generate the slat and flap positions. Further, RBF is utilized to generate the metamodel of the system, and the sensitivity analysis indices are computed. The sensitivity analysis results show that the slat overhang influenced the overall performance of the three-element airfoil. Overall, the forward elements influenced the trailing elements' performance considerably, while the reverse is minimum. Significantly, the flap positions do not affect the slats, while it affects the main-element performance slightly. The total effects indicate the interaction effects, and it is found that the lift and moment coefficients of the flap are influenced mainly by the interaction effects among the others.

ACKNOWLEDGEMENTS

The authors would like to acknowledge the Department of Science and Technology, Government of India, for financial support for this research (Grant no. SPR/2020/000353).

REFERENCES

Anderson, W. K., Bonhaus, D. L., McGhee, R. J., & Walker, B. S. (1995). Navier-Stokes computations and experimental comparisons for multielement airfoil configurations. *Journal of Aircraft*, 32(6), 1246–1253. https://doi.org/10.2514/3.46871

Balaji, R., Bramkamp, F., Hesse, M., & Ballmann, J. (2006). Effect of flap and slat riggings on 2-D high-lift aerodynamics. *Journal of Aircraft*, 43(5), 1259–1271.

Burhenne, S., Tsvetkova, O., Jacob, D., Henze, G. P., & Wagner, A. (2013). Uncertainty quantification for combined building performance and cost-benefit analyses. *Building and Environment*, 62, 143–154. https://doi.org/10.1016/j.buildenv.2013.01.013

Chen, W., Jin, R., & Sudjianto, A. (2005). Analytical Variance-Based Global Sensitivity Analysis in Simulation-Based Design Under Uncertainty. *Journal of Mechanical Design*, 127(5), 875. https://doi.org/10.1115/1.1904642

Chin, V., Peters, D., Spaid, F., & Mcghee, R. (1993, July). Flowfield measurements about a multi-element airfoil at high Reynolds numbers. *23rd Fluid Dynamics, Plasmadynamics, and Lasers Conference*. https://doi.org/10.2514/6.1993-3137

Deck, S., & Laraufie, R. (2013). Numerical investigation of the flow dynamics past a three-element aerofoil. *Journal of Fluid Mechanics*, 732, 401–444.

Grendysa, W., & Olszański, B. (2021). Optimization of the three-segment aerofoil arrangement based on wind tunnel tests and numerical analysis. *Proceedings of the Institution of Mechanical Engineers, Part G: Journal of Aerospace Engineering*, 235(1), 117–137.

Hardy, R. L. (1971). Multiquadric equations of topography and other irregular surfaces. *Journal of Geophysical Research*, 76(8), 1905–1915. https://doi.org/10.1029/JB076i008p01905

Kim, S., Alonso, J., & Jameson, A. (2002, January). Design optimization of high-lift configurations using a viscous continuous adjoint method. *40th AIAA Aerospace Sciences Meeting & Exhibit.* https://doi.org/10.2514/6.2002-844

Li, D. (2013). *Multi-Objective Design Optimization for High-Lift Aircraft Configurations Supported By Surrogate Modeling* (Issue December) [Cranfield University]. https://files.core.ac.uk/pdf/23/20338966.pdf

Lin, J. C., & Dominik, C. J. (1997). Parametric Investigation of a High-Lift Airfoil at High Reynolds Numbers. *Journal of Aircraft, 34*(4), 485–491. https://doi.org/10.2514/2.2217

Martin, J. D., & Simpson, T. W. (2005). Use of Kriging Models to Approximate Deterministic Computer Models. *AIAA Journal, 43*(4), 853–863. https://doi.org/10.2514/1.8650

Mullur, A. A., & Messac, A. (2005). Extended Radial Basis Functions: More Flexible and Effective Metamodeling. *AIAA Journal, 43*(6), 1306–1315. https://doi.org/10.2514/1.11292

Nakayama, A., Kreplin, H. P., & Morgan, H. L. (1990). Experimental investigation of flowfield about a multielement airfoil. *AIAA Journal, 28*(1), 14–21. https://doi.org/10.2514/3.10347

Nakayama, H., Kim, H.-J., Matsushima, K., Nakahashi, K., & Takenaka, K. (2006). Aerodynamic Optimization of Multi-Element Airfoil. *44th AIAA Aerospace Sciences Meeting and Exhibit,* January, 1–15. https://doi.org/10.2514/6.2006-1051

Oakley, J. E., & O'Hagan, A. (2004). Probabilistic sensitivity analysis of complex models: a Bayesian approach. *Journal of the Royal Statistical Society: Series B (Statistical Methodology), 66*(3), 751–769. https://doi.org/10.1111/j.1467-9868.2004.05304.x

Pagani, C. C., Souza, D. S., & Medeiros, M. A. F. (2017). Experimental investigation on the effect of slat geometrical configurations on aerodynamic noise. *Journal of Sound and Vibration, 394,* 256–279.

Pehlivanoglu, Y. V., & Yagiz, B. (2012). Aerodynamic design prediction using surrogate-based modeling in genetic algorithm architecture. *Aerospace Science and Technology, 23*(1), 479–491. https://doi.org/10.1016/j.ast.2011.10.006

Mohamed Abubacker Siddique, P. M., & Raj, L. P. (2023). Sensitivity analysis of geometric parameters on the aerodynamic performance of a multi-element airfoil. *Aerospace Science and Technology, 132,* 108074.

Pomeroy, B., Williamson, G., & Selig, M. (2012). Experimental Study of a Multielement Airfoil for Large Wind Turbines. *30th AIAA Applied Aerodynamics Conference,* June, 835–852. https://doi.org/10.2514/6.2012-2892

Prince Raj, L., Lee, J. W., & Myong, R. S. (2019). Ice accretion and aerodynamic effects on a multi-element airfoil under SLD icing conditions. *Aerospace Science and Technology, 85,* 320–333. https://doi.org/10.1016/j.ast.2018.12.017

Prince Raj, L., Yee, K., & Myong, R. S. (2020). Sensitivity of ice accretion and aerodynamic performance degradation to critical physical and modeling parameters affecting airfoil icing. *Aerospace Science and Technology, 98.* https://doi.org/10.1016/j.ast.2019.105659

Rogers, S. E. (1994). Progress in high-lift aerodynamic calculations. *Journal of Aircraft, 31*(6), 1244–1251. https://doi.org/10.2514/3.46642

Rumsey, C. L., Gatski, T. B., Ying, S. X., & Bertelrud, A. (1998). Prediction of High-Lift Flows Using Turbulent Closure Models. *AIAA Journal, 36*(5), 765–774. https://doi.org/10.2514/2.435

Saltelli, A. (2002). Making best use of model evaluations to compute sensitivity indices. *Computer Physics Communications*, *145*(2), 280–297. https://doi.org/10.1016/S0010-4655(02)00280-1

Saltelli, A., & Homma, T. (1996). Importance measures in global sensitivity analysis of model output. *Reliab. Eng. Sys. Safety*, *52*, 1–17.

Smith, A. M. O. (1975). High-lift aerodynamics. *Journal of Aircraft*, *12*(6), 501–530. https://doi.org/10.2514/3.59830

Sobol, I. M. (2011). Global sensitivity indices for nonlinear mathematical models and their Monte Carlo estimates. *Mathematics and Computers in Simulation*, *4*(2), 989–999. https://doi.org/10.4271/2011-01-1529

Soulat, L., Fosso Pouangué, A., & Moreau, S. (2016). A high-order sensitivity method for multi-element high-lift device optimization. *Computers & Fluids*, *124*, 105–116. https://doi.org/10.1016/j.compfluid.2015.10.013

Spaid, F., & Lynch, F. (1996). High Reynolds number, multi-element airfoil flowfield measurements. *34th Aerospace Sciences Meeting and Exhibit*, *37*(3). https://doi.org/10.2514/6.1996-682

Sudret, B. (2008). Global sensitivity analysis using polynomial chaos expansions. *Reliability Engineering & System Safety*, *93*(7), 964–979. https://doi.org/10.1016/j.ress.2007.04.002

Valarezo, W., Dominik, C., Mcghee, R., Goodman, W., & Paschal, K. (1991, September). Multi-element airfoil optimization for maximum lift at high Reynolds numbers. *9th Applied Aerodynamics Conference*. https://doi.org/10.2514/6.1991-3332

Valarezo, W. O. (1992, July). High-lift testing at high Reynolds numbers. *AIAA 17th Aerospace Ground Testing Conference, 1992*. https://doi.org/10.2514/6.1992-3986

Valarezo, W. O., Dominik, C. J., & Mcghee, R. J. (1992). Reynolds and Mach number effects on multielement airfoils. *The Fifth Symposium on Numerical and Physical Aspects of Aerodynamic Flows*. https://ntrs.nasa.gov/citations/19930018257

van Dam, C. P. (2002). The aerodynamic design of multi-element high-lift systems for transport airplanes. *Progress in Aerospace Sciences*, *38*(2), 101–144. https://doi.org/10.1016/S0376-0421(02)00002-7

Werbos, P. J. (2005). Applications of advances in nonlinear sensitivity analysis. In *System Modeling and Optimization* (pp. 762–770). Springer-Verlag. https://doi.org/10.1007/BFb0006203

Wild, J. (2013). Mach and Reynolds number dependencies of the stall behavior of high-lift wing-sections. *Journal of Aircraft*, *50*(4), 1202–1216.

Zhang, X., & Zerihan, J. (2003). Aerodynamics of a double-element wing in ground effect. *AIAA Journal*, *41*(6), 1007–1016.

Chapter 7

Design optimization and sensitivity analysis in aerospace engineering

Mrityunjai Verma and Varun Pratap Singh
School of Advanced Engineering, UPES, Dehradun, India

7.1 INTRODUCTION

The field of aerospace engineering has experienced a continuous evolution, pushing the boundaries of human achievement in flight and space exploration. A pivotal aspect of this advancement is the intricate interplay between various components and variables that govern aerospace systems. Sensitivity analysis, as a fundamental analytical tool, has emerged as a critical driver in deciphering these intricate relationships. This chapter delves into the dynamic landscape of sensitivity analysis within the realm of aerospace engineering, unravelling its multifaceted significance and far-reaching implications. In the pursuit of designing and optimizing aerospace systems, comprehending the impact of different parameters becomes paramount. The quest for enhanced aerodynamic efficiency, structural integrity, and operational robustness necessitates a profound understanding of the sensitivities between system variables. Sensitivity analysis is the key to unlocking this understanding, enabling engineers and researchers to identify influential factors and predict the outcomes of design modifications. From a historical perspective, sensitivity analysis has been an instrumental tool in aerospace engineering, allowing for the refinement of aircraft and spacecraft designs, propulsion systems, and navigation strategies [1].

The scope of this chapter encompasses a comprehensive exploration of sensitivity analysis in the context of aerospace engineering applications. It traverses the spectrum from classical methodologies to cutting-edge techniques, shedding light on both conventional and contemporary practices. The chapter delves into the domains of design optimization, uncertainty quantification, and system-level analysis, elucidating how sensitivity analysis intertwines with each of these facets. Furthermore, the chapter extends its purview to include emerging and unconventional areas such as electric propulsion systems, UAVs, and space exploration missions. By delving into this diverse array of applications, the chapter aims to offer a holistic view of sensitivity analysis's far-reaching impact within the aerospace domain [2].

DOI: 10.1201/9781003465171-7

The primary objectives of this chapter are to provide readers with a comprehensive overview of sensitivity analysis, including its historical underpinnings and its evolution in aerospace engineering. It also examines established sensitivity analysis techniques, offering insights into their strengths, weaknesses, and limitations when applied to aerospace systems. Additionally, the chapter explores the future of sensitivity analysis in aerospace engineering, including the integration of computational methods such as machine learning and artificial intelligence. It also delves into various applications of sensitivity analysis within contemporary aerospace engineering, addressing challenges inherent in aerospace systems. In essence, this chapter aims to be a guiding compass, illuminating the path forward for researchers, engineers, and enthusiasts in harnessing the power of sensitivity analysis to conquer the intricate challenges of aerospace engineering [3].

7.2 SENSITIVITY ANALYSIS IN WIND TUNNEL TESTING AND EXPERIMENTAL TECHNIQUES

Wind tunnel testing is a critical tool in aerospace engineering for evaluating the aerodynamic performance of aircraft and spacecraft designs. In this chapter, we delve into the application of sensitivity analysis in wind tunnel testing, a methodology crucial for enhancing the precision and efficiency of these experiments. Understanding how variations in parameters such as airspeed, angle of attack, or model geometry affect aerodynamic forces and moments is essential. Sensitivity analysis aids in precisely identifying and varying these parameters during the experiments, allowing for comprehensive data collection and error reduction. It also guides the design of experiments, helping researchers choose appropriate measurement techniques and configurations. Moreover, it assists in post-test data analysis by accessing data quality, identifying uncertainties, and establishing error bounds, ensuring the reliability and accuracy of experimental results. The chapter explores innovative experimental techniques guided by sensitivity analysis, such as adaptive testing protocols and integration with computational fluid dynamics simulations. It also discusses future trends, including the integration of artificial intelligence and machine learning algorithms, showcasing how sensitivity analysis is evolving to advance aerospace technology. Through case studies and discussions, the chapter underscores the profound impact of sensitivity analysis on aerospace research, ultimately contributing to the development of more efficient and optimized aerospace designs [4].

7.3 SENSITIVITY ANALYSIS IN AEROSPACE ENGINEERING

The field of aeronautical engineering is characterized by its continuous quest for innovation and efficiency, which requires a deep understanding of the intricate interplay of variables inside complex systems. Sensitivity analysis is an important and fundamental tool that provides a structured framework for comprehending the intricate interconnections inherent in these systems. The purpose of this part is to offer a thorough examination of sensitivity analysis, emphasizing its significance and many applications in the domain of aeronautical engineering [5].

7.3.1 Definition and importance of sensitivity analysis

Sensitivity analysis is a fundamental aspect of quantitative engineering analysis, involving a methodical evaluation of how variations in input parameters impact the behaviour and performance of the output in a specific system. This procedure enables a systematic understanding of the quantitative influence that changes in parameters exert on important output reactions. The importance of sensitivity analysis is in its capacity to reveal key aspects that have a substantial impact on the outcomes of a system. Through the identification of these crucial aspects, engineers and researchers are able to strategically deploy resources, prioritize design revisions, and maximize the overall performance of the system [6].

The primary objective of sensitivity analysis is to assess the impact of fluctuations in input parameters on the results of a mathematical model. This analytical methodology aids in identifying the most suitable data points for conducting thorough evaluations of a project's return on investment (ROI). In addition, sensitivity analysis serves as a vital tool for engineers in facilitating the creation of more reliable and robust designs. This is achieved by the methodical assessment of regions of uncertainty inherent in the design's structure [7].

7.4 ROLE IN DESIGN OPTIMIZATION

The utilization of sensitivity analysis within the field of aerospace engineering has significant importance in the process of designing and assessing intricate aircraft systems. The utilization of this analytical methodology enables engineers and academics to quantitatively assess the impact of fluctuations in input parameters on the outcomes of complex aeronautical models and simulations. In a discipline that prioritizes accuracy and dependability, sensitivity analysis functions as a potent instrument for comprehending the connections between design factors and system responses. Sensitivity analysis offers significant insights in several applications, such as refining the aerodynamic profile of an airplane, fine-tuning the trajectory of a space mission, or boosting the efficiency of a propulsion system. Aerospace professionals can enhance design efficiency, reduce costs, and improve overall safety and performance by conducting systematic assessments of how alterations in parameters such as airfoil shape, propulsion thrust, or structural materials impact crucial performance metrics like lift, drag, fuel consumption, or structural integrity. Moreover, sensitivity analysis has significant importance in situations characterized by inherent uncertainty and unpredictability, enabling aeronautical engineers to properly consider and address these aspects. Sensitivity analysis plays a crucial role in directing the optimization of aeronautical systems within a dynamic and demanding sector [8].

7.5 UNCERTAINTY QUANTIFICATION AND RISK ASSESSMENT

The aircraft environment is characterized by a high level of unpredictability, which can be attributed to several factors such as manufacturing tolerances, operational situations, and external influences. Sensitivity analysis is an essential approach for assessing the impact of these unknown factors on the performance of a system. This quantification helps with risk assessment since it sheds light on the possibility for outcomes to vary depending on a number of factors (Figure 7.1).

In the context of aerospace systems, uncertainty analysis can be further segmented into two bifurcations, which can be referred to as "sensitivity characterization" (SC) and "sensitivity propagation" (SP). Sensitivity characterization involves quantifying uncertainties associated with various aspects of the model inputs. Here, "inputs" encompass a wide range of parameters within the model, derived from real-world data. These parameters encompass model variables, boundary and initial conditions, loading conditions, as well as geometric and material properties. The primary sources of uncertainty in these inputs typically arise from measurement imprecision (e.g., the inability to precisely measure certain quantities) and inherent natural variability (e.g., variations in physiological metrics across a population). The objective of sensitivity characterization is to establish probability distributions that describe the uncertainty associated with each input parameter. This task is often data-driven and can be particularly

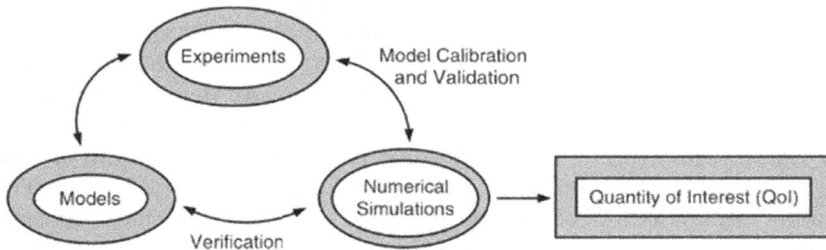

Figure 7.1 Procedures and methodologies involved in establishing the quantification of uncertainty.

challenging for intricate aerospace models with numerous parameters, where even estimating mean values can be demanding [9].

The second stage, sensitivity propagation, involves the process of transmitting the identified input uncertainties through the model to assess the resulting uncertainty in crucial model outputs. This phase can pose significant computational challenges, especially in aerospace systems, as it often necessitates a substantial number of simulations to be executed [10].

7.6 SYSTEM-LEVEL ANALYSIS AND DECISION-MAKING

Aerospace systems are complex arrangements of interconnected subsystems, wherein modifications to a single component can have far-reaching effects on the entire system. Sensitivity analysis expands its scope to the system level, enabling engineers to gain a comprehensive understanding of the overall consequences of differences in parameters. This knowledge plays a crucial role in making informed decisions and allocating resources effectively. In the context of aerospace systems, sensitivity analysis plays a significant role as a technique for determining particular requirements in the decision-making process. In the realm of complex systems, the identification of factors that exert the most substantial influence on system performance or outcomes is sometimes a formidable task. Sensitivity analysis facilitates the identification of essential factors by evaluating the impact of variations in each parameter on the outcomes. This information provides guidance to decision-makers in directing their attention and allocating resources towards the most impactful elements of the system [11].

Risk assessment is a critical process that necessitates the inclusion of sensitivity analysis. Decision-makers have the ability to evaluate the possible risks associated with various scenarios by comprehending how uncertainty in input parameters propagates through a system and impact outcomes. This empowers individuals to make well-informed decisions that take into consideration uncertainties and effectively address potential negative outcomes. In the context of system-level analysis, it is common to encounter various design choices and trade-offs that necessitate optimization. Sensitivity analysis plays a crucial role in the optimization process as it provides insights into the influence of parameter modifications on performance metrics or targets. This information may be utilized by decision-makers to ascertain the most favourable design or operating circumstances that align with their objectives. The utilization of sensitivity analysis aids in the efficient allocation of resources. Organizations may effectively prioritize investments, manage funds, and allocate staff by identifying characteristics that have a large impact. This enables them to focus resources on areas that will provide the most influence on system performance and results [12].

Scenario analysis is a crucial aspect of decision-making, since it necessitates the consideration of several scenarios and potential consequences.

Sensitivity analysis facilitates the process of scenario design by examining the impact of varying parameter values or assumptions on outcomes. This practice facilitates the preparation for various potential future scenarios and enables the formulation of resilient decision-making strategies. Continuous improvement is a crucial aspect in dynamic systems. Iterative utilization of sensitivity analysis enables the evaluation of the effects of modifications over a period of time. This enables decision-makers to enhance strategies, procedures, and designs in response to the availability of fresh data or the evolution of conditions [13].

The utilization of sensitivity analysis offers a transparent means of communicating the potential consequences of actions to stakeholders, hence facilitating effective engagement. The utilization of this approach aids in effectively illustrating the resilience of a plan or system, hence fostering productive discourse by showcasing the inherent compromises and uncertainties associated with it. The present scenario under examination involves the utilization of an unmanned aerial vehicle (UAV) in the context of search and rescue missions. The assessment of sensitivity analysis involves evaluating the influence of variations in key factors such as UAV propulsion efficiency, sensor accuracy, and communication range on the overall mission's level of achievement. By comprehending the interdependence of these elements, engineers have the ability to enhance the design of the UAV in order to guarantee consistent and dependable performance in crucial situations [14].

Sensitivity analysis serves as a fundamental tool for gaining engineering understanding in the context of aeronautical applications. This section has thoroughly examined the importance of the subject matter, encompassing the identification of crucial variables, the enhancement of designs, the measurement of uncertainties, and the provision of guidance for decisions at the system level. In the ever-changing aerospace industry, sensitivity analysis is an essential technique that guides engineers in effectively managing complexity and driving advancements in aerospace engineering [15].

7.7 CURRENT METHODOLOGIES AND LIMITATIONS

Many different approaches to sensitivity analysis have been developed in the field of aeronautical engineering. These methods are the backbone upon which complicated interactions may be understood and system performance can be optimized. Various approaches are discussed in detail, along with the benefits and drawbacks of each (Figure 7.2).

7.8 PARAMETRIC SENSITIVITY ANALYSIS

A fundamental method, parametric sensitivity analysis attempts to quantify the effect of modest changes to individual input parameters on system outputs. This approach includes repeatedly changing one parameter while

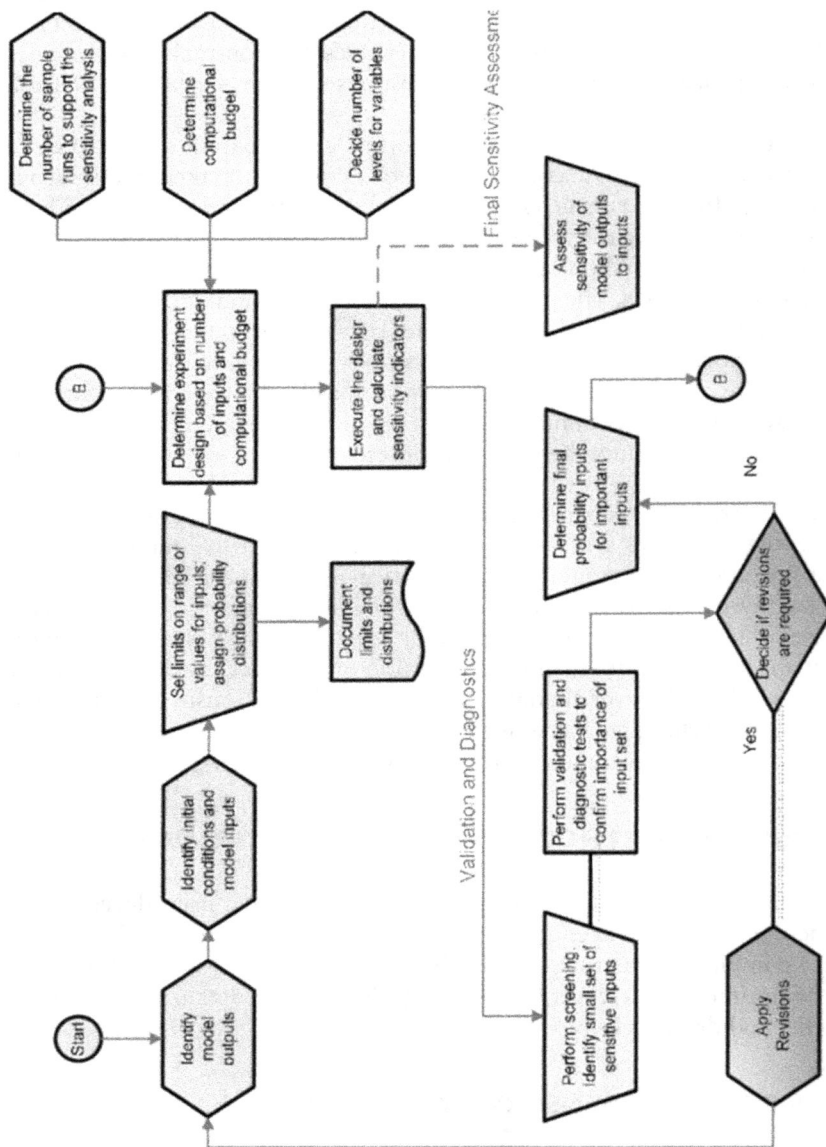

Figure 7.2 Process diagram of conducting sensitivity analysis.

holding all others fixed and then analysing the subsequent shifts in output responses. Parametric sensitivity analysis is a straightforward method for learning more about the linear connections between variables and outcomes. As an example of aero-elastic modelling in aeronautical engineering, think of a simulated airplane wing. Parametric sensitivity analysis shows how changes in factors like material qualities or wing form affect wing deflection and stress distribution. However, it only works when parameter interactions are linear, which rules out more complicated systems where nonlinear effects may play a role [16].

7.9 GRADIENT-BASED SENSITIVITY ANALYSIS

Gradient-based sensitivity analysis relies on the principles of calculus and is especially suitable for situations in which continuous and differentiable functions are widespread. By calculating partial derivatives of output responses with respect to input parameters, engineers gain insight into the rate of change of outputs as parameters shift. This technique is invaluable when precision and accuracy are crucial. In the context of propulsion system optimization, gradient-based sensitivity analysis enables engineers to determine how changes in fuel composition affect combustion efficiency and thrust. By calculating derivatives, the optimal fuel mixture can be pinpointed, facilitating improvements in engine performance. However, this method assumes differentiability and is less effective when dealing with nonlinear or discontinuous systems [17].

7.10 GLOBAL SENSITIVITY ANALYSIS

Global sensitivity analysis goes beyond examining isolated parameter variations and instead considers the entire parameter space's influence on output responses. This holistic approach acknowledges that system behaviour might change non-monotonically due to interactions among multiple parameters. By exploring parameter space comprehensively, engineers can capture intricate relationships and interactions. For instance, in spacecraft trajectory optimization, global sensitivity analysis might reveal that multiple parameters—such as initial velocity, launch angle, and gravitational forces—simultaneously impact the spacecraft's trajectory. By understanding how these factors collectively influence the trajectory, engineers can formulate optimal launch strategies that account for a range of scenarios [18].

7.11 CHALLENGES IN TRADITIONAL APPROACHES

While these methodologies offer valuable insights, they are not without limitations. One common challenge is the curse of dimensionality, particularly prevalent in high-dimensional parameter spaces. As aerospace systems become more intricate, traditional sensitivity analysis methods struggle to efficiently explore and quantify sensitivities in these spaces. This can lead to impractical computational costs and hinder accurate analysis.

In addition, many aerospace systems have complex nonlinear interactions, which are not captured by conventional sensitivity analysis methods since they presume linear correlations. Emergent behaviours that have a major effect on system performance may be missed by these approaches. As aeronautical engineering incorporates more complicated designs and cutting-edge technology, the limitations of conventional approaches become more apparent [19].

In conclusion, the variety of sensitivity analysis approaches in aeronautical engineering is highlighted by the methods described in this section. Parametric, gradient-based, and global sensitivity assessments all have their uses, but they have drawbacks as well, such as linearity assumptions and difficulties in high-dimensional spaces, which emphasize the need for new methods that can handle the complexities of today's aerospace systems [20].

7.12 FUTURE TRENDS IN SENSITIVITY ANALYSIS

Sensitivity analysis in aeronautical engineering is on the cusp of a revolutionary new era, one that will be propelled by cutting-edge tools and groundbreaking approaches. This section deconstructs the emerging tendencies in sensitivity analysis and explains how they are changing the way aerospace systems are analysed and optimized (Figure 7.3).

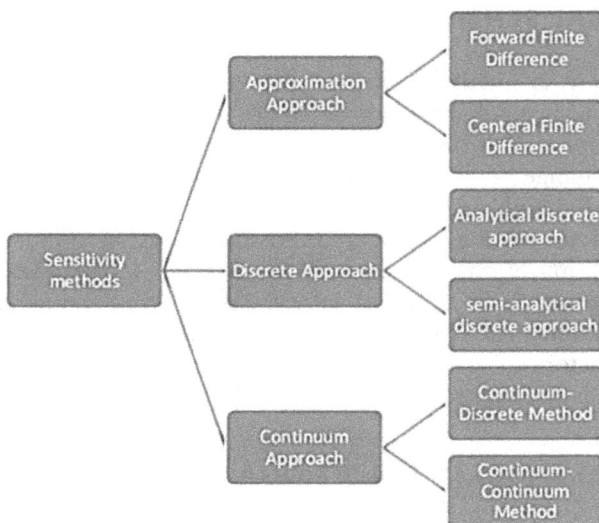

Figure 7.3 Techniques and methodologies involved in sensitivity analysis.

7.13 INTEGRATION OF MACHINE LEARNING AND AI

The integration of machine learning (ML) and artificial intelligence (AI) with sensitivity analysis represents a significant advancement in aerospace engineering. These technologies possess the capability to revolutionize the field by efficiently handling extensive and intricate datasets, uncovering subtle patterns, and capturing nonlinear relationships that might be elusive to conventional analytical methods. This integration holds great promise for enhancing our understanding of complex system behaviours and optimizing aerospace designs [16].

To illustrate the potential impact, let's consider the optimization of a hypersonic vehicle's aerodynamics. Traditional methods may struggle to decipher the multifaceted interactions between numerous design parameters. However, by incorporating ML algorithms, aerospace engineers gain the ability to analyse massive datasets containing aero-elastic simulations, sensor measurements, and real-time flight data. The AI-powered system can discern intricate and nonlinear dependencies among these parameters, shedding light on emergent phenomena critical for precise performance predictions. This not only streamlines the design process but also contributes to the development of more efficient and capable aerospace systems. The marriage of sensitivity analysis with ML and AI represents a transformative step towards addressing the complexities and challenges inherent in modern aerospace engineering, ultimately leading to safer, more efficient, and innovative aerospace solutions [21].

7.14 DATA-DRIVEN SENSITIVITY ANALYSIS TECHNIQUES

Conventional sensitivity analysis techniques, often based on mathematical models, encounter constraints when confronted with complex system dynamics or the absence of a clearly defined mathematical framework. The utilization of empirical data in data-driven sensitivity analysis approaches has emerged as an effective approach to address the limitations mentioned, allowing for the identification of previously undisclosed sensitivities and correlations.

In the field of aerospace materials development, there has been significant progress in the application of data-driven sensitivity analysis. Currently, engineers possess the capability to effectively utilize a vast amount of experimental data pertaining to material characteristics, structural behaviour, and thermal conductivity. By utilizing advanced machine learning methods, researchers are able to unveil subtle relationships between the aforementioned material qualities and the overall performance of aeronautical systems. The utilization of a data-driven methodology not only improves our understanding of intricate aerospace materials but also enables us to make informed choices on the selection of materials for certain aircraft applications [22].

For example, in the process of developing aerospace components that are subjected to high levels of stress, such as turbine blades or spacecraft heat shields, engineers have the ability to examine a comprehensive dataset containing material attributes and performance parameters. Machine learning algorithms have the capability to discern intricate associations among variables such as tensile strength, thermal expansion coefficients, and structural stability. Equipped with this acquired information, engineers possess the capacity to make informed judgements based on facts regarding material selections that aim to maximize performance, decrease weight, improve durability, and minimize expenses. The aerospace industry's use of data-driven sensitivity analysis reflects its dedication to utilizing new technology in order to make better informed and efficient engineering choices, resulting in the development of safer and more sophisticated aircraft systems [23].

7.15 HYBRID APPROACHES: PHYSICS-BASED AND DATA-DRIVEN

The fusion of physics-based models and data-driven techniques represents a pivotal advancement in the realm of sensitivity analysis, offering the potential to revolutionize aerospace engineering practices. Hybrid approaches harmonize the precision of physics-based understanding with the adaptability and empirical grounding of data-driven methods, delivering a holistic perspective on the intricate dynamics of aerospace systems [24].

For instance, let's delve into the complexities of optimizing a re-entry vehicle's heat shield design. This critical component must withstand the extreme conditions of re-entering Earth's atmosphere, balancing the imperatives of thermal protection and structural integrity. In this scenario, a hybrid approach can be employed, simultaneously leveraging computational fluid dynamics (CFD) simulations to capture the intricate physics of fluid flow and heat transfer while integrating empirical data gleaned from numerous real-world re-entry missions. This synergistic methodology empowers engineers to navigate the delicate equilibrium between the underlying physics and the inherent variability introduced by actual re-entry scenarios. Consider the extensive dataset compiled from past missions that includes measurements of heat shield temperature, structural deformation, and material behaviour under varying conditions. Machine learning algorithms can analyse this wealth of empirical data, revealing nuanced correlations between design parameters, operational factors, and performance metrics. This holistic approach allows engineers to craft heat shield designs that are not only grounded in a deep understanding of the physics at play but also attuned to the practical challenges posed by unpredictable re-entry scenarios [25].

In an era where aerospace missions grow increasingly complex and diversified, hybrid sensitivity analysis approaches emerge as a potent tool, facilitating the design of aerospace systems that excel in both theoretical rigor and real-world adaptability. This innovative convergence of disciplines exemplifies the aerospace industry's commitment to harnessing the full spectrum of available knowledge and technologies to drive progress in aerospace engineering, thus ultimately advancing safety, efficiency, and mission success rates.

7.16 ADVANCEMENTS IN SENSITIVITY ANALYSIS ALGORITHMS

Due to the increasing complexity of aircraft systems, sensitivity analysis techniques are constantly emerging in aeronautical engineering. These algorithmic advances are a major departure from current methods and will efficiently solve aeronautical engineering's complex problems. They excel in understanding nonlinear connections, high-dimensional parameter spaces, and complex system dynamics. A notable advancement in this field is surrogate-based optimization. This revolutionary strategy uses restricted high-fidelity simulations to build surrogate models. Surrogate models efficiently approximate complicated aeronautical systems for fast sensitivity analysis. Surrogate-based optimization is ideal for exploring large design spaces in propulsion systems and spaceship configurations by reducing resource-intensive simulations. This paradigm change in sensitivity analysis techniques allows aeronautical engineers to evaluate complex design decisions and their far-reaching effects. These sophisticated algorithms are essential for engineering solutions to stay practical and optimum in an increasingly complex and linked aerospace landscape as aerospace systems push technical boundaries [26] (Figure 7.4).

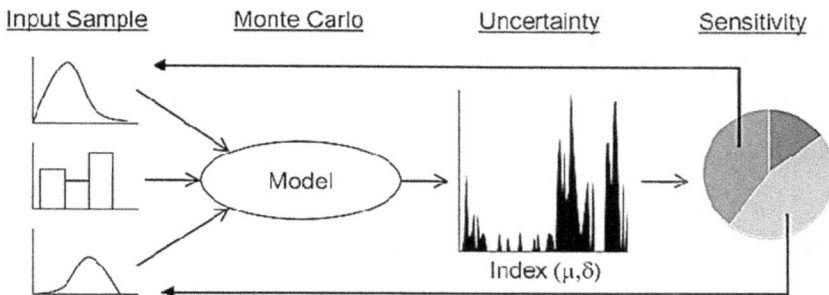

Figure 7.4 Methods to quantify uncertainty and identifying risks.

When integrated with machine learning, Monte Carlo algorithms may explore large parameter sets and quantify uncertainty. This is critical when external factors like atmospheric conditions cause system behaviour uncertainty. A hybrid technique using Monte Carlo simulations and ML algorithms can reveal aircraft system resilience under different scenarios. The sensitivity analysis trends promise unprecedented accuracy and efficiency. Aircraft engineers use machine learning, data-driven methods, hybrid techniques, and sophisticated algorithms to optimize designs, robust performance, and better understand complex behaviours in modern aircraft systems [27].

7.17 APPLICATION IN EMERGING AEROSPACE AREAS

Sensitivity analysis, once rooted in traditional aerospace domains, has expanded its horizons to encompass emerging areas that are at the forefront of shaping the future of aerospace engineering. This section provides insights into the specific applications of sensitivity analysis within these burgeoning fields, highlighting how this methodology is effectively addressing complex challenges and catalysing innovation.

7.18 SENSITIVITY ANALYSIS IN ELECTRIC PROPULSION SYSTEMS

With the promise of greater efficiency and reduced environmental impact, electric propulsion technologies are ushering in a new age in aircraft propulsion. Optimization of such systems relies heavily on sensitivity analysis, which guarantees the best possible thrust-to-power ratios, propellant consumption rates, and other performance parameters. Electric propulsion makes use of sensitivity analysis to assess how changes to factors like power input, propellant composition, and magnetic field intensity influence the ability to generate thrust. Engineers can determine the most effective setups, taking into account real-world considerations like power limitations and the overall mass of the spacecraft. Electric propulsion systems are essential for long-term space missions, satellite station-keeping, and interplanetary exploration, and this knowledge directs their development [28].

7.19 UAV DESIGN AND OPERATION OPTIMIZATION

UAVs, or unmanned aerial vehicles, have quickly become indispensable tools in fields as varied as surveillance and remote sensing. When applied to UAV design and operational optimization, sensitivity analysis is a crucial component in controlling critical performance metrics, including range, battery life, and mission success. Consider the engineering of a solar-powered UAV for use in ecological research. The impact of key factors, such as wing aspect ratio, solar cell efficiency, and payload weight, on the UAV's endurance and flying characteristics is evaluated by sensitivity analysis. Engineers may create UAV configurations with the ideal combination of energy economy, data gathering, and mission endurance by isolating the most crucial design features [29].

7.20 SENSITIVITY ANALYSIS IN SPACE EXPLORATION MISSIONS

The effectiveness of space exploration missions and the data gleaned from them is influenced by a tangled web of factors. Sensitivity analysis is a powerful framework for investigating how different factors influence spacecraft paths, the efficiency of scientific instruments, and the dependability of communication networks. Imagine an expedition sent to investigate a faraway celestial body like a comet. Understanding how changes in parameters like propulsion thrust, trajectory correction manoeuvres, and communication latency affect the spacecraft's arrival time and closeness to the target may be determined with the use of a sensitivity analysis. Engineers may ensure accurate rendezvous with the celestial body and more easily acquire high-fidelity scientific data by quantifying these sensitivity factors and then methodically optimizing mission plans [30].

7.21 ROLE IN SUSTAINABLE AEROSPACE TECHNOLOGIES

The aerospace sector is now undergoing a significant shift towards sustainability, driven by the urgent need to decrease environmental impacts and improve operational effectiveness. The employment of sensitivity analysis has become a significant factor in the development of sustainable aircraft technologies. It offers a quantitative approach to evaluating the potential impacts of design decisions on important aspects such as emissions, energy consumption, and material usage [31].

The present analysis contemplates the advancement of a future airplane design that is meticulously constructed to significantly mitigate carbon emissions. The utilization of sensitivity analysis allows for the examination of the many effects that result from changes in parameters related to aerodynamic profiles, propulsion system efficiency, and the incorporation of lightweight materials. Engineers are able to design airplanes that effectively reduce environmental impact and meet rigorous performance criteria by identifying and focusing on the most significant characteristics. The utilization of sensitivity analysis in new aerospace sectors highlights its inherent flexibility and versatility as a decision-support tool. In the domains of electric propulsion, UAV improvement, space exploration, and the development of sustainable aerospace technologies, sensitivity analysis remains an essential tool for aerospace engineers. It assists in guiding them towards innovative solutions that effectively address the distinctive challenges posed by these dynamic and ever-changing fields [32].

7.22 CHALLENGES AND CONSIDERATIONS

Although sensitivity analysis is a powerful tool in the field of aeronautical engineering, it is not immune to the difficulties associated with understanding intricate systems. This section explores the various complex issues and factors that engineers need to take into account while performing sensitivity analysis to handle the complexities of aeronautical systems.

7.22.1 High-dimensional parameter spaces

Contemporary aircraft systems frequently incorporate a diverse array of characteristics, with the capacity to exert influence on system behaviour. The task of doing sensitivity analysis becomes more challenging when dealing with parameter spaces that have a high number of dimensions. Conventional approaches encounter challenges in effectively navigating extensive design areas, resulting in computational inefficiencies and constrained precision.

The present study examines an aircraft design characterized by a multitude of factors that have influence on aerodynamic performance. The process of manoeuvring through the many factors inside a given area, such as wing form, engine placement, and control surface deflections, can be quite complex. To tackle this difficulty, it is necessary to employ creative methodologies such as surrogate modelling and dimensionality reduction approaches in order to effectively investigate and evaluate parameter spaces that include a high number of dimensions [33].

7.22.2 Dealing with complex and nonlinear relationships

Aerospace systems frequently exhibit intricate interactions and nonlinear behaviours that traditional sensitivity analysis techniques might struggle to capture. As systems become more complex, identifying and quantifying these nonlinear relationships become increasingly critical. Failing to account for these complexities can lead to inaccurate sensitivity results and suboptimal designs. Consider the dynamics of a spacecraft during a complex orbital transfer manoeuvre. The interaction between propulsion thrust, gravitational forces, and atmospheric drag introduces nonlinearities that traditional linear sensitivity analysis techniques might not adequately capture. Advanced methods like finite difference approximations and gradient-free optimization can help address these nonlinear challenges [34].

7.22.3 Computational costs and efficiency

Sensitivity analysis often relies on simulations or mathematical models that demand substantial computational resources. Performing numerous simulations to quantify sensitivities can incur high computational costs, limiting the speed at which analyses can be conducted and hampering real-time decision-making.

For instance, in the analysis of thermal management systems for spacecraft, conducting multiple simulations to assess how variations in coolant flow rates impact temperature distribution can be computationally intensive. Strategies like adaptive sampling, surrogate modelling, and parallel computing can alleviate computational burdens, enabling more efficient sensitivity analysis (Figure 7.5).

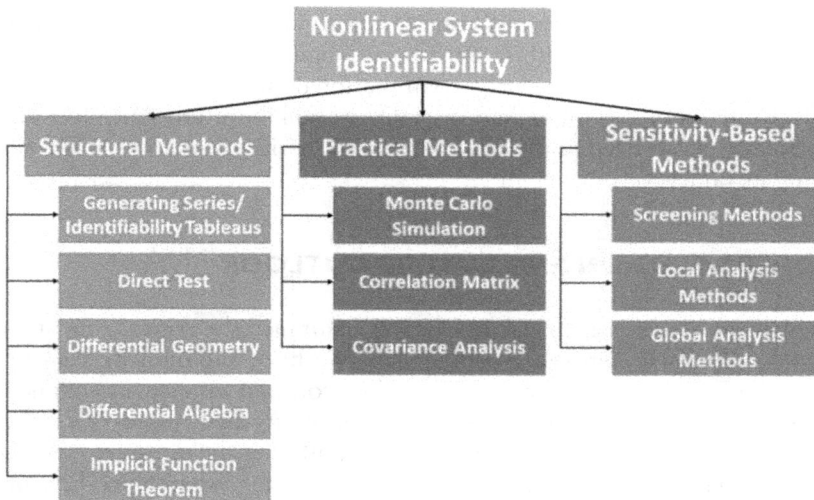

Figure 7.5 Nonlinear system identifiability—methodologies and sub-methodologies.

7.22.4 Uncertainty quantification and robustness analysis

Aerospace systems are exposed to uncertainties arising from factors like manufacturing variations, environmental conditions, and operational scenarios. Sensitivity analysis must address these uncertainties to provide reliable insights. Neglecting uncertainty quantification can lead to incomplete understanding and suboptimal design choices. Consider the design of a satellite's communication system. Uncertainty in solar radiation and atmospheric conditions can impact communication performance. Sensitivity analysis should incorporate uncertainty propagation techniques, such as Monte Carlo simulations, to assess the range of potential outcomes under various scenarios. Robustness analysis then ensures that design decisions are resilient across these uncertainties [35].

7.22.5 Ethical considerations in sensitivity analysis

As sensitivity analysis guides engineering decisions with far-reaching implications, ethical considerations become paramount. Decisions influenced by sensitivity analysis can impact human safety, environmental sustainability, and societal well-being. Engineers must carefully weigh these considerations, adhering to ethical principles and considering potential consequences [36].

For example, in autonomous drone navigation for medical supply delivery, sensitivity analysis guides decisions regarding flight paths and operational parameters. Ethical considerations must encompass issues like avoiding populated areas, minimizing noise pollution, and ensuring equitable access to medical resources [37].

In essence, the challenges and considerations in sensitivity analysis underscore its complexity within aerospace engineering. Addressing high-dimensional spaces, nonlinearity, computational efficiency, uncertainty, and ethical dimensions requires interdisciplinary collaboration and innovative methodologies to harness the full potential of sensitivity analysis while mitigating its limitations.

7.23 CONCLUSION AND FUTURE OUTLOOK

In conclusion, this chapter has thoroughly examined the critical role of sensitivity analysis in aerospace engineering. We have explored its historical significance, versatile applications, and its profound impact on enhancing aerospace system efficiency and reliability. Sensitivity analysis has emerged as an essential tool for aerospace professionals, offering the means to optimize designs, quantify uncertainties, and make informed system-level decisions. From parametric sensitivity analysis to advanced global sensitivity analysis techniques, we have dissected the methodologies that empower aerospace engineers to decipher intricate relationships and drive innovation. Looking ahead, the integration of cutting-edge technologies such as machine learning and artificial intelligence promises to revolutionize sensitivity analysis, thus enabling deeper insights and more efficient optimization. Emerging areas such as electric propulsion, UAVs, and space exploration missions will continue to rely on sensitivity analysis to shape the future of aerospace engineering. However, we must not overlook the challenges ahead—managing high-dimensional parameter spaces, addressing computational efficiency, and navigating ethical considerations. These challenges demand innovative solutions and collaborative efforts from the aerospace community. In summary, sensitivity analysis remains the North Star for aerospace engineers, guiding us through the complex challenges of this ever-evolving field. It empowers us to push the boundaries of human achievement in flight and space exploration, steering aerospace engineering towards new heights of excellence. As we continue to explore the frontiers of the skies and beyond, sensitivity analysis will remain an unwavering ally, propelling us to new horizons of aerospace engineering achievement. Researchers have studied aeronautical engineering sensitivity analysis fundamentals, methods, and applications. We found its role in design optimization, uncertainty quantification, and system-level analysis. The combination of machine learning, data-driven methods, and hybrid approaches has improved accuracy and efficiency. Our work in electric propulsion, UAV optimization, space exploration, and sustainability shows sensitivity analysis's versatility. Researchers have seen this methodology's complicated environment by handling high-dimensional spaces, nonlinearities, and uncertainties.

Sensitivity analysis has major aerospace engineering consequences. It helps engineers optimise designs, make decisions, and improve system performance. As aircraft systems advance, sensitivity analysis assures flexibility and provides a solid foundation for resolving complexity and uncertainties. It shapes aircraft innovation from electric propulsion to UAVs and space exploration. Sensitivity analysis in aircraft engineering offers several options for research and development. Hybrid methods that combine physical insights with data-driven tactics can help researchers and engineers solve complicated problems. Surrogate modelling, dimensionality reduction, and uncertainty quantification can improve high-dimensional parameter spaces and nonlinear system sensitivity assessments. As aeronautical technology progress, ethical issues will need further study to assure fairness. Untapped potential exists in integrating sensitivity analysis with quantum computing and sustainability research. Further research can create algorithms that adapt to changing aerospace systems, handle real-time data, and deliver quick insights.

In conclusion, sensitivity analysis stands as a guiding star in the constellation of aerospace engineering tools. Its journey—from unravelling fundamental relationships to steering futuristic innovations—is one of continuous growth and adaptation. The strides taken in this field are a testament to the indomitable spirit of aerospace engineering, harnessing the power of analysis to conquer the complexity of the skies and beyond. The journey continues, beckoning us towards uncharted territories and new frontiers of knowledge.

REFERENCES

[1] O. Baysal and M. E. Eleshaky, "Aerodynamic design optimization using sensitivity analysis and computational fluid dynamics," *AIAA J.*, vol. 30, no. 3, pp. 718–725, 1992.

[2] A. Kumar, V. P. Singh, C. S. Meena, and N. Dutt, *Thermal Energy Systems: Design, Computational Techniques, and Applications*, 1st ed. CRC Press, 2023.

[3] J. Jodei, M. Ebrahimi, and J. Roshanian, "Multidisciplinary design optimization of a small solid propellant launch vehicle using system sensitivity analysis," *Struct. Multidiscip. Optim.*, vol. 38, pp. 93–100, 2009.

[4] G. Grossir, B. Van Hove, S. Paris, P. Rambaud, and O. Chazot, "Free-stream static pressure measurements in the Longshot hypersonic wind tunnel and sensitivity analysis," *Exp. Fluids*, vol. 57, pp. 1–13, 2016.

[5] J. C. Newman III, A. C. Taylor III, R. W. Barnwell, P. A. Newman, and G. J.-W. Hou, "Overview of sensitivity analysis and shape optimization for complex aerodynamic configurations," *J. Aircr.*, vol. 36, no. 1, pp. 87–96, 1999.

[6] A. Kundu, A. Kumar, N. Dutt, V. P. Singh, and C. S. Meena, "Chapter 7: Modelling and Simulation of Thermal Energy System for Design Optimization," in *Thermal Energy Systems: Design, Computational Techniques, and Applications*, A. Kumar, N. Dutt, V. Singh, and C. Meena, Eds. CRC Press, 2023, pp. 103–137.

[7] M. Oberguggenberger, J. King, and B. Schmelzer, "Classical and imprecise probability methods for sensitivity analysis in engineering: A case study," *Int. J. Approx. Reason.*, vol. 50, no. 4, pp. 680–693, 2009.

[8] K. Maute, M. Nikbay, and C. Farhat, "Sensitivity analysis and design optimization of three-dimensional non-linear aeroelastic systems by the adjoint method," *Int. J. Numer. Methods Eng.*, vol. 56, no. 6, pp. 911–933, 2003.

[9] A. Saxena, A. Prajapati, G. Pant, C. Meena, A. Kumar, and V. Singh, "Water Consumption Optimization of Hybrid Heat Pump Water Heating System," in *Lecture Notes in Mechanical Engineering- Recent Advances in Mechanical Engineering Select Proceedings of FLAME 2022*, A. Shukla, B. Sharma, A. Arabkoohsar, and P. Kumar, Eds. Springer Nature Singapore, 2023, pp. 721–732.

[10] E. Alyanak, R. Grandhi, and H.-R. Bae, "Gradient projection for reliability-based design optimization using evidence theory," *Eng. Optim.*, vol. 40, no. 10, pp. 923–935, 2008.

[11] V. P. Singh, C. S. Meena, A. Kumar, and N. Dutt, "Double Pass Solar Air Heater: A Review," *Int. J. Energy Resour. Appl.*, vol. 1, no. 2, pp. 22–43, 2022, https://doi.org/10.56896/IJERA.2022.1.2.009

[12] T. Sivasakthivel, V. Verma, R. Tarodiya, C. S. Meena, V. P. Singh, and R. Kumar, "Chapter 11: Analysis of Optimum Operating Parameters for Ground Source Heat Pump System for Different Cases of Building Heating and Cooling Mode Operations," in *Thermal Energy Systems: Design, Computational Techniques, and Applications*, 1st ed., A. Kumar, N. Dutt, V. Singh, and C. Meena, Eds. CRC Press, 2023, pp. 183–207.

[13] H.-P. Wan, W.-X. Ren, and M. D. Todd, "Arbitrary polynomial chaos expansion method for uncertainty quantification and global sensitivity analysis in structural dynamics," *Mech. Syst. Signal Process.*, vol. 142, p. 106732, 2020.

[14] S. Dabetwar, N. N. Kulkarni, M. Angelosanti, C. Niezrecki, and A. Sabato, "Sensitivity analysis of unmanned aerial vehicle-borne 3D point cloud reconstruction from infrared images," *J. Build. Eng.*, vol. 58, p. 105070, 2022.

[15] V. P. Singh and S. Jain, "Economic analysis of a large scale solar updraft tower power plant," *Sustain. Energy Technol. Assessments*, vol. 58, p. 103325, 2023, https://doi.org/10.1016/j.seta.2023.103325

[16] V. Chabridon, "Reliability-oriented sensitivity analysis under probabilistic model uncertainty–Application to aerospace systems." Université Clermont Auvergne [2017-2020], 2018.

[17] J. E. V. Peter and R. P. Dwight, "Numerical sensitivity analysis for aerodynamic optimization: A survey of approaches," *Comput. Fluids*, vol. 39, no. 3, pp. 373–391, 2010.

[18] A. A. Pohya, K. Wicke, and T. Kilian, "Introducing variance-based global sensitivity analysis for uncertainty enabled operational and economic aircraft technology assessment," *Aerosp. Sci. Technol.*, vol. 122, p. 107441, 2022.

[19] A. Kundu, A. Kumar, V. P. Singh, C. S. Meena, and N. Dutt, "Chapter 1: Introduction to Thermal Energy Resources and Their Smart Applications," in *Thermal Energy Systems:Design, Computational Techniques, and Applications*, 2023, pp. 1–15.

[20] K. Chauhan and V. P. Singh, "Proceedings Prospect of biomass to bioenergy in India: An overview," *Mater. Today Proc.*, no. xxxx, 2023, https://doi.org/10.1016/j.matpr.2023.01.419

[21] A. R. Singh, S. K. Singh, and V. P. Singh, "Process parameters optimization of carbon nano tube based catalytic transesterification of algal oil," *Mater. Today Proc.*, no. xxxx, 2023, https://doi.org/10.1016/j.matpr.2023.01.418

[22] V. P. Singh and G. Dwivedi, "Technical Analysis of a Large - Scale Solar Updraft Tower Power," *Energies*, vol. 16, no. 1, pp. 1–29, 2023, doi: https://doi.org/10.3390/en16010494

[23] S. L. Brunton et al., "Data-driven aerospace engineering: reframing the industry with machine learning," *AIAA J.*, vol. 59, no. 8, pp. 2820–2847, 2021.

[24] S. Jain, N. Kumar, V. P. Singh, S. Mishra, and N. K. Sharma, "Transesterification of Algae Oil and Little Amount of Waste Cooking Oil Blend at Low Temperature in the Presence of NaOH," *Energies*, vol. 16, no. 1, pp. 1–12, 2023, https://doi.org/10.3390/ en16031293

[25] M. G. Kapteyn, D. J. Knezevic, D. B. P. Huynh, M. Tran, and K. E. Willcox, "Data-driven physics-based digital twins via a library of component-based reduced-order models," *Int. J. Numer. Methods Eng.*, vol. 123, no. 13, pp. 2986–3003, 2022.

[26] V. P. Singh et al., "Nanomanufacturing and Design of High-Performance Piezoelectric Nanogenerator for Energy Harvesting," in *Nanomanufacturing and Nanomaterials Design: Principles and Applications*, 1st ed., S. Singh, S. K. Behura, A. Kumar, K. Verma, Ed. 2022, pp. 241–272.

[27] J. Sobieszczanski-Sobieski, "Sensitivity analysis and multidisciplinary optimization for aircraftdesign-Recent advances and results," *J. Aircr.*, vol. 27, no. 12, pp. 993–1001, 1990.

[28] A. Datta, A. Kumar, A. Kumar, A. Kumar, and V. P. Singh, "Advanced materials in biological implants and surgical tools," in *Advanced Materials for Biomedical Applications*, 1st ed., N. D. Ashwani Kumar, Yatika Gori, Avinash Kumar, Chandan Swaroop Meena, Ed. 2022, pp. 283–298.

[29] R. K. Shubham Srivastava, Deepti Verma, Shreya Thusoo, Ashwani Kumar, Varun Pratap Singh, "Nanomanufacturing for Energy Conversion and Storage Devices," in *Nanomanufacturing and Nanomaterials Design: Principles and Applications*, 2022, pp. 165–174.

[30] N. Dutt, A. Binjola, A. J. Hedau, A. Kumar, V. P. Singh, and C. S. Meena, "Comparison of CFD Results of Smooth Air Duct with Experimental and Available Equations in Literature," *Int. J. Energy Resour. Appl.*, vol. 1, no. 1, pp. 40–47, 2022, https://doi.org/10.56896/IJERA.2022.1.1.006

[31] L. Ribeiro, O. Saotome, R. d'Amore, and R. de Oliveira Hansen, "High-Speed and High-Temperature Calorimetric Solid-State Thermal Mass Flow Sensor for Aerospace Application: A Sensitivity Analysis," *Sensors*, vol. 22, no. 9, p. 3484, 2022.

[32] V. P. Singh, A. Karn, G. Dwivedi, T. Alam, and A. Kumar, "Experimental Assessment of Variation in Open Area Ratio on Thermohydraulic Performance of Parallel Flow Solar Air Heater," *Arab. J. Sci. Eng.*, vol. 41, no. 12, pp. 1–17, 2022, https://doi.org/10.1007/s13369-022-07525-7

[33] V. P. Singh, S. Jain, and J. M. L. Gupta, "Analysis of the effect of variation in open area ratio in perforated multi-V rib roughened single pass solar air heater- Part A," *Energy Sources, Part A Recover. Util. Environ. Eff.*, vol. 44, no. Jan, pp. 1–21, 2022, https://doi.org/10.1080/15567036.2022.2029976

[34] V. P. Singh, S. Jain, and J. M. L. Gupta, "Performance assessment of double-pass parallel flow solar air heater with perforated multi-V ribs roughness — Part B," *Exp. Heat Transf.*, vol. 00, no. 00, pp. 1–18, 2022, https://doi.org/10.1080/08916152.2021.2019147

[35] V. P. Singh et al., "Heat transfer and friction factor correlations development for double pass solar air heater artificially roughened with perforated multi-V ribs," *Case Stud. Therm. Eng.*, vol. 39, no. September, p. 102461, 2022, https://doi.org/10.1016/j.csite.2022.102461

[36] V. P. Singh, S. Jain, and J. M. L. Gupta, "Analysis of the effect of perforation in multi-v rib artificial roughened single pass solar air heater: - Part A," *Exp. Heat Transf.*, pp. 1–20, Oct. 2021, https://doi.org/10.1080/08916152.2021.1988761

[37] R. Yondo, E. Andrés, and E. Valero, "A review on design of experiments and surrogate models in aircraft real-time and many-query aerodynamic analyses," *Prog. Aerosp. Sci.*, vol. 96, pp. 23–61, 2018.

Chapter 8

Modeling of flows through porous media

Sunil Kumar, Maria King, and Robert G. Hardin

Texas A&M University, College Station, TX, USA

8.1 BASICS OF FLOW THROUGH POROUS MEDIA

This section presents basic definitions relevant to flow through porous media.

8.1.1 Porosity

The porosity (ϕ) of a porous media is defined as the ratio of the pore volume to the bulk volume of the porous medium.

$$\phi = {V_p}\big/{V_b} \tag{8.1}$$

Here, V_b and V_p are the bulk volume and pore volume, respectively. The porosity varies between zero and unity depending on the porous medium material properties. Porosity value zero signifies that fluid will not pass through the porous medium material, while unity signifies that fluid will pass through the porous medium without any resistance. However, the porosity for most of the materials falls between zero and unity. For flow in porous media particularly in compacted mediums is interconnected and dead pore spaces coexist. Interconnected spaces allow flow to pass, while the dead pore spaces do not allow flow to take place [1].

8.1.2 Darcy's law and permeability

Darcy's law was introduced by Henry Darcy. It stands as the cornerstone in groundwater science, governing fluid movement through porous materials. Like Ohm's law in electrostatics, it linearly links the fluid flow rate to hydraulic head difference (often proportional to pressure difference) using hydraulic conductivity. Darcy's law is one of the first mathematical

DOI: 10.1201/9781003465171-8

relationships proposed for water flow through saturated porous media. The integral form of Darcy's law is given by [2]:

$$Q = \frac{kA}{\mu} \frac{\Delta P}{L} \qquad (8.2)$$

Here Q is the volumetric flow rate, k is the permeability, A is the cross-sectional area of the porous sample normal to flow, L is the length of the sample in flow direction, ΔP is the pressure drop, and μ is the viscosity of the fluid. In equation (8.2), the flow resistance can be identified as $\left(R = \frac{\mu L}{kA}\right)$. For flow through porous media, the permeability is primarily influenced by the pore structure, which is the inherent characteristic of a porous medium enabling the passage of the fluid. The measurements are performed for various flow rates of the fluid as a function of pressure drop. Permeability is obtained by fitting the data points.

8.1.3 Pore size and pore size distribution

The distribution of the pore size is arbitrary. Given the pore size distribution function $\alpha(D)$ with respect to pore size D, the selection of the specific pore size is determined as [3]

$$\int_0^\infty \propto (D) dD = 1 \qquad (8.3)$$

For measuring the pore size and distribution, the mercury porosimeter approach is adopted. Mercury porosimetry involves measuring the mercury volume that enters a sample under varying pressures. Pore size is then determined using Laplace's capillarity equation, considering a capillary tube pore model. This method, along with adsorption isotherms, is commonly used for pore size distribution analysis. In general, pores are more like irregular particles and are interconnected with necks or throats between them. The sizes of both the pore bodies and throats impact macroscopic properties.

8.1.4 Specific surface

The interior surface area of the voids is the particular surface of a porous material. It is essential in a range of porous media applications including adsorption, catalysis, and ion-exchange processes. Specific surface is an essential element in models of porous media conductivity or permeability for a porous media flow. Two porous materials with the same porosity, for example, may have varying permeabilities due to differing specific surface areas. The average pore size and permeability of a medium with a larger specific surface area are smaller [4].

8.1.5 Hydraulic conductivity

The ease with which a fluid may pass through the pores of porous material or fracture networks is measured by hydraulic conductivity, a fundamental property. The degree of saturation, inherent permeability of the material, and physical characteristics of the fluid, such as density and viscosity, all have a role. Hydraulic conductivity is a measure of a soil's capacity to convey water under hydraulic gradient conditions. It may be calculated as the volume flow to hydraulic gradient ratio.

Hydraulic conductivity can be determined using an experimental approach that utilizes hydraulic experiments using Darcy's law. The experimental approach adopts laboratory tests and field tests (including both small- and large-scale field tests). The laboratory tests (constant head and falling head) methods are discussed next.

8.1.5.1 Constant head method for hydraulic conductivity prediction

The constant head method is typically used for granular soil, allowing water to pass through the soil under a steady state head condition. The measurements are performed over a period. Assume that a flow is taking place in a circular pipe of cross-sectional area A, with porous medium length L with hydraulic head h. Let us assume that the experiment was initiated to achieve steady state and measurements are performed. Thereby knowing the volume ΔV collected in time Δt.

Therefore, the volumetric flow rate is

$$Q = \frac{\Delta V}{\Delta t} \tag{8.4}$$

Using Darcy's law of permeability and $\Delta P = \rho g h$

$$Q = \frac{\Delta V}{\Delta t} = -\frac{kA}{\mu}\frac{\Delta P}{L} = -\frac{kA}{\mu}\frac{\rho g h}{L} \tag{8.5}$$

Hence

$$\frac{\Delta V}{\Delta t} = -\frac{k\rho g A}{\mu L} h \tag{8.6}$$

where hydraulic conductivity

$$K = \frac{k\rho g}{\mu} \tag{8.7}$$

8.1.5.2 *Falling head method for hydraulic conductivity prediction*

In this approach, the porous samples are saturated under specific head conditions. Water is allowed to flow without additional water from any external source. As a result, the pressure head declines with the flow of water through the porous material. This approach is beneficial for both fine- and coarse-grained porous mediums.

The Darcy's law from equation 8.6 indicates that

$$\frac{\Delta V}{\Delta t} = -K \frac{A}{L} h \tag{8.8}$$

The decrease in volume is related to falling head by $\Delta V = \Delta h A$ and considering $\Delta t \to 0$. Equation 8.8 can be rewritten as follows:

$$\frac{dh}{dt} = -\frac{K}{L} h \tag{8.9}$$

Then equation 8.9 can be integrated to provide:

$$h(t) = h_i e^{-\frac{K}{L}(t-t_i)}$$

Assume that the head drops from h_i to h_f in a time Δt and plugging $h(t_f) = h_f$. The thermal conductivity can be written as:

$$K = \frac{L}{\Delta t} \ln \frac{h_f}{h_i} \tag{8.10}$$

8.2 FUNDAMENTALS OF CAPILLARITY IN POROUS MEDIA

8.2.1 Laplace's equation

Capillarity is the phenomena of fluids absorbing into tiny holes or porous rocks and materials as a result of surface energy via capillary pressure. The deformable interfaces produced between two phase fluids in contact with each other, such as a liquid and a vapor, are caused by capillary pressure. Capillarity and its relevance have been extensively acknowledged in a variety of sectors, including civil engineering, soil science, hydrology, carbon storage, and petroleum engineering. The basic equation of capillarity is Laplace's equation describing a very important phenomenon, i.e., the existence of pressure difference Δp at a point between two sides of the curved fluid surface or interface.

$$\Delta P = \sigma \left(\frac{1}{R_1} + \frac{1}{R_2} \right) \tag{8.11}$$

where σ is the surface tension and R_1 and R_2 are the radius of the curvature at point P. It can be concluded from equation 8.11 that for a plane surface, the two radii are infinite and therefore the pressure difference across a plane surface is zero.

8.2.2 Surface wettability and contact angle

In a porous medium, the immiscible fluids that are present concurrently contend for occupancy of the pore surface. The relative wettability of the pore surface by the two immiscible fluids determines how the interface between the immiscible fluids is configured. A porous medium's relative wettability is determined by the angle of contact (θ) between two fluids is what defines them. It is useful to think about what might happen to a drop of liquid positioned on a flat surface (see Figure 8.1) [4].

The contact angle θ is defined as the angle between the tangent to the liquid–solid boundary constructed at a point on the three-phase line of contact and tangent to the gas–liquid boundary constructed at the same point. The force balance at the point on the three-phase line of contact along the liquid–solid boundary is expressed by Young's equation.

$$\sigma_{sg} - \sigma_{sl} = \sigma_{lg}\cos\theta \tag{8.12}$$

Here σ_{sg}, σ_{lg}, and σ_{sl} are the surface tension of the solid, liquid, and interfacial tension between the liquid and the solid. The contact angle of a gas-liquid-solid or a liquid-liquid-solid system may vary between 0° and 180°. Based on the wetting behavior, the fluids can be classified into two categories known as wetting and non-wetting fluids. The wetting fluids make contact angle between $0° \leq \theta \leq 90°$, while non-wetting fluids make contact angle between $90° \leq \theta \leq 180°$. In the pores of the porous media, the curved interfaces between the two immiscible fluids may take different shapes and directions of the curvature.

8.2.3 Capillary pressure and capillary rise

In the porous media, the two fluids which are immiscible separate from each other by the curved interface, across which there is pressure difference, which is known as capillary pressure [5–7]. The capillary pressure is

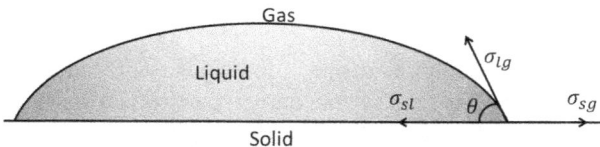

Figure 8.1 Contact angle of liquid on solid surface.

balanced at equilibrium at any point of the contact between wetting and non-wetting phases.

$$P_c = \Delta P = P_{nw} - P_w \tag{8.13}$$

Here, P_c is the capillary pressure, ΔP is the pressure difference, P_{nw} is the pressure by the non-wetting phase, and P_w is the pressure by the wetting phase. At equilibrium, $P_{nw} > P_w$, therefore, capillary pressure is always positive.

The wetting phase can be distinguished from the non-wetting phase by its capacity to selectively diffuse over capillary walls. A fluid's "wettability" is influenced by its surface tension, the forces that cause it to occupy the least amount of space feasible and its contact angle. By adjusting the capillary surface qualities (such as roughness and hydrophilicity), one may regulate the "wettability" of a fluid. However, in oil-water systems, water generally represents the wetting phase, whereas in gas-oil systems, oil typically represents the wetting phase. No matter the system, a pressure differential develops at the resultant curved contact between the two fluids.

The capillary pressure formulae are obtained from the pressure connection between two fluid phases in an equilibrium capillary tube, which is defined as force in an upward and downward direction.

Force in upward direction = interfacial tension of the fluid acting along the perimeter of the capillary tube

Force in downward direction = (density gradient difference) × (cross section area) × (height of capillary rise in the tube)

Young-Laplace equation is used for the force in the upward direction.

$$P_c = \Delta P = \frac{2\sigma \cos\theta}{r_c} \tag{8.14}$$

where σ is the interfacial tension, r is the effective radius of the interface, and θ is the wetting angle of the liquid on the surface of the capillary. The force down formula for capillary pressure is seen as:

$$P_c = \Delta P = \frac{\pi r^2 h \left(\Gamma_w - \Gamma_{nw} \right)}{\pi r^2} = h \left(\Gamma_w - \Gamma_{nw} \right) \tag{8.15}$$

Here h is the height of capillary rise, Γ_w and Γ_{nw} are the density gradient of the wetting and non-wetting phases.

This phenomenon has applications in multiple area(s) such as petrochemical engineering, soil science, pharmaceuticals, materials science, microfluidics, geology and hydrology, enhanced oil recovery, agriculture, and construction (Figure 8.2).

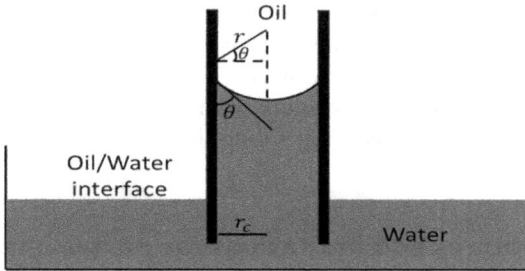

Figure 8.2 Capillary rise of water.

The capillary rise of capillarity is a phenomenon in which the liquid spontaneously rises or falls in a narrow space such as a thin tube or in the voids of a porous material. The rise of the liquid in the capillary tube is due to surface tension. This is because the adhesive force of the liquid (capillary action) is greater than the cohesive force between liquids (surface tension) and this will lead to a rise in the liquid in the capillary tube. As a result, positive and negative rises are observed. The capillary rise method is generally used to calculate surface tension of unknown fluids. At equilibrium, the pressure drop across the interface is equal to the hydraulic pressure drop in the liquid column

$$\Delta P = \rho g h = \frac{2\sigma \cos\theta}{r_c} \tag{8.16}$$

Here ρ is density and g is the gravitational constant.

$$h = \frac{2\sigma \cos\theta}{\rho g r_c} \tag{8.17}$$

As r is in the denominator, the liquid rises further, making the space the thinner. The height of the column also rises with lighter liquid and lower gravity. Capillary tubes are thin cylindrical tubes with extremely tiny diameters. The liquid inside the capillary is seen to either rise or sink in relation to the surrounding liquid level when these tiny tubes are submerged in a liquid. Such tubes are known as capillary tubes, and this phenomenon is known as the capillary action.

8.3 MODELING OF FLOW IN POROUS MEDIUM

Porous media modeling is used in a variety of single phase and multiphase problems, flowing through packed beds, filter papers, flow distributors, and tube banks. While simulating porous media, dedicated cell zones are selected where porous media model is applied and pressure loss across the media is determined with the help of inputs provided for simulations. The "porous jump," a 1D approximation of the porous media model, may be used to simulate a thin membrane with well-known velocity/pressure-drop properties. The porous jump model is applied to a face zone, not a cell zone.

8.3.1 Assumptions and limitations for porous media modeling

For part of the model defined as "porous," the porous media model integrates an experimentally calculated flow resistance. The porous medium model is an extra momentum sink in the governing momentum equations. As a result, it is important to understand the following modeling presumptions and limitations [8]:

1. The porous modeling code employs and presents a superficial velocity inside the porous medium, based on the volumetric flow rate, to maintain continuity of the velocity vectors across the porous medium interface since the volume obstruction that is physically present is not represented in the model.
2. The effect of the porous medium on the turbulence field is approximated.
3. For both single and multiphase flows, the porosity is assumed isotropic and can vary spatially and temporally.
4. Superficial velocity formulation and physical velocity formulations are available for multiphase porous media.
5. For each phase, the momentum resistance and heat source terms are calculated separately.
6. Porous modeling assumes thermal equilibrium between the porous media and multiphase flows.
7. The solid temperature is determined using phase temperatures.
8. Porous media modeling in a moving frame of reference utilizes relative reference frame or the absolute reference frame. This is achieved through the application of a relative velocity resistance formulation, which ensures precise prediction of source terms.

8.3.2 Momentum equations for porous media

The superficial velocity porous is utilized in the porous media models for both single-phase and multiphase flows. Volumetric flow rate in the porous region is used for the superficial phase or mixture velocities calculation. By including a momentum source element to the common fluid flow equations, porous media are represented. Viscosity loss term and inertial loss term made up the source term [8].

$$S_i = -\left(\sum_{j=1}^{3} D_{ij} \mu v_j + \sum_{j=1}^{3} C_{ij} \frac{1}{2} \rho |v| v_j \right) \tag{8.18}$$

where $|v|$ is the velocity magnitude, S_i is the source term, and D and C are the required diagonal matrices. The source term acts as a momentum sink that helps create a pressure gradient in porous cells, resulting in a pressure drop that is proportional to the fluid velocity inside the cell. For homogenous porous media, equation 8.18 can be written as:

$$S_i = -\left(\frac{\mu}{\alpha} v_i + \frac{1}{2} C_2 \rho |v| v_i \right) \tag{8.19}$$

where α is the permeability and C_2 is the inertial resistance factor. The source terms can also be modeled as a power law of the velocity magnitude:

$$S_i = -C_0 |v|^{C_1} = C_0 |v|^{(C_1-1)} v_i \tag{8.20}$$

C_0 and C_1 are user defined empirical coefficients.

8.3.3 Darcy's Law in porous media

The pressure drop in laminar flows through porous media is normally proportional to velocity; therefore, C_2 can be considered zero. Darcy's Law results from the porous medium model [8].

$$\nabla p = -\frac{\mu}{\alpha} \vec{v} \tag{8.21}$$

The pressure drops in each coordinate direction within the porous region can be shown by the following equation:

$$\Delta p_x = \sum_{j=1}^{3} \frac{\mu}{\alpha_{xj}} v_j \Delta n_x, \Delta p_y = \sum_{j=1}^{3} \frac{\mu}{\alpha_{yj}} v_j \Delta n_y, \Delta p_z = \sum_{j=1}^{3} \frac{\mu}{\alpha_{zj}} v_j \Delta n_z \tag{8.22}$$

where $1/\alpha_{ij}$ are the entries in D in equation 8.22, v_j are velocity components in the x, y, and z direction, and Δn_x, Δn_y, and Δn_z are the actual thicknesses of the porous medium in the x, y, and z direction.

8.3.4 Inertial losses in porous media

The constant C_2 corrects the inertial losses in the porous material at high velocities. The pressure drop may be calculated as a function of the dynamic head using this constant since it serves as a loss coefficient per unit length along the flow direction. By only accounting of inertial losses when modeling a perforated plate or a bank of tubes:

$$\nabla p = -\sum_{j=1}^{3} C_{2ij} \left(\frac{1}{2} \rho v_j |v| \right) \tag{8.23}$$

In three dimensions, equation 8.23 can be rewritten as follows:

$$\Delta p_x \approx \sum_{j=1}^{3} C_{2xj} \Delta n_x \frac{1}{2} \rho v_j |v|,$$

$$\Delta p_y \approx \sum_{j=1}^{3} C_{2yj} \Delta n_y \frac{1}{2} \rho v_j |v|, \tag{8.24}$$

$$\Delta p_z \approx \sum_{j=1}^{3} C_{2yj} \Delta n_z \frac{1}{2} \rho v_j |v|$$

Here, Δn_x, Δn_y, and Δn_z are the thicknesses of the porous medium in the simulation zone.

8.3.5 Energy equation in porous media

The standard energy equation is solved in the porous media region with modifications to the conduction flux and transient terms only. The conduction flux uses an effective conductivity in the porous medium, and the transient term includes thermal inertia of the solid region on the medium [9].

$$\frac{\partial}{\partial t} \left(\gamma \rho_f E_f + (1-\gamma) \rho_s E_s \right) + \nabla \cdot \left(\vec{v} \left(\rho_f E_f + p \right) \right) \tag{8.25}$$

$$= \nabla \cdot \left[k_{eff} \nabla T - \left(\sum_i h_i J_i \right) + \left(\bar{\bar{\tau}} . \vec{v} \right) \right] + S_f^h$$

where E_f is the total fluid energy, E_s is the total solid medium energy, γ is the porosity of the medium, k_{eff} is the effective thermal conductivity of the medium, and S_f^h is the fluid enthalpy source term. The effective isotropic thermal conductivity in the porous medium is computed as the average of the fluid and solid conductivity.

$$k_{eff} = \gamma k_f + (1-\gamma) k_s \tag{8.26}$$

Here, γ is the medium porosity, k_f is the fluid thermal conductivity, and k_s is the solid medium thermal conductivity.

8.3.6 Porous media modeling based on physical velocity for single and multiphase systems

In general, the porous mediums are exposed to the flows for different purposes. The porous modeling is performed based on velocities [8]. The superficial velocity is calculated as

$$\vec{v}_s = \gamma \vec{v}_p \tag{8.27}$$

where $\vec{v}_s$ is the superficial velocity, while $\vec{v}_p$ is the physical velocity, and γ is the porosity of the porous media. It is important to note that superficial velocities inside and outside the porous zone are equal. To improve accuracy in simulating flows through porous media, the actual or physical velocity across the entire flow field must be determined, rather than relying on superficial velocity.

8.3.6.1 Single-phase porous media modeling

Let us assume a general scalar ϕ and physical velocity $\vec{v}_p$, then the governing equation in an isotropic porous media has the following form:

$$\frac{\partial (\gamma \rho \phi)}{\partial t} + \nabla \cdot (\gamma \rho \vec{v}_p \phi) = \nabla \cdot (\gamma \Gamma \nabla \phi) + \gamma S_\phi \tag{8.28}$$

The volume-averaged continuity and momentum conservation equations are as follows, supposing isotropic porosity and single-phase flow:

$$\frac{\partial (\gamma \rho)}{\partial t} + \nabla \cdot (\gamma \rho \vec{v}_p) = 0 \tag{8.29}$$

$$\frac{\partial (\gamma \rho \vec{v}_p)}{\partial t} + \nabla \cdot (\gamma \rho \vec{v}_p \vec{v}_p) = -\gamma \nabla p + \nabla \cdot (\gamma \vec{\tau}) + \gamma \vec{B}_f - \left(\frac{\mu}{\alpha} + \frac{C_2 \rho}{2} |\vec{v}_p| \right) \vec{v}_p \tag{8.30}$$

The above equation includes both viscous and inertial drag forces imposed on the fluid.

8.3.6.2 Multiphase porous media modeling

The porous media exposed to multiphase flow are solved using physical velocity porous media formulations, including both porous and non-porous regions. Let us assume that a general scalar ϕ in the qth phase is ϕ_q. The governing equation for the isotropic porous medium can be written as:

$$\frac{\partial\left(\gamma\alpha_q\rho_q\phi_q\right)}{\partial t}+\nabla\cdot\left(\gamma\alpha_q\rho_q\vec{v}_q\phi_q\right)=\nabla\cdot\left(\gamma\Gamma_q\nabla\phi_q\right)+\gamma S_{\phi,q} \tag{8.31}$$

where γ is the porosity, ρ_q is the phase density, α_q is the phase volume fraction, $\vec{v}_q$ is the phase velocity vector, $S_{\phi,q}$ is the source term, and Γ_q is the diffusion coefficient. This general scalar equation applies to all transport equations in the Eulerian multiphase model. In multiphase flows, for isotropic porosity the governing continuity, momentum, and energy equations can be written as:

8.3.6.2.1 The continuity equation

$$\frac{\partial\left(\gamma\alpha_q\rho_q\right)}{\partial t}+\nabla\cdot\left(\gamma\alpha_q\rho_q\phi_q\right)=\gamma\sum_{p=1}^{n}\left(\dot{m}_{pq}-\dot{m}_{qp}\right)+\gamma S_q \tag{8.32}$$

Here mass transfer from phase p to phase q is m_{pq}, mass transfer from phase q to phase p is m_{qp}, and S_q is the source term.

8.3.6.2.2 The momentum equation

The fluid-fluid momentum balance for phase q is shown in equation 8.33, where $\vec{R}_{pq}$ is the interaction force among phases, $\vec{F}_q$ is the external force on the body, $\vec{F}_{lift,q}$ is the lift force, $\vec{v}_{pq}$ is the interphase velocity, and τ is the stress tensor. The momentum equation for the phase can be written as:

$$\frac{\partial\left(\gamma\alpha_q\rho_q\vec{v}_q\right)}{\partial t}+\nabla\cdot\left(\gamma\alpha_q\rho_q\vec{v}_q\vec{v}_q\right)=-\gamma\alpha_q\nabla p+\nabla\cdot\left(\gamma\bar{\bar{\tau}}_q\right)+\gamma\alpha_q\rho_q\vec{g}+\gamma\sum_{p=1}^{n}$$

$$\left(\vec{R}_{pq}+\dot{m}_{pq}\vec{v}_{pq}-\dot{m}_{qp}\vec{v}_{qp}\right)+\gamma\left(\vec{F}_q+\vec{F}_{lift,q}+\vec{F}_{vm,q}\right)+\alpha_q\left(\frac{\mu}{k}+\frac{C_2}{2}\left|\vec{v}_q\right|\right)\vec{v}_q \tag{8.33}$$

The last term in the above equation is known as the sink term for porous media simulations The permeability (k) and inertial resistance factor (C_2) are known to be function of $(1-\gamma)$. Here $\bar{\bar{\tau}}_q$ is the qth phase stress-strain tensor and $\vec{F}_{vm,q}$ is the virtual mass force.

$$\bar{\bar{\tau}}_q=\alpha_q\mu_q\left(\nabla v_q^T\right)+\alpha_q\left(\lambda_q-\frac{2}{3}\mu_q\right)\nabla\cdot\vec{v}_q\hat{I} \tag{8.34}$$

Here, μ_q and λ_q are the shear and bulk viscosity of the phase q and p is the pressure shared by all phases. Equation 8.33 must be closed with appropriate expressions for the interphase force $\vec{R}_{pq}$. The force depends on the friction, pressure, cohesion, and other effects and is subjected to the conditions that $\vec{R}_{pq} = -\vec{R}_{qp}$ and $\vec{R}_{qq} = 0$.

$$\sum_{p=1}^{n} \vec{R}_{pq} = \sum_{p=1}^{n} K_{pq}\left(\vec{v}_p - \vec{v}_q\right) \tag{8.35}$$

where $K_{pq}\ (= K_{qp})$ is the interphase momentum exchange coefficient.

8.3.6.2.3 The energy equation

To satisfy the conservation of energy in Eulerian multiphase applications, a separate enthalpy equation can be written for each phase as:

$$\frac{\partial\left(\alpha_q \rho_q h_q\right)}{\partial t} + \nabla.\left(\alpha_q \rho_q \vec{u}_q h_q\right) = \alpha_q \frac{\partial p_q}{\partial t} + \bar{\bar{\tau}}_q : \nabla \vec{u}_q - \nabla.\vec{q}_q$$

$$+ S_q + \sum_{p=1}^{n}\left(Q_{pq} + \dot{m}_{pq}h_{pq} - \dot{m}_{qp}h_{qp}\right) \tag{8.36}$$

$$\frac{\partial\left(\gamma\alpha_q \rho_q h_q\right)}{\partial t} + \nabla.\left(\gamma\alpha_q \rho_q \vec{v}_q h_q\right) = -\gamma\alpha_q \frac{\partial p_q}{\partial t} + \gamma\bar{\bar{\tau}}_q : \nabla \vec{v}_q$$

$$- \nabla.\left(\vec{q}_q\right) + \gamma S_q + \gamma\sum_{p=1}^{n}\left(Q_{pq} + \dot{m}_{pq}h_{pq} - \dot{m}_{qp}h_{qp}\right) + Q_{sp} \tag{8.37}$$

where h_q is the specific enthalpy of the qth phase, $\vec{q}_q$ is the heat flux, S_q is the source term that includes sources of the enthalpy (e.g., due to the chemical reaction of radiation), Q_{pq} is the intensity of the heat exchange between p^{th} and q^{th} phases, and h_{pq} is the interphase enthalpy (the enthalpy of the vapor at the temperature of the droplets, in the case of evaporation). The heat exchange between phases must comply with the local balance conditions $Q_{pq} = -Q_{qp}$ and $Q_{qq} = 0$.

where Q_{sp} is the heat transfer between solids surface and phase q in the porous medium. Assuming only convective heat transfer Q_{sp} can be written as:

$$Q_{sp} = (1-\gamma)\alpha_q h_{q,eff}\left(T_s - T_q\right) \tag{8.38}$$

$h_{q,\,eff}$ is the effective convective heat transfer coefficient, T_s is the solid surface temperature and T_q is the phase q temperature. Q_{sp} is associated with phase q temperature via heat conduction equation:

$$\frac{\partial\left(\rho_s h_s\right)}{\partial t} + \nabla.\left(\vec{v}_s \rho_s h_s\right) = \nabla.\left(k_s \nabla T\right) - \sum_{p=1}^{n}\left(Q_{sp}\right) \tag{8.39}$$

If it is assumed that the overall heat transfer between multiphase fluid and solids is in equilibrium, then

$$\sum_{p=1}^{n}\left(Q_{sp}\right) = 0 \tag{8.40}$$

Therefore,

$$T_s = \frac{\sum_{p=1}^{n} \alpha_q h_{q,eff} T_p}{\sum_{p=1}^{n} \alpha_q h_{q,eff}} \tag{8.41}$$

The model's ability to anticipate the rise in velocity in porous zones is limited by the fact that the surface velocity values within and outside of a porous region are the same.

8.4 EXAMPLE CASE STUDY FOCUSED ON DRYING OF COTTON IN PNEUMATIC CONVEYER BELT

The flow through porous material and its effect can be well understood with the help of an example case study. The presented case study focuses on drying of cotton in a pneumatic conveyer belt system.

8.4.1 Motivation for cotton drying study

Field-harvested cotton seeds have a natural moisture content that must be carefully controlled during the ginning process. The creation of clumps caused by too much moisture prevents the separation of the fiber into separate locks, which might ultimately stop the ginning process altogether. On the other hand, a lack of moisture makes cotton stick to metallic surfaces, creating static charges. Consequently, a carefully monitored drying process is essential for ginning facilities. Therefore, controlled drying of cotton is necessary for machinery systems.

8.4.2 Background of research work relevant to cotton drying

More than 20% of the variable costs for cotton ginning processes are connected to power costs [10]. A major portion of this cost is attributed to the operation of centrifugal fans used for pneumatic conveying, as they consume more than half of the electricity [11, 12], reduction of which will not only reduce electric bill but also decrease the carbon footprint thereby critically supporting reduced emissions [13, 14]. When no material is being conveyed, the pressure drop is proportional to the square of the fan speed, and the energy consumption of these fans is inversely correlated with their speed. However, the pressure drop rises when porous materials are present for conveyance due to material resistance, deviating from the square of the fan speed for a given mass flow rate. According to research on horizontal conveying in a 20.3 cm (8 in) pipe, air velocity is needed for efficient conveying to decrease with mass flow ratio and moisture content [15]. According to [16], the recommended conveying air velocities for seed cotton now vary from 17.8 to 25.4 m/s. The pressure drop per unit length decreases with lower air velocities, reaches a minimum, and then increases with further decreases in velocity, according to an experiment conducted by [17] to characterize pressure drop per unit length at various air velocities and mass flow ratios for various materials. Hardin [18] constructed models for the lowest conveying velocity and pressure drop per unit length and validated this trend for seed cotton.

Moisture is a key component of the ginning process and is naturally present in seed cotton that has been gathered from fields. The effective cleaning process in the gin is hampered by excessive moisture that exceeds the crucial threshold of 7% fiber moisture content (w.b.) [19]. Additionally, it makes it harder to separate the cotton into locks, making the ginning process more challenging. In extreme circumstances, this may result in equipment jams, possible gin component damage, or even a total shutdown of ginning operations. Excess moisture causes the ginned lint to twist tightly, forming "spindle twists," which detract from the finished product's aesthetic attractiveness. On the other side, very low fiber moisture content below 5% causes static charges to build up on the fibers, which makes them attach to oppositely charged parts of the gin system, frequently in the vicinity of the condenser rollers [18]. For processing systems to be effective, it is essential to comprehend how moisture is absorbed and lost [20]. To attain the necessary moisture levels, efficient ginning, and high-quality lint production, precise drying techniques must be used.

Local environmental factors inside the conveyor belt have an impact on the drying of cotton wads, causing differences in temperature and humidity. The emergence of strata with varying temperatures is commonly referred to as thermal stratification in terms of temperature variation [21–23]. For some systems, thermal stratification is advantageous [24], but for some critical systems, performance may be declining [25–30]. This may lead to regional variations in drying speeds and uneven moisture distribution throughout the cotton wads. Even though it is normal practice to conduct studies to evaluate the drying effectiveness of agricultural products, the expense and practical constraints prevent a thorough investigation of all drying techniques and characteristics. Numerical simulation technologies provide a practical substitute and a successful method for critical parametric analysis. These numerical methods have been extensively used to simulate the drying of many different products, including soybeans [31], particle evaporation in hot airflow [32], vacuum drying of woods [33], iron ore pellets [34], viscoelastic fluid [35]. This has given researchers important new insights into drying procedures. Numerical simulations help researchers understand drying processes in a cost-effective and comprehensive manner, opening the way for improved and optimized drying procedures in a variety of industries.

8.4.3 Drying of porous material in a conveyer belt system

A typical pneumatic conveying system used for drying of cotton wads is shown in Figure 8.3. The details of the system can be found in the research paper by Hardin [18].

The conveying system with transparent acrylic viewing spool is shown in Figure 8.3. With this setup, cotton flow may be easily observed. A 0.203-meter-diameter conduit allows the carrying air and cotton to be discharged into a settling chamber. This setup is intended for monitoring and regulating the cotton conveying process effectively. The cotton wad and schematic of the conveyer belt system are shown in Figure 8.4.

The main goal of this study is to determine how differences in relative air speed and air temperature impact the drying of cotton wads. As a result, the

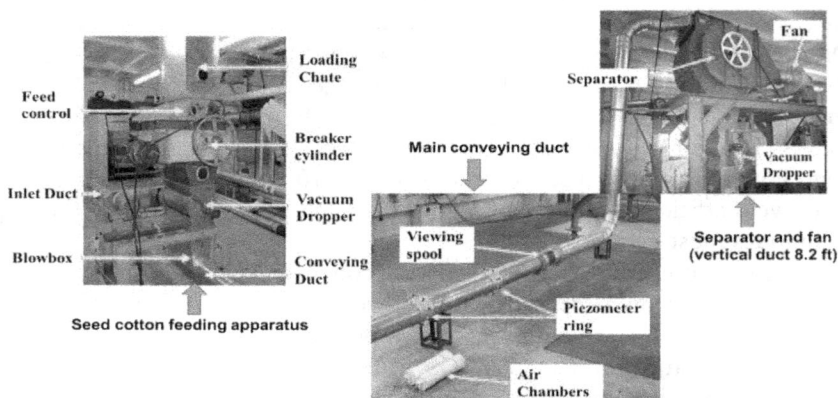

Figure 8.3 Standard pneumatic conveying system at USDA ARS, Mississippi [18].

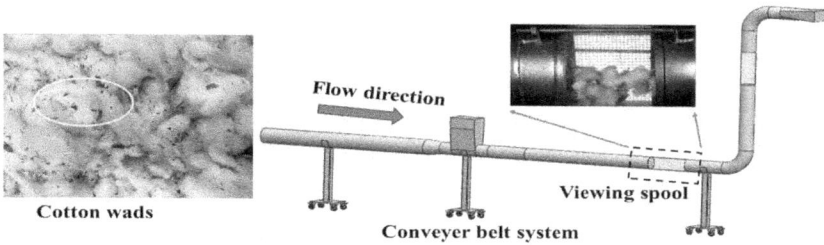

Figure 8.4 Cotton wads and schematic of the conveyer belt system.

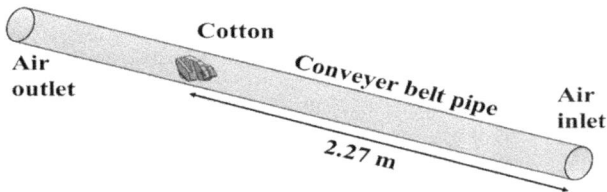

Figure 8.5 Cotton wad in a conveyer belt pipe.

horizontal portion of the conveying system was the focus of the CFD simulations. The stationary cotton wad in Figure 8.5 has the following dimensions: 0.23 meters in length, 0.131 meters in breadth, and 0.084 meters in height. This cotton wad is positioned within a straight pipe that is 3.5 meters long and 0.2032 meters (equal to 8 inches) in diameter. The cotton wad serves as the focus point for this study's analysis of the drying process since it is situated 2.27 meters from the pipe's entrance.

8.4.3.1 Meshing of the geometry of system

Figure 8.6 shows a mesh representation of the pipe in the conveyance system. Smaller-sized meshes were specially designed at the entrance, outflow, next to the wall, and at the air-cotton contact to account for abrupt gradients and interface phenomena. A total of 260,630 polyhedral cells were employed in the simulation.

8.4.4 Numerical methods

The interaction between hot air and moisture in the cotton wads during the drying process causes the moisture to turn into vapor and subsequently depart with the hot air. While cotton is regarded as a porous media in the simulation, hot air is viewed as a continuous phase. Air, water vapor, and liquid water are the three different phases that are considered in this simulation. During the drying process, the air phase, which includes the vapor, moves through the porous zone, while the liquid water phase stays static

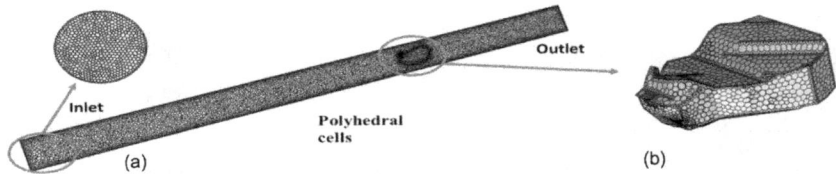

Figure 8.6 Meshed (a) horizontal pipe and (b) cotton wad.

inside the cotton. ANSYS FLUENT 2019R3 was employed to carry out thorough CFD simulations, in accordance with the instructions provided in the ANSYS Fluent User Guide [9].

8.4.4.1 Modeling assumptions and critical equations

Cotton drying is a multiphase problem. The multiphase porous media equations 8.31–8.41 were applied to simulate the drying of cotton in the conveyer belt system. The cotton does not have any heat or fluid generation source. Therefore, respective source terms were eliminated. The following presumptions were made for cotton drying study:

1. A single stationary cotton wad was considered for drying.
2. Cotton wad has uniform porosity throughout the volume.
3. The moisture was evenly distributed throughout its porous medium.
4. Thermal equilibrium was considered for all the phases.
5. Heated conveyance air was presumed to interact with the moisture (water) in the cotton, leading to the initiation of evaporation upon interaction.
6. The pipe walls are adiabatic.

8.4.4.2 Boundary conditions for simulations

The cotton-air interface was set to contact areas, the pipe walls were assumed to be adiabatic with standard roughness and no slip boundary conditions, and the pipe inlet and outlet were both set to velocity inlet and pressure outlet.

8.4.5 Cotton moisture calculation

Cotton wad moisture content is typically calculated on a wet basis. This wet basis moisture content (Mw) is stated as the percentage expressing the ratio of the weight of water (Ww) to the total weight of the material (Wt).

$$M_w = \left(\frac{W_w}{W_t} \right) \times 100 = \frac{W_w}{W_w + W_d} \qquad (8.42)$$

where W_d is the dry weight of the material.

In this study, the cotton wad had a total envelope volume of 1.1612 10⁻³ m³, a cotton density of 80 kg/m³, and a porosity of 95%. The dry weight of the 95% porous cotton was found to be 4.6448 grams. A simulated cotton wad was augmented with 0.35 grams of water, which was uniformly dispersed in accordance with the cotton wads' uniform porosity, at a seed cotton moisture content of 7%.

8.4.6 Heat transfer and evaporation (moisture removal) model

8.4.6.1 Heat transfer model

Along with the flow energy equations solved for the multiphase flow, the Ranz-Marshall correlation was used for solving the heat transfer coefficient h_{pq} between air and water phases. Here κ_q is thermal conductivity, Nu is the Nusselt number, d_p is the particle diameter, Re is the Reynolds number, and Pr is the Prandtl number.

$$h_{pq} = \frac{6\kappa_q \alpha_p \alpha_q \mathrm{Nu}_p}{d_p^{\,2}} \tag{8.43}$$

$$\mathrm{Nu} = 2.0 + 0.6\,\mathrm{Re}^{1/2}\,\mathrm{Pr}^{1/3} \tag{8.44}$$

8.4.6.2 Evaporation model

The momentum equation for the evaporation model used by ANSYS contains a mass transfer expression from the liquid state to the gas state (evaporation). Mass transfer from the gas to liquid state (condensation) is zero in the present case of cotton drying.

$$\frac{\partial}{\partial t}\left(\alpha.\rho_v\right) + \nabla.\left(\alpha.\rho_v \vec{V}_v\right) = \dot{m}_{l \to v} - \dot{m}_{v \to l} \tag{8.45}$$

where v is the vapor index, α is the volume fraction of the vapor, ρ_v is the vapor density, $\vec{V}_v$ is the velocity of the gas state, and $\dot{m}_{l \to v} - \dot{m}_{v \to l}$ is the evaporation rate. Mass transfer is defined as:

$$\dot{m}_{l \to v} = coeff.\alpha_l \rho_l \frac{\left(T - T_{sat}\right)}{T_{sat}} \tag{8.46}$$

8.4.7 Cotton drying CFD simulations

Modeling of cotton drying in a conveyer pipe was performed using pressure-based, transient simulations. The model employed an Eulerian approach with a multi-fluid volume of fluid model to account for air-water and air-vapor fluids. The air-water interaction was represented using a dispersed technique. The major phase in the simulations was described as moist air (air-vapor mixture), while the secondary phase was the water-liquid content of the porous cotton medium. Evaporation-condensation models were employed for mass transfer between the primary and secondary phases while the Ranz-Marshall correlation was used for heat transfer. The SST k-w (shear stress transport k-w) turbulence model with low Reynolds number modification was utilized. The SST k-w turbulence model is excellent for modeling both forced and natural circulation flow systems, according to Kumar et al [36]. The temperature of the incoming air was changed to provide the required 50% relative humidity for the simulations. Cotton's airflow was believed to be laminar due to its low average velocity and assumed 95% porosity. An inertial resistance of 76 m^{-1} and viscous resistance of 6990000 m^{-2} were set in the direction of the air flow. The simulations were regarded to have converged when they reached the limit of 10^{-5}. The relative viscosity in the cotton zone was calculated using Brinkman equations.

The drying process was examined using relative air speeds of 0.1, 1.0, 3.0, and 5.0 m/s. The phrase "relative velocity" is used to characterize the difference in velocity between the air and seed cotton in the conveying system. For each relative air speed, four inlet air temperatures were taken into account: 20°C, 66°C, 121°C, and 177°C (equivalent to 68°F, 150°F, 250°F, and 350°F, respectively), with corresponding relative humidity values of 50%, 2%, 0%, and 0%. These temperature ranges are within the recommended range for gin drying. Cotton with moisture values of 7% was considered. A complete investigation of the drying dynamics under various conditions is possible.

8.4.8 Results and discussion

The effects of temperature fluctuations and relative air velocity on the drying process are examined in the following sections.

8.4.8.1 Mesh independency test

Three mesh sizes (205,390, 260,630, and 310,948 cells) were used in simulations for a total of 60 seconds to ensure mesh independence. According to the findings, mesh sizes of 260,630 and 310,948 cells performed similarly, with a moisture content differential of under 5% (Figure 8.7a). The mesh

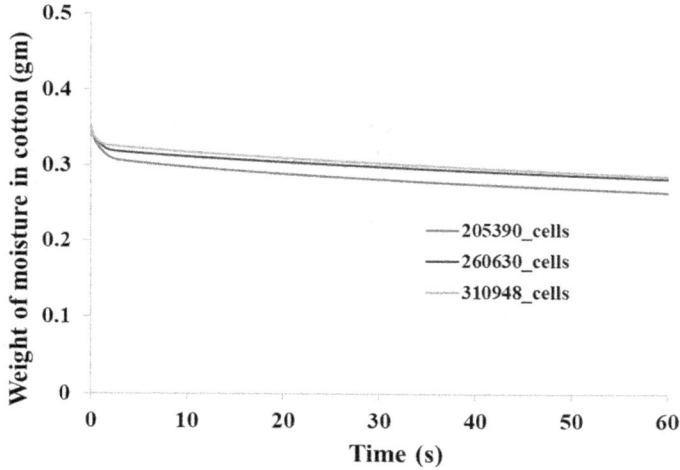

Figure 8.7 Weight of the moisture present in cotton wad during the drying process.

with 260,630 cells was consequently chosen for future research owing to its balance between accuracy and computing efficiency.

8.4.8.2 Characteristic properties of the conveying system

Figure 8.8 illustrates key characteristics of the conveyance system, including variations in air velocity, and temperature. To elaborate briefly, air velocity rises as it flows in the narrow passage around the cotton wad. Temperature changes in the cotton zone indicate an initial surface temperature increase upon contact with hot air, followed by a gradual rise inward. Notably, the contours also offer insights into the effective width of the area, providing valuable information about the system's dynamics.

Velocity Variation (m/s)

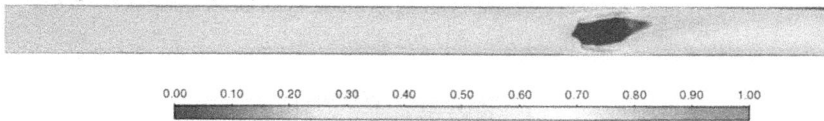

Temperature Variation (°C)

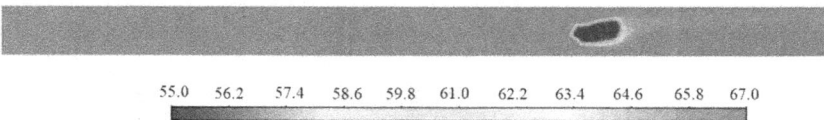

Figure 8.8 Velocity and temperature variation of conveying air.

8.4.8.3 Effect of relative air speed on cotton drying

The speed of the air relative to cotton is a critical factor in moisture removal. Greater relative velocities cause speeding up the evaporation of water from the cotton. Moreover, higher relative velocities effectively remove the water vapor phase. Figure 8.9 demonstrates the impact of relative air velocity on moisture removal from cotton wads with moisture levels of 7%. Figure 8.9a–d depict moisture loss in cotton when air at temperatures of 20°C, 66°C, 121°C, and 177°C (68°F, 150°F, 250°F, and 350°F) is blown at relative velocities of 0.5 m/s, 1.0 m/s, 3.0 m/s, and 5.0 m/s. In all scenarios, higher relative air velocities at fixed temperatures result in more rapid moisture reduction in the cotton. Furthermore, increased air temperatures contribute to even faster moisture removal from the cotton wads.

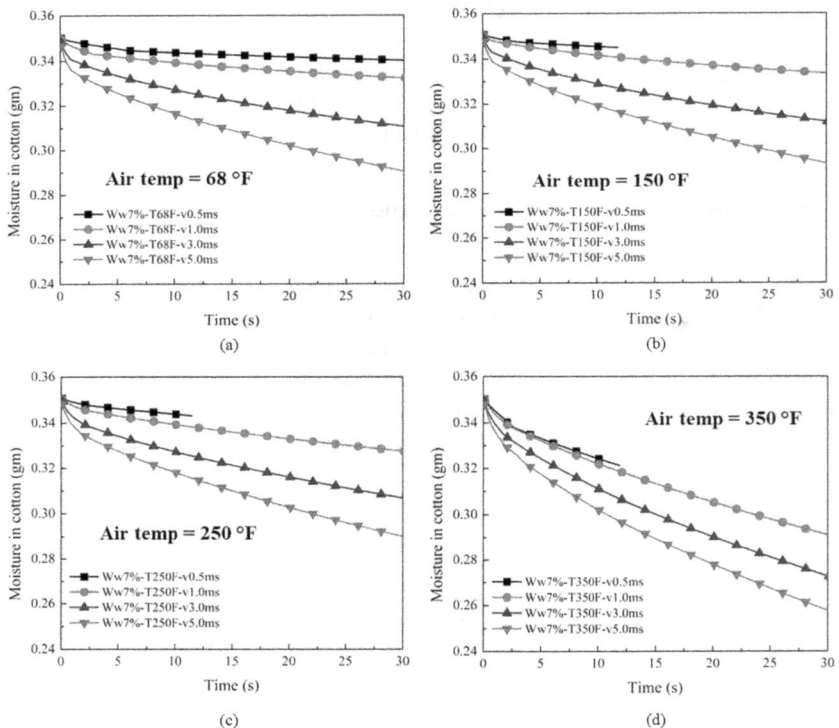

Figure 8.9 Loss of moisture from cotton with 7% moisture by blowing air at a relative velocity of 0.5, 1.0, 3.0, and 5.0 m/s and air temperature of (a) 20°C (68°F), (b) 66°C (150° F), (c) 121°C (250°F), and (d) 177°C (350°F).

8.4.8.4 Comparison of moisture loss during cotton drying

Figures 8.10a and b compare the evaporation of moisture in cotton wads with 7% moisture levels at 10 and 30 seconds, respectively. This comparison considers various air temperatures (68°F, 150°F, 250°F, and 350°F) and air velocities (0.5 m/s, 1.0 m/s, 3.0 m/s, and 5.0 m/s). The temperature of the conveying air significantly impacts moisture removal efficiency. Higher air temperatures are expected to enhance moisture removal. However, it is noted that up to 250°F, moisture removal isn't significantly affected at any given air velocity. Increasing air velocity alone is enough to improve moisture removal. Conversely, at 350°F, heating the air substantially enhances moisture removal. A higher moisture level will be relatively slow in evaporating due to thermal inertia and would use more heat energy to raise the temperature of the trapped moisture, making less moist cotton dry faster under similar conditions. It has been observed that when the inlet air speed equals or exceeds 1.0 m/s, 50% of the expected moisture is removed within the initial 10 seconds, compared to the total moisture removed at 30 seconds. This delay in moisture removal is due to the time required for heat energy to penetrate and heat the trapped moisture while moving from the cotton wad's outer surface to its center.

8.4.9 Conclusion from the case study

From this study, several conclusions emerge, such as the moisture within the cotton changing from water to vapor due to the surrounding blowing air, leading to its removal. The moisture removal is enhanced by higher relative air speeds at a constant temperature, with the most effective moisture removal occurring with elevated air temperature and relative velocity. Moreover, an increase in air temperature up to 250°F does not significantly impact moisture removal, whereas higher air temperatures (350°F) improve

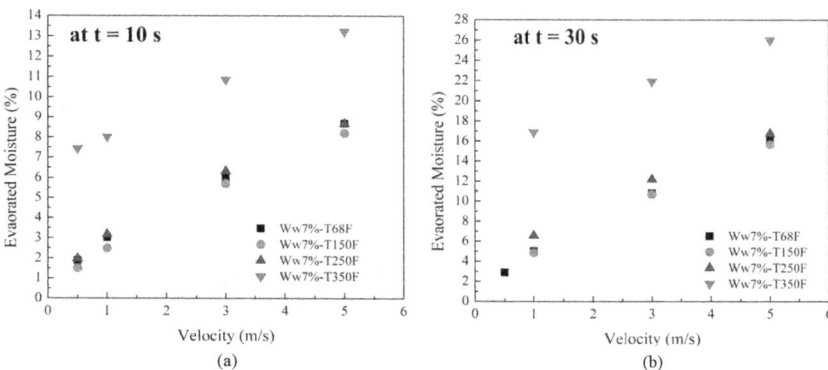

Figure 8.10 Comparison of moisture loss during cotton drying at (a) 10s and (b) 30s.

removal at all velocities. Furthermore, half of the expected moisture removal occurs within the initial 10 seconds at any operating parameter compared to 30 seconds. The cotton wad with initial lower moisture will dry early primarily due to heat energy being utilized to raise trapped moisture's temperature, in the case of cotton with high moisture.

8.5 LEARNING FROM THIS CHAPTER

This chapter provides fundamental definitions essential for comprehending the flow characteristics within porous media. The discussion on modeling flow through porous media offers an extensive grasp of these flows, which can be adapted to various types of porous materials with necessary adjustments. Furthermore, the chapter includes a real-world case study that centers on the drying process of cotton using a pneumatic conveyor belt. The study's outcomes are valuable for achieving precise cotton drying and optimizing the performance of cotton ginning facilities.

ACKNOWLEDGMENT

This research work presented in this paper was funded by USDA ARS, USA.

REFERENCES

[1] R.J. Goldstein, W.E. Ibele, S. V. Patankar, T.W. Simon, T.H. Kuehn, P.J. Strykowski, K.K. Tamma, J.V.R. Heberlein, J.H. Davidson, J. Bischof, F.A. Kulacki, U. Kortshagen, S. Garrick, V. Srinivasan, K. Ghosh, R. Mittal, Heat transfer—A review of 2004 literature, *Int. J. Heat Mass Transf.* 53 (2010) 4343–4396. http://doi.org/10.1016/j.ijheatmasstransfer.2010.05.004

[2] A. Verruijt, Darcy's law, *Theory Appl. Transp. Porous Media.* 30 (2018) 49–58. https://doi.org/10.1007/978-3-319-61185-3_6

[3] A. Fouda, H El Deeb, H Abou Taleb, A. Ragab, Determination of Pore Size, Porosity and Pore Size Distribution of Woven Structures by Image Analysis Techniques, *J. Text. Sci. Eng.* 7 (2017). https://doi.org/10.4172/2165-8064.1000314

[4] Multiphase Flow Handbook (2005). https://doi.org/10.1201/9781420040470

[5] Capillary pressure, (n.d.). https://en.wikipedia.org/wiki/Capillary_pressure (accessed September 5, 2023).

[6] P. Glover, *Fluid Saturation and Cappilary Pressure*, (n.d.).

[7] Capillary pressure I MOOSE, (n.d.). https://mooseframework.inl.gov/modules/porous_flow/capillary_pressure.html (accessed September 5, 2023).

[8] Fluent Theory Guide, (n.d.). https://ansyshelp.ansys.com/account/secured?returnurl=/Views/Secured/corp/v201/en/flu_th/flu_th.html

[9] ANSYS Fluent User Guide, (n.d.). http://www.pmt.usp.br/academic/martoran/notasmodelosgrad/ANSYS Fluent Users Guide.pdf.

[10] T.D. Valco, J.M. Fannin, R.A. Isom, The Cost of Ginning Cotton – 2010 Survey Results (2012) 2010–2013.

[11] R.G. Hardin, P.A. Funk, Electricity use patterns in cotton gins, *Appl. Eng. Agric.* 28 (2012) 841–849. https://doi.org/10.13031/2013.42471

[12] P. Funk, R.G. Hardin, P. Funk, R.G. Hardin IV, Cotton gin electrical energy use trends and 2009 audit results, *Appl. Eng. Agric.* 28 (2012) 503–510. https://doi.org/10.13031/2013.42078

[13] S. Kumar, K. Rathore, Renewable energy for sustainable development goal of clean and affordable energy, *Int. J. Mater. Manuf. Sustain. Technol.* 2 (2023) 01–15. https://doi.org/10.56896/IJMMST.2023.2.1.001

[14] K. Rathore, S. Kumar, Net zero emissions for our future generations through renewables: A brief review, *Int. J. Energy Resour. Appl.* 2 (2023) 39–56. https://doi.org/10.56896/IJERA.2023.2.1.004

[15] G. J. Mangialardi Jr., Conserving energy-reduce volume of air to transport and dry seed cotton, *Texas Cotton Ginners Journal and Yearbook* 45, no. 1, 1977.

[16] B.J.S. Baker, E.P. Columbus, R.C. Eckley, *Pneumatic and mechanical handling systems*, Washington, DC, 1994.

[17] F.A. Zenz, Two-phase fluid-solid flow, *Ind. Eng. Chem.* 41 (2002) 2801–2806. https://doi.org/10.1021/IE50480A032

[18] R.G. Hardin, Pneumatic conveying of seed cotton: Minimum velocity and pressure drop, *Trans. ASABE.* 57 (2014) 391–400. https://doi.org/10.13031/trans.57.10437

[19] J.B. Cocke, *Effect of seed cotton moisture level at harves on ginned lint*, 1974.

[20] G.L. Barker, R.V. Baker, J.W. Laird, A cotton processing quality model, *Agric. Syst.* 35 (1990) 1–20.

[21] S. Kumar, P.K. Vijayan, U. Kannan, M. Sharma, D.S. Pilkhwal, Experimental and computational simulation of thermal stratification in large pools with immersed condenser by, *Appl. Therm. Eng.* 113 (2017) 345–361. https://doi.org/10.1016/j.applthermaleng.2016.10.175

[22] J. Fan, S. Furbo, Buoyancy driven flow in a hot water tank due to standby heat loss, *Sol. Energy.* 86 (2012) 3438–3449. http://dx.doi.org/10.1016/j.solener.2012.07.024

[23] S. Kumar, R.B. Grover, P.K. Vijayan, U. Kannan, Numerical investigation on the effect of shrouds around an immersed isolation condenser on the thermal stratification in large pools, *Ann. Nucl. Energy.* 110 (2017) 109–125. https://doi.org/10.1016/j.anucene.2017.06.022

[24] A. Gil, M. Medrano, I. Martorell, A. Lázaro, P. Dolado, B. Zalba, L.F. Cabeza, State of the art on high temperature thermal energy storage for power generation. Part 1—Concepts, materials and modellization, *Renew. Sustain. Energy Rev.* 14 (2010) 31–55. http://dx.doi.org/10.1016/j.rser.2009.07.035

[25] S. Kumar, R.B. Grover, H. Yadav, P.K. Vijayan, U. Kannan, A. Agrawal, R.B. Grover, U. Kannan, H. Yadav, A. Agrawal, Experimental investigations on thermal stratification in a large pool of water with immersed isolation condenser, *Int. Conf. Nucl. Eng. Proceedings, ICONE.* 3 (2020) 1–7. https://doi.org/10.1115/ICONE2020-16647

[26] M. Colombo, M. Fairweather, Study of nuclear reactor external vessel passive cooling using computational fluid dynamics, *Nucl. Eng. Des.* 378 (2021) 111186. https://doi.org/10.1016/J.NUCENGDES.2021.111186

[27] S. Kumar, P.K. Vijayan, M. Sharma, D.S. Pilkhwal, U. Kannan, A.K. Nayak, Experimental and numerical analysis of thermal stratification in large water pool, (2015). http://inis.iaea.org/Search/search.aspx?orig_q=RN:47073982 (accessed January 12, 2021).

[28] T. Khurana, B. Prasad, K. Ramamurthi, S. Murthy, T. Kumar Khurana, B. Prasad, K. Ramamurthi, S. Murthy, Thermal stratification in ribbed liquid hydrogen storage tanks, *Int. J. Hydrogen Energy*. 31 (2006) 2299–2309. https://doi.org/10.1016/j.ijhydene.2006.02.032

[29] S. Kumar, A. Thyagarajan, D. Banerjee, Experimental Investigation of Thermal Energy Storage (TES) platform leveraging phase change materials in a chevron plate heat exchanger, *Int. Mech. Eng. Congr. Expo.*, 2022: pp. 1–7. https://doi.org/10.1115/IMECE2022-96226

[30] S. Kumar, A. Thyagarajan, D. Banerjee, An investigation on chevron plate heat exchanger filled with organic phase change material, in: *Proc. ASME 2020 Heat Transf. Summer Conf. HT2020* July 13-15, 2020, Virtual, Online, American Society of Mechanical Engineers (ASME), 2020: pp. 1–6. https://doi.org/10.1115/HT2020-9122

[31] S. Rafiee, A. Keyhani, A. Mohammadi, Soybean seeds mass transfer simulation during drying using finite element method, *World Appl. Sci.* 4 (2008) 284–288.

[32] L. Huang, K. Kumar, Mujumdar, Arun S., Computational fluid dynamic simulation of droplet drying in a spray dryer *Computational Fluid Dynamic Simulation of Droplet Drying in a 1*. Mechanical Engineering Department, National University of Singapore, Kent Ridge Crescent, Singapore, (2015).

[33] F. Nadi, G.H. Rahimi, R. Younsi, T. Tavakoli, Z. Hamidi-Esfahani, Numerical simulation of vacuum drying by Luikov's Equations, *Dry. Technol.* 30 (2012) 197–206. https://doi.org/10.1080/07373937.2011.595860

[34] A. Ljung, Drying of Iron Ore Pellets -Analysis with CFD, (2008).

[35] M.K. Awasthi, N. Dutt, A. Kumar, S. Kumar. Electrohydrodynamic capillary instability of Rivlin–Ericksen viscoelastic fluid film with mass and heat transfer. *Heat Transfer*. 2023; 1–19. doi:10.1002/htj.22944

[36] S. Kumar, R.B. Grover, H. Yadav, P.K. Vijayan, U. Kannan, A. Agrawal, Experimental and numerical investigation on suppression of thermal stratification in a water-pool: PIV measurements and CFD simulations, *Appl. Therm. Eng.* 138 (2018) 686–704. https://doi.org/10.1016/j.applthermaleng.2018.04.070

Chapter 9

Numerical analysis of turbulent flow in tubes with longitudinal fin using Al$_2$O$_3$-water nanofluids

Anand Kumar Solanki
Gayatri Vidya Parishad College of Engineering, Visakhapatnam, India

Ankit Rajkumar Singh
Indian Institute of Technology Bombay, Mumbai, India

Nitesh Dutt
College of Engineering Roorkee, Uttarakhand, India

9.1 INTRODUCTION

Nowadays, heat exchangers are widely used in heating, ventilation, air-conditioning (HVAC) system, refrigeration industry, process industries, and power plants [1]. Nowadays, the effective design of the heat exchangers is important for the industries to obtain a maximum heat transfer with a smaller pressure drop. As such, an extended surface inside the tube has been widely used to enhance the heat transfer by increasing the surface area of the tube and the turbulence intensity on the flow. The development of extended surfaces, such as integral inner-fin tubing, and the augmentation of heat transfer coefficients on extended surfaces, by shaping or interrupting the surfaces, is of particular interest. Moreover, by using nanofluids, heat transfer rates are also improved. Nanofluids have extremely high thermal conductivity and good heat transfer properties. Khdher et al. [2] performed experimental and numerical studies to carry out the heat transfer coefficient and pressure drop of Al$_2$O$_3$/water nanofluid in circumferentially ribbed tubes. Under the same operating condition, the enhanced tubes show a 92–621% heat transfer coefficient increment compared to the smooth tube. Li et al. [3] conducted the numerical simulations to find the performance of dimpled enhanced tubes. They showed the effect of dimple shape, depth, and pitch on the Nusselt number and the friction factor. They found that the ellipsoidal dimpled tube showed better performance compared to spherical and cone dimpled tubes. Qi et al. [4] numerically and experimentally investigated the heat transfer of TiO$_2$-water nanofluids and deionized water in corrugated tubes with the variation of the Reynolds number and nanoparticle mass fraction. They observed that the corrugated tubes show a significant effect on the heat transfer and friction factor. Abdolbaqi et al. [5] have initiated

DOI: 10.1201/9781003465171-9

studies on the BioGlycol/water mixture-based TiO_2 nanofluid flowing in the flat tubes under different operating temperatures of 30°C, 50°C, and 70°C. They found that the heat transfer of the fixed solid volume increases with an increase of temperature. Baba et al. [6] discussed on the heat transfer characteristics of Fe_3O_4–water nanofluid in the tube with multiple internal longitudinal fins. They investigated that at the fixed solid volume fraction of 4% Fe_3O_4–water nanofluid, the internal longitudinal fins provided 80–90% more heat transfer compared to the plain tube. Kumait et al. [7] investigated the heat transfer of TiO_2 nanofluid in the tube with integral rib under the turbulent flow condition with consideration of constant heat flux. Also, they developed the correlation to predict the friction factor and a Nusselt number in the enhanced tubes. Rabbani et al. [8] attempted to demonstrate the heat transfer and the pressure drop of MgO/water-EG laminar nanofluid flowing in the metal foams filled in the copper tubes. They investigated that the heat transfer rate of metal foams filled in the copper tubes raised intensely, but also pressure drop in the same tube enhanced greatly. Karimi et al. [9] presented the numerical simulation of alumina/water nanofluid flowing inside the double tube heat exchanger having a twisted tape. They found that the Reynolds number, nanofluid concentration, and pitch ratio had a significant effect on the Nusselt number and friction factor.

Kristiawan et al. [10] experimentally studied the heat transfer and pressure drop of TiO_2/water nanofluids in a helical microfin tube. Moreover, they suggested a correlation to predict the heat transfer and pressure drop in the helical micro-fin tube for the TiO_2/water nanofluids. Wang et al. [11] presented the numerical simulation to carry out the heat exchange efficiency of silica-water nanofluids in the round tube and triangular tube with porous twisted tape. They found that performance of the triangular tube is well compared to that of the round tube. E. Kumar et al. [12] investigated the influence of micro-fin in the helically coiled tube upon heat transfer. Also, they presented the effect of coil pitch and coil diameter on the heat transfer and pressure drop. Solanki et al. [13] numerically analyzed the effect of the solid volume fraction, the Reynolds number, and the aspect ratio of the flattened tube on heat transfer and pressure drop of Al_2O_3-water nanofluids. They found that both heat transfer and pressure drop enhanced with aspect ratio of the flattened helically coiled tubes. Singh et al. [14] showed the thermal-hydraulic characteristics of the micro-fin flattened straight tubes. They reported that the value of friction factor declined with the increase of the number of fins in the tube. Islam et al. [15] numerically investigated the heat transfer efficiency of non-uniform wall corrugations under laminar condition for hybrid nanofluid which is the mixture of graphene nanoplatelets and multi-walled carbon nanotubes with water. They found that the non-uniform wall corrugations showed a better performance compared to uniform corrugation UWC.

As per the aforementioned literature, a number of research studies have been done on the heat transfer and pressure drop of various types of nanofluids in different geometries of the tubes. In this chapter, the heat transfer coefficient and pressure drop characteristics of the Al_2O_3-Water nanofluid inside the longitudinal fin tube are numerically examined by ANSYS FLUENT. The studies were carried out at different solid volume fractions of nanofluid ranging from 0% to 3%, with the Reynolds number varying from 10,000 to 20,000. The effect of the fin, Reynolds number, and solid volume fraction of the nanofluid on the heat transfer coefficient is discussed.

9.1.1 Physical model

The geometry of the tube with the longitudinal integral fin is shown in Figure 9.1. In this study, the diameter of the tube is 26 mm, the length of the tube is 1,000 mm, and the number of the longitudinal integral fins is 4.

9.1.2 Governing equations and boundary conditions

In the current work, the heat transfer coefficient and pressure drop of Al_2O_3-water based nanofluid are numerically examined with the Reynolds number varying from 10,000 to 20,000. The following assumptions for Al_2O_3-water based nanofluid in a tube are taken: (1) three-dimensional flow is considered as a steady-state. (2) The size and shape of nanoparticles are uniform and spherical, respectively. (3) The working fluid is a Newtonian homogeneous and incompressible. (4) The effects of viscous heating and radiation are ignored. (5) No-slip condition between water and the nanoparticles phase

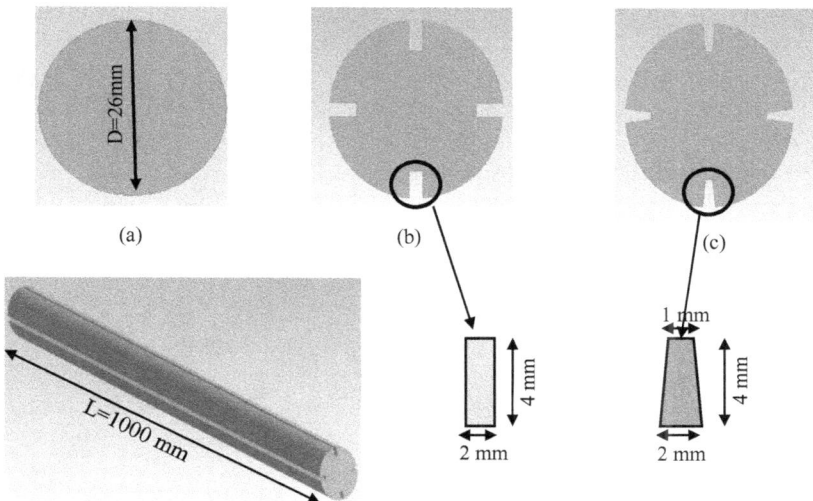

Figure 9.1 Geometry of tubes (a) smooth circular tube, (b) tube with rectangular integral fins, and (c) tube with trapezoidal integral fins.

is considered. (6) Water and the nanoparticles phase are in thermal equilibrium. As per the above assumption, the governing equations are as follows:

Continuity equation:

$$\frac{\partial}{\partial x_i}\left(\rho_{nf}u_i\right) = 0 \tag{9.1}$$

where u_i is the axial velocity in the tube and ρ_{nf} is the density of the nanofluid.

Momentum equation:

$$\frac{\partial}{\partial x_j}\left(\rho_{nf}u_iu_j\right) = -\frac{\partial P}{\partial x_i} + \frac{\partial}{\partial x_j}\left[\mu_{nf}\left(\frac{\partial u_i}{\partial x_j}+\frac{\partial u_j}{\partial x_i}\right)\right] + \frac{\partial}{\partial x_j}\left(-\rho_{nf}\overline{u_i'u_j'}\right) \tag{9.2}$$

where μ_{nf}, u', and u_j are the nanofluid viscosity, fluctuating velocity, and axial velocity components. The term $\rho_{nf}\overline{u_i'u_j'}$ is the turbulent shear stress or the Reynolds stress.

Energy equation:

$$\frac{\partial}{\partial x_i}\left(\rho_{nf}u_iT\right) = \frac{\partial}{\partial x_j}\left[\left(\frac{\mu_{nf}}{\text{Pr}}+\frac{\mu_{t,nf}}{\text{Pr}}\right)\frac{\partial T}{\partial x_j}\right] \tag{9.3}$$

where T and Pr are the temperature and Prandtl number. The subscript t refers to the turbulent flow.

For the numerical simulation in the present study, the standard $k-\varepsilon$ turbulence model is applied:

$$\left(-\rho_{nf}\overline{u_i'u_j'}\right) = \mu_{t,nf}\left(\frac{\partial u_i}{\partial x_j}+\frac{\partial u_j}{\partial x_i}\right) \tag{9.4}$$

The turbulent kinematic viscosity is:

$$\mu_{t,nf} = \rho_{nf}C_\mu\frac{k^2}{\varepsilon} \tag{9.5}$$

Turbulent kinetic energy k and energy dissipation ε equation are:

$$\frac{\partial\left(\rho_{nf}ku_i\right)}{\partial x_i} = \frac{\partial}{\partial x_j}\left[\left(\mu_{nf}+\frac{\mu_{t,nf}}{\sigma_k}\right)\frac{\partial k}{\partial x_j}\right] + \Gamma_k - \rho_{nf}\,\varepsilon \tag{9.6}$$

$$\frac{\partial\left(\rho_{nf}\varepsilon u_i\right)}{\partial x_i} = \frac{\partial}{\partial x_j}\left[\left(\mu_{nf}+\frac{\mu_{t,nf}}{\sigma_\varepsilon}\right)\frac{\partial\varepsilon}{\partial x_j}\right] + C_{1\varepsilon}\frac{\varepsilon}{k}\Gamma_k + C_{2\varepsilon}\rho_{nf}\frac{\varepsilon^2}{k} \tag{9.7}$$

where Γ_k and ε are the turbulence kinetic energy generation and its dissipation rate, respectively. For the realizable k-e turbulence model, the following empirical constants are applied: $C_{1\varepsilon} = 1.44$, $C_{2\varepsilon} = 1.90$, $\sigma_k = 1.0$, $\sigma_\varepsilon = 1.2$.

All governing equations using the CFD software ANSYS FLUENT were solved with a control volume scheme. The geometry of the tube with the longitudinal integral fin and its meshing were created using ANSYS WORKBENCH and ANSYS Mesh package, respectively. The SIMPLE method was used with pressure velocity coupling. The second-order upwind scheme was used for discretizing the moment, turbulent, and energy equations. Moreover, the convergence criteria of 10^{-5} are applied for the continuity, momentum, and energy equation.

9.1.3 Boundary conditions

The following boundary conditions are used in all of the above governing equation: The inlet temperature of the working fluid is 298.15 K, and the inlet velocity is taken on the basis of the Reynolds number. The tube wall temperature remains constant at 353.15 K. At the outlet of the tube, the gauge pressure is set as zero. Moreover, the inlet turbulence intensity is taken by

$$I = 0.16\left(\text{Re}_l\right)^{-\frac{1}{8}} \tag{9.8}$$

9.1.4 Grid independence

Unstructured grids were applied in the computational domain with local grid refinement given in the boundary layer, as shown in Figure 9.2. Near the tube wall, 15 inflation layers were provided to resolve the boundary region because the temperature and velocity showed large gradients.

In order to check grid independence, Figure 9.3 and Table 9.1 present the relationship between the Nusselt number and the friction factor with

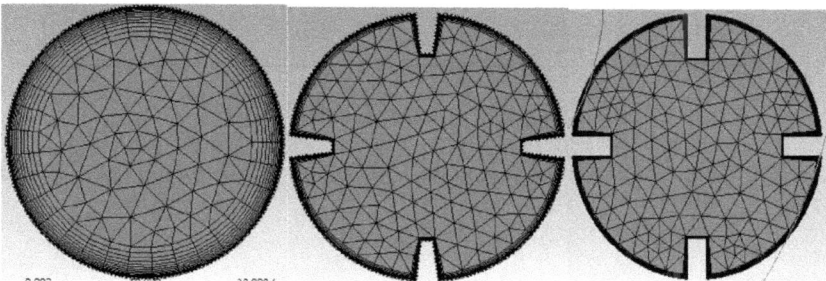

Figure 9.2 Meshing of smooth straight tube and tube with integral fins.

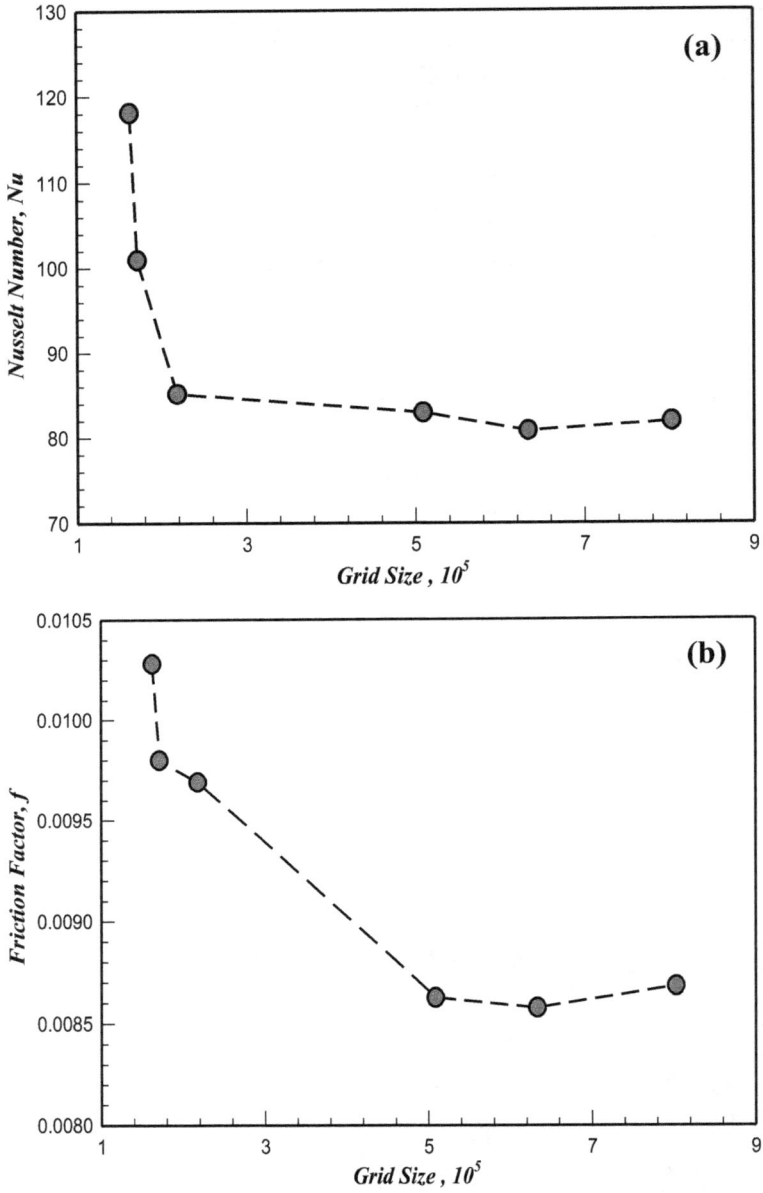

Figure 9.3 Grid independence test in Re = 10000 (a) Nusselt number (b) friction factor.

Table 9.1 Grid independent test

Number of Grid	Nusselt Number (Nu)	Friction Factor (f)
162,572	118.11	0.05653
171,321	100.95	0.04554
217,983	85.146	0.03909
508,830	82.92	0.03325
633,001	80.74	0.03252
802,981	81.84	0.03406

different sizes of grids. It is found from Figure 9.3 that the Nusselt number and the friction factor decrease with an increase of grid elements. The difference of the Nusselt number and the friction factor between 633,001 and 802,981 grid elements is less than 1.5%. Therefore, in order to ensure the solution accuracy and convergent time, the grid element 633,001 is applied to perform the following calculations.

9.1.5 Model validation

First, as shown in Figure 9.4, the numerical Nusselt number results of smooth straight tubes were compared with the data predicted by the renowned correlation given by Dittus-Boeltier equation [16] and Gnielinski equation [17] with maximum turbulent flow of water. It can be apparently seen that the numerical data was well predicted by Gnielinski equation [17], followed by Dittus-Boeltier equation [16] within 19% maximum deviation. Similarly, numerical friction factor results were compared with correlation suggested by Blausius correlation [18] and Petukhov correlation [19]. It was found that the maximum deviation of numerical data falls within 6%. This suitable agreement of the numerical Nusselt number and the friction factor with predicted data certifies the simulation work.

9.1.6 Data reduction

The Nusselt number is calculated by the following equation:

$$\mathrm{Nu} = \frac{hd_h}{k} \tag{9.9}$$

where h is the average heat transfer co-efficient of the fluid, d_h is the hydraulic diameter of the tube, and k is the thermal conductivity of working fluid.
    The friction factor is given by:

$$f = \frac{2\Delta P d_h}{\rho u^2 L} \tag{9.10}$$

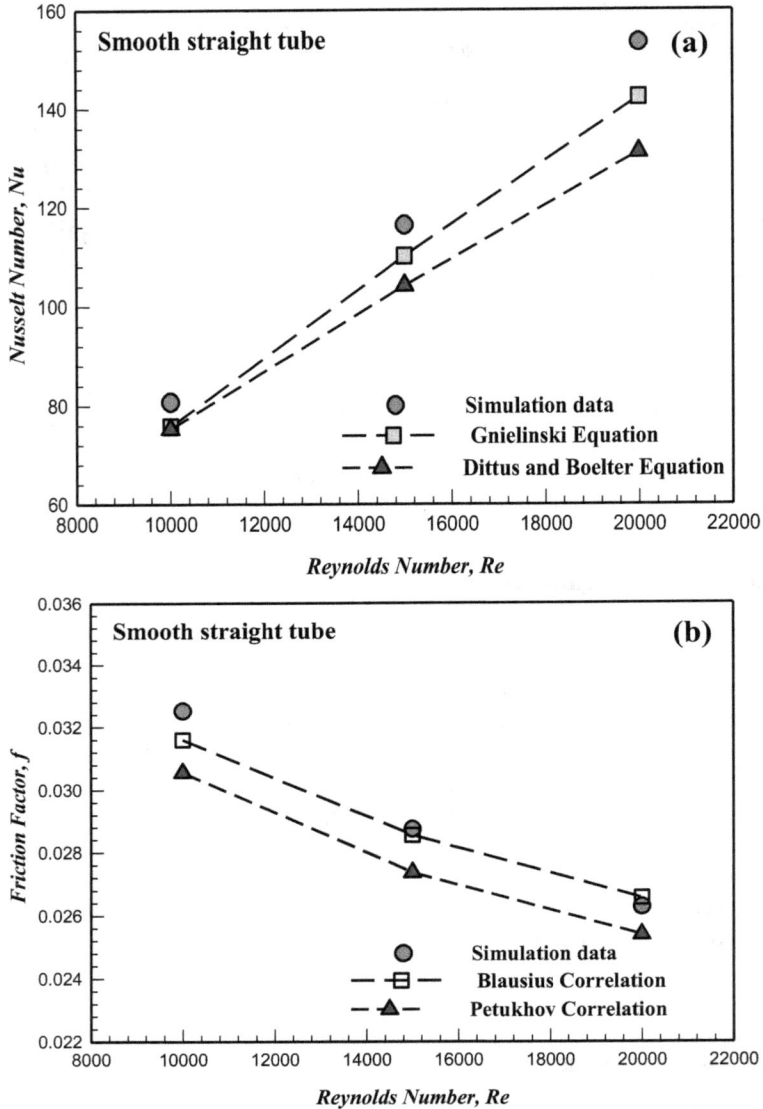

Figure 9.4 Comparison of the numerical results in smooth straight tube with the correlations for (a) Nusselt number (b) Friction factor.

where ΔP is the the pressure drop between the inlet pressure and outlet of tube and u is the mean velocity of the working fluid.

9.1.7 Thermophysical property of nanofluid

The following correlation and model was used to find the thermophysical properties of Al_2O_3-water nanofluids.

The density of nanofluid was taken from Pak and Cho equation [20].

$$\rho_{eff} = \rho_b\left(1 - \phi_{np}\right) + \rho_{np}\phi_{np} \tag{9.11}$$

where ρ_{eff}, ρ_b, ϕ_{np} and ρ_{np} are the effective density of nanofluid, the density of the base fluid, the volumetric concentration of nanoparticles and the density of nanoparticles, respectively.

The thermal conductivity of Al_2O_3-water nanofluids can be calculated by using the following correlation which was suggested by Maïga et al. [21]:

$$k_{nf} = \left(1 + 2.72\varphi + 4.97\varphi^2\right)k_{bf} \tag{9.12}$$

where φ, k_{nf}, and k_{bf} are the solid volume fraction, thermal conductivity of the nanofluid, and thermal conductivity of base fluid, respectively.

The specific heat (C_{peff}) of the nanofluid was calculated from the following equation given by Xuan and Roetzel [20]:

$$c_{peff} = \frac{\left(1 - \phi_{np}\right) \times \rho_b c_{pb} + \left(\phi_{np} \times \rho_{np}c_{pnp}\right)}{\rho_{eff}} \tag{9.13}$$

where c_{peff}, c_{pnp}, c_{pb}, and ϕ_{np} are the effective specific heat, the specific heat of nanoparticles, the specific heat of base fluid, and the volumetric concentration of nanoparticles.

For determining the dynamic viscosity [9] of the nanofluid, the following correlation is used:

$$\mu_{nf} = \left(1 + 1.73\varphi + 123\varphi^2\right)\mu_{bf} \tag{9.14}$$

where μ_{nf} and μ_{bf} are the viscosity of the nanofluid and base fluid, respectively. The value of the thermophysical properties of Al_2O_3-water nanofluids are summarized in Table 9.2.

Table 9.2 Thermophysical properties of Al_2O_3–water nanofluid

Nanofluids	Density, ρ (kg/m³)	Specific heat, Cp, (J/kg K)	Viscosity, μ, (mPa/s)	Thermal conductivity, k (W/mK)
Al_2O_3-water ($\varphi = 0$)	997.1	4183	0.895	0.605
Al_2O_3-water ($\varphi = 0.02$)	1051.5	3949	0.964	0.639
Al_2O_3-water ($\varphi = 0.03$)	1078.7	3841	1.034	0.657

9.2 RESULTS AND DISCUSSIONS

In this study, the effect of the Reynolds number (10,000, 15,000, and 20,000) and solid volume fraction of Al_2O_3-water nanofluid (0–3%) on the heat transfer coefficient and pressure drop inside a smooth straight tube and tube with internal rectangular and trapezoidal longitudinal fins have been examined. The results obtained from the tube with internal rectangular and trapezoidal longitudinal fin tubes were also compared with the results from those smooth straight tubes at the same operating condition.

9.2.1 Effects of solid volume fraction

Figure 9.5 plots the correlation between the heat transfer coefficient and solid volume fraction of Al_2O_3-water nanofluids. In this, the effect of the solid volume fraction of nanofluids on the heat transfer coefficient is presented for different tubes. It can be seen that the solid volume fraction shows a significant effect on the heat transfer coefficient. With an increase of the percentage of the particle in the base fluid, the heat transfer coefficient is increased for all tubes. For example, the solid volume fraction of the 3% Al_2O_3-water nanofluid shows 4–6% higher heat transfer coefficient compared to the water (0% SVF) in the case of the trapezoidal integral fin, as shown in Figure 9.5. The similar effects for the smooth straight tube and the rectangular fin tube have been found. The reason behind this is that the thermal conductivity of the nanofluid increases with an increase of the amount of the nanoparticle in the base fluid. Therefore, higher solid volume fraction nanofluid gives more heat transfer.

The relationship between the pressure drop and the Reynolds number for various solid volume fractions of nanofluid in a tube is revealed in Figure 9.6. It is apparently seen from Figure 9.6 that the pressure drop in the tube has been increased with an increase of the solid volume fraction and Reynolds numbers. This phenomenon is mainly caused by increasing the viscosity of nanofluid when the percentage of the solid volume fraction increased. For instance, the solid volume fraction of the 3% Al_2O_3-water nanofluid presents 9–24% higher pressure drop compared to the water (0% SVF) in the case of the rectangular integral fin, as shown in Figure 9.6.

9.2.2 Comparison of tubes

Figure 9.7 illustrates the temperature contours for the smooth straight tube, rectangular integral fin tube and trapezoidal integral fin tube. It can be seen that at the same Reynolds number, the rectangular integral fin tube shows a higher outlet temperature compared to the smooth straight tube and the rectangular integral fin tube at the fixed solid volume fraction of nanofluids.

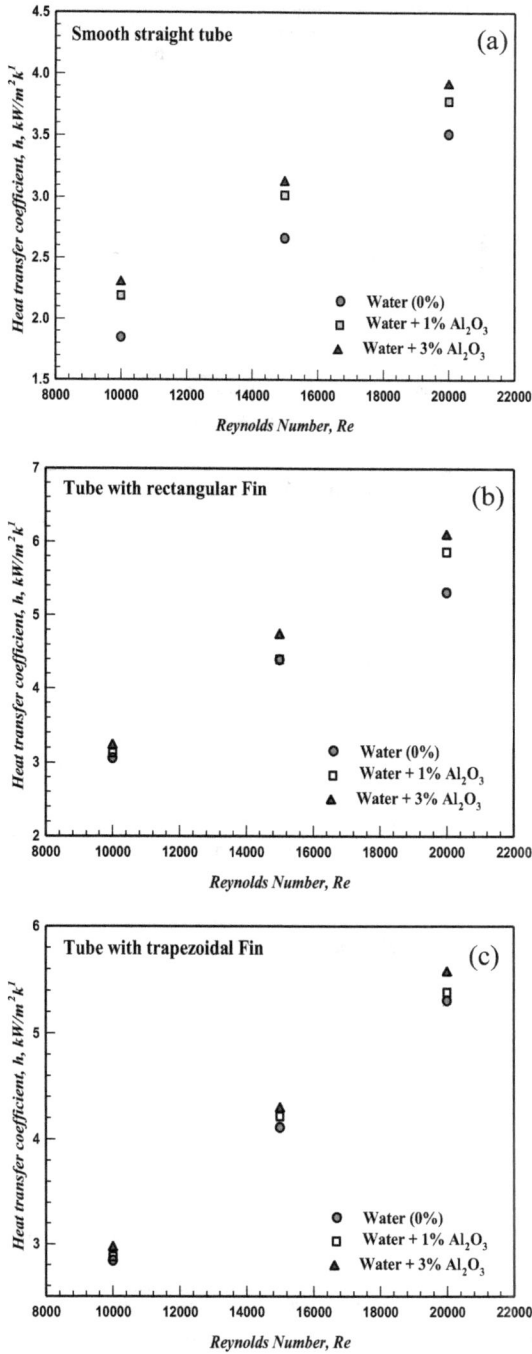

Figure 9.5 Effect of solid volume fraction on the heat transfer coefficient (a) smooth straight tube, (b) tube with rectangular fin, and (c) tube with trapezoidal fin.

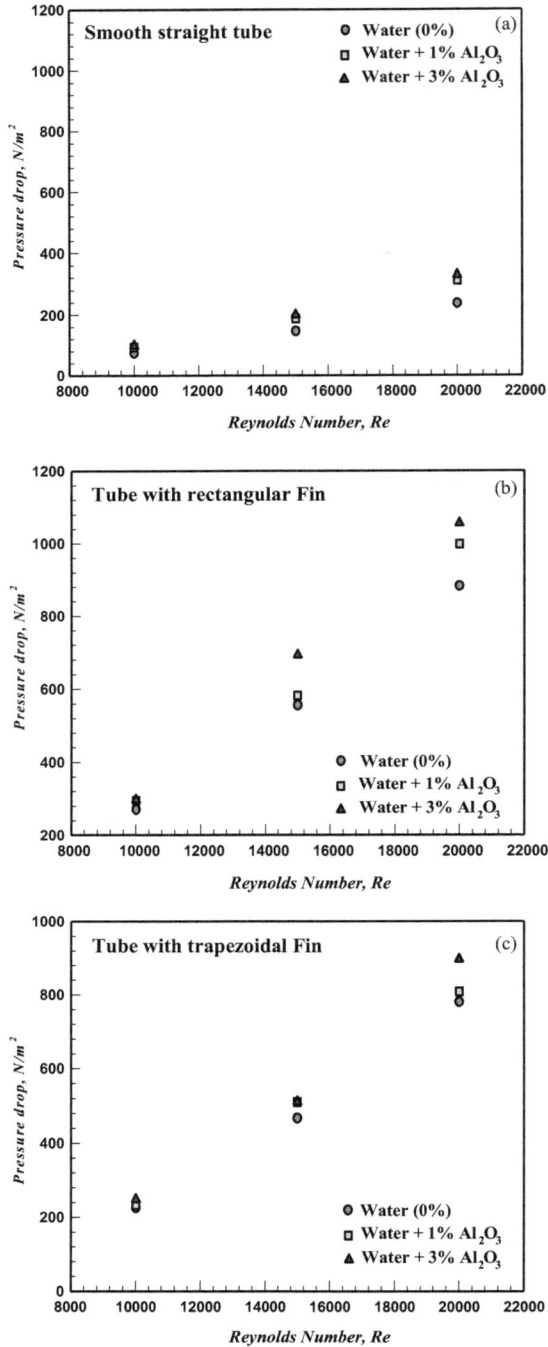

Figure 9.6 Effect of solid volume fraction on the pressure drop (a) smooth straight tube, (b) tube with rectangular fin (c) and tube with trapezoidal fin.

Tubes	Circular Smooth tube	Tube with trapezoidal integral fin	Tube with rectangular integral fin
Temperature TC 3.531e+002 3.481e+002 3.431e+002 3.381e+002 3.331e+002 3.281e+002 3.231e+002 3.181e+002 3.131e+002 3.081e+002 3.031e+002 2.981e+002 [K]	**Re=20000, Water+ 3% Al₂O₃, tube outlet temperature**		
	Re=10000, Water+ 3% Al₂O₃, tube outlet temperature		

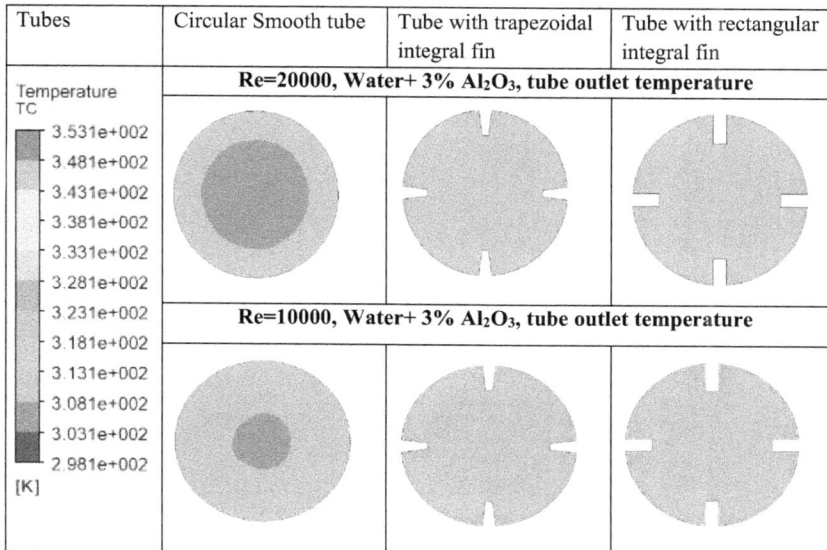

Figure 9.7 Cross-section view of outlet temperature contours for tubes.

Figure 9.8 shows the effect of integral fin on the heat transfer of the nano-fluid at the same operating conditions. The results indicate that the tube with the rectangular fin shows a higher heat transfer coefficient compared to the smooth straight tube and the tube with the trapezoidal fin in the percentage of 40–55% and 8–10%, respectively, as shown in Figure 9.8. This is because the tube with the rectangular integral fin has a higher surface area than the other tubes. Similarly, Figure 9.9 presents a comparison of tubes for the pressure drop of nanofluids. It can be seen from Figure 9.9 that the tube with rectangular integral fin produces more pressure drop than the smooth straight tube and the tube having a trapezoidal fin in the percentage of 200–240% and 17–35%, respectively.

9.3 CONCLUSIONS

In the present work, heat transfer coefficient and pressure drop of Al_2O_3-water-based nanofluid are numerically examined with the Reynolds number varying from 10,000 to 20,000 in the tube with integral rectangular and trapezoidal fins. The following results are obtained.

1. The rectangular fin shows a higher heat transfer coefficient compared to the smooth straight tube and the tube with trapezoidal fin in the percentage of 40–55% and 8–10%, respectively.

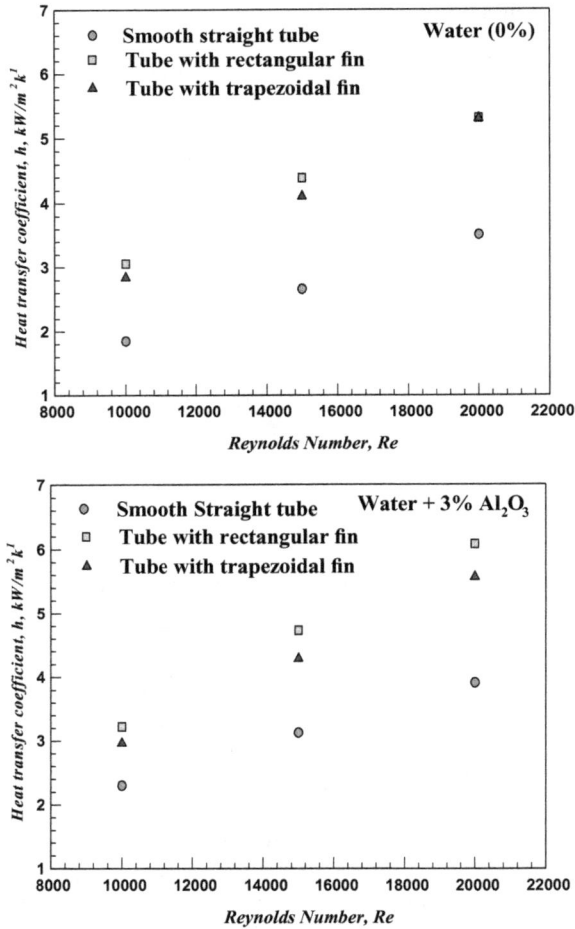

Figure 9.8 Heat transfer coefficient comparison for tubes.

2. The tube with rectangular integral fin produces more pressure drop than the smooth straight tube and the tube having trapezoidal fin in the percentage of 200–240% and 17–35%, respectively.
3. With an increase of the percentage of the particle in the base fluid, the heat transfer coefficient is increased for all tubes. For example, the solid volume fraction of the 3% Al_2O_3-water nanofluid shows 4–6% higher heat transfer coefficient compared to the water (0% SVF) in the case of the trapezoidal integral fin.
4. The solid volume fraction of the 3% Al_2O_3-water nanofluid presents 9–24% higher pressure drop compared to the water (0% SVF) in the case of the rectangular integral fin.

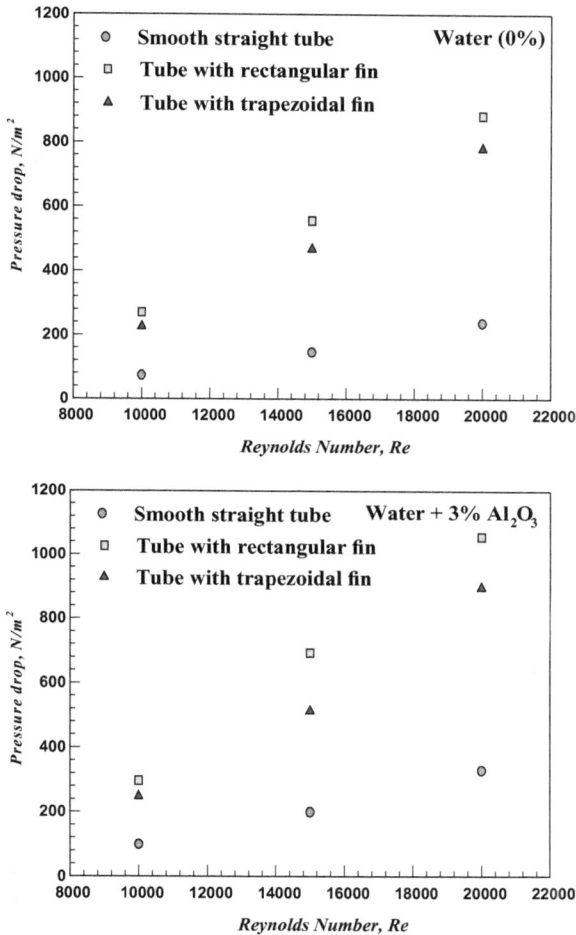

Figure 9.9 Pressure drop comparison for tubes.

REFERENCES

[1] R.L. Webb, A.E. Bergles, G.H. Junkhan, Bibliography of U.S. patents on augmentation of convective heat and mass transfer-II, 1983.

[2] A.B.M. Khdher, N.A.C. Sidik, R. Mamat, W.A.W. Hamzah, Experimental and numerical study of thermo-hydraulic performance of circumferentially ribbed tube with Al₂O₃ nanofluid, *Int. Commun. Heat Mass Transf.* (2015). https://doi.org/10.1016/j.icheatmasstransfer.2015.10.003

[3] M. Li, T.S. Khan, E. Al Hajri, Z.H. Ayub, Geometric optimization for thermal-hydraulic performance of dimpled enhanced tubes for single phase flow, *Appl. Therm. Eng.* (2016). https://doi.org/10.1016/j.applthermaleng.2016.04.141

[4] C. Qi, Y.L. Wan, C.Y. Li, D.T. Han, Z.H. Rao, Experimental and numerical research on the flow and heat transfer characteristics of TiO2-water nanofluids in a corrugated tube, *Int. J. Heat Mass Transf.* (2017). https://doi.org/10.1016/j.ijheatmasstransfer.2017.08.098

[5] M.K. Abdolbaqi, R. Mamat, N.A.C. Sidik, W.H. Azmi, P. Selvakumar, Experimental investigation and development of new correlations for heat transfer enhancement and friction factor of BioGlycol/water based TiO2 nanofluids in flat tubes, *Int. J. Heat Mass Transf.* (2017). https://doi.org/10.1016/j.ijheatmasstransfer.2016.12.024

[6] M.S. Baba, A.V.S.R. Raju, M.B. Rao, Heat transfer enhancement and pressure drop of Fe3O4 -water nanofluid in a double tube counter flow heat exchanger with internal longitudinal fins, *Case Stud. Therm. Eng.* (2018). https://doi.org/10.1016/j.csite.2018.08.001

[7] A.A.R. Al Kumait, T.K. Ibrahim, M.A. Abdullah, Experimental and numerical study of forced convection heat transfer in different internally ribbed tubes configuration using TiO2 nanofluid, *Heat Transf. - Asian Res.* (2019). https://doi.org/10.1002/htj.21457

[8] P. Rabbani, A. Hamzehpour, M. Ashjaee, M. Najafi, E. Houshfar, Experimental investigation on heat transfer of MgO nanofluid in tubes partially filled with metal foam, *Powder Technol.* (2019). https://doi.org/10.1016/j.powtec.2019.06.037

[9] A. Karimi, A.A.A.A. Al-Rashed, M. Afrand, O. Mahian, S. Wongwises, A. Shahsavar, The effects of tape insert material on the flow and heat transfer in a nanofluid-based double tube heat exchanger: Two-phase mixture model, *Int. J. Mech. Sci.* 156 (2019) 397–409. https://doi.org/10.1016/j.ijmecsci.2019.04.009

[10] B. Kristiawan, A.I. Rifa'i, K. Enoki, A.T. Wijayanta, T. Miyazaki, Enhancing the thermal performance of TiO2/water nanofluids flowing in a helical microfin tube, *Powder Technol.* (2020). https://doi.org/10.1016/j.powtec.2020.08.020

[11] Y. Wang, C. Qi, Z. Ding, J. Tu, R. Zhao, Numerical simulation of flow and heat transfer characteristics of nanofluids in built-in porous twisted tape tube, *Powder Technol.* (2021). https://doi.org/10.1016/j.powtec.2021.07.066

[12] E. Pavan Kumar, A. Kumar Solanki, M. Mohan Jagadeesh Kumar, Numerical investigation of heat transfer and pressure drop characteristics in the micro-fin helically coiled tubes, *Appl. Therm. Eng.* 182 (2021) 116093. https://doi.org/10.1016/j.applthermaleng.2020.116093

[13] A.K. Solanki, B.K. Ram, V.S. Siddharth, Numerical analysis on heat transfer enhancement of Al_2O_3 nanofluid in the flattened conically coiled tubes, *Heat Transf.* (2022). https://doi.org/10.1002/htj.22591

[14] A.R. Singh, A.K. Solanki, Numerical study on thermal-hydraulic characteristics of flattened microfin tubes, *Chem. Prod. Process Model.* (2023). https://doi.org/10.1515/cppm-2022-0005

[15] R.M. Insiat Islam, N.M.S. Hassan, M.G. Rasul, P. V. Gudimetla, M.N. Nabi, A.A. Chowdhury, Effect of non-uniform wall corrugations on laminar convective heat transfer through rectangular corrugated tube by using graphene nanoplatelets/MWCN hybrid nanofluid, *Int. J. Therm. Sci.* (2023). https://doi.org/10.1016/j.ijthermalsci.2023.108166

[16] F.W. Dittus, L.M.K. Boelter, Heat transfer in automobile radiators of the tubular type, *Int. Commun. Heat Mass Transf.* 12 (1985) 3–22. https://doi.org/10.1016/0735-1933(85)90003-X

[17] V. Gnielinski, New Equations for Heat and Mass Transfer in Turbulent Pipe and Channel Flow, *Int. Chem. Eng.* (1976).

[18] D.A.   P.R.H.   Blasius,   Aehnlichkeitsgesetz   bei   Reibungsvorgangen   in Flüssigkeiten, *Forschungsheft.* 131 (1913) 1–41.

[19] B.S. Petukhov, Heat Transfer and Friction in Turbulent Pipe Flow with Variable Physical Properties, *Adv. Heat Transf.* (1970). https://doi.org/10.1016/S0065-2717(08)70153-9

[20] K. Khanafer, K. Vafai, A critical synthesis of thermophysical characteristics of nanofluids, *Int. J. Heat Mass Transf.* 54 (2011) 4410–4428. https://doi.org/10.1016/j.ijheatmasstransfer.2011.04.048

[21] S. El Bécaye Maïga, S.J. Palm, C.T. Nguyen, G. Roy, N. Galanis, Heat transfer enhancement by using nanofluids in forced convection flows, *Int. J. Heat Fluid Flow.* 26 (2005) 530–546. https://doi.org/10.1016/j.ijheatfluidflow.2005.02.004

Chapter 10

Gravity-modulated thermal instability in Oldroyd-B nanofluids

Shilpee, B.S. Bhadauria, and Ismail

Babasaheb Bhimrao Ambedkar University, Lucknow, India

10.1 INTRODUCTION

"Nanofluids" is a the term that has so much popularity in the field of fluids research nowadays. The range from 1 to 100 nm size of particles is known as nanoparticles. It is a combination of metal oxides like CuO, Al_2O_3, and silica used as materials for the nanoparticles and base fluids that comprise oils, organic fluids such as glycols, polymeric solutions, bio-fluids, and lubricants. The mixture of nanoparticles in base fluids is termed nanofluids having its properties and specially used for increasing the heat-transfer rate. In 1995 Choi [1] was the first to use the term nanofluids. However, later Buongiorno [2] was the first (2006) to obtain the governing equations for convective transport in nanofluids. Using these equations for nanofluids, many authors such as Kuznetsov and NieldKuznetsov [3], Nield [4] and Nield [5], and Bhadauria [6] examined the problem of thermal instability in nanofluids. With passing days, it has been used in many areas like factories, laboratories, etc.

Most often the term "porous medium" is characterized by porosity. In simple words, it is defined as a rigid body having a large number of pores that are typically filled with fluids. Some of the naturally occurring porous materials are rock and soil, some biological tissues, zeolite, and many more. Many properties of pores lie in the fact by considering they are porous media. Based on the results of the experiment, Henary-Darcy was the first to observe and formulate a law known as Darcy's law which gives an equation that describes the flow of a fluid through a porous medium. He was the man who published a treatise in 1856 [7]. Darcy's law is used in many applications like petroleum engineering, coffee brewing, and many more. Horton and Rogers (1945) examined the onset of convection in a fluid saturated by the porous layer based on Darcy's-law. Later, Sheikholeslami [8] used the Darcy model for the motion of the nanofluid in a permeable medium under the influence of an electric field. In 1901, Forchheimer worked on this, and there were several issues on Darcy's law and many extensions were needed. Frochheimer [9] in his original paper showed that the coarse sand and the

DOI: 10.1201/9781003465171-10

hydraulic conductivity shrink when hydraulic gradient increases. Later in 1949 came Brinkman [10], who was interested in modifying Darcy's law as the article says, "A calculation of viscous force exerted by a flowing fluid on a dense swarm of particles." Going through Darcy's law where gravitational forces have been ignored, Brinkman noticed that no definition of viscous force was described and hence considered it as inappropriate. Working through this he took Navier-Stokes equation for viscous incompressible fluid yielding.

$$div.v_F = 0$$

where v_F is the seepage velocity.

He suggested combining both equations, as due to partial differential equations involved in Navier-Stokes equation, it is not possible to form rational boundary equations. Concerning Brinkman's equation, he assumes that there is a difference between the two viscosity parameters that is obvious.

In the last few years, non-Newtonian fluids are becoming more important due to their application in engineering fields and industrial fields. Non-Newtonian fluids comprise fluids such as colloidal solutions, paints, suspensions, clay coating cosmetic products, and different polymer solutions.

Delayed response (delay of the elasticity) and the time needed for initial stress can lead to relaxation and retardation phenomena. This is because of the consistent behavior of the viscoelastic fluid on the thermodynamic particle. Study of these terms led to the model of another fluid given by James G. Oldroyd. The Oldroyd-B model was first used by James G. Oldroyd. This model is used to describe the flow of fluids that is viscoelastic fluids. This model is regarded as an extension of the upper-convected Maxwell model. The model can be written as

$$\left(1+\lambda_1 \frac{\partial}{\partial t}\right)\tau_{ij} = \mu\left(1+\lambda_2 \frac{\partial}{\partial t}\right)e_{ij} \tag{10.1}$$

where λ_1 is the relaxation time and λ_2 is the retardation time.

Fetecau [11] had demonstrated Oldroyd-B-fluid flow over a plate. This new result gave the idea and seeks the attention of many researchers. Vieru [12] investigated the influence of Oldroyd-B-fluid due to constantly accelerating plate. Hayat [13] demonstrated the MHD flow of an Oldroyd-B-fluid in a porous channel. Azeem [14] showed the repercussion due to stretching sheets of an Oldroyd-B-fluid. Sheu [15] investigated the convection in an Oldroyd-B type viscoelastic nanofluid layer using linear stability for both the porous and nonporous medium, and he analyzed that oscillatory convection is possible in both cases for top and heavy arrangements of nanoparticles. Kumar [16] extended and studied the nonlinear stability analysis, while Srivastava [17] took into consideration the revised model which is embedded in the porous medium. Rana [18] had investigated the electro-thermal convection in the shear-thinning viscoelastic fluid layer in the porous medium.

The term "gravity modulation" is observed when a vessel containing liquid is made to vibrate in a vertical form and a pattern having standing waves is observed at a surface (free). Faraday in 1831 was the first to study the waves. Later Matthissein performed experiments where he observed that vibrations were synchronous. Benjamin [19] examined the stability of the surface of a liquid in vertical periodic motion. Wadih [20] examined the effect of gravity modulation in a long vertical cylinder with natural convection. Gresho [21] studied the effect of gravity modulation and the stability of the fluid layer heating from below. Murray [22] examined the repercussion of gravity modulation on solutal convection having directional solidification. Recently there are many papers Thomas [23] that studied the repercussions of gravity modulation in a Darcy-Brinkman layer of a porous medium. Zhao [24] studied chaotic Darcy-Brinkman convection in a fluid layer saturated porous layer under gravity modulations. Many applications are there for the effect of gravity modulation, for example, Stergiou [25] on the dynamics of radial reaction and Smith [26] experimental application in the inverted pendulum.

Electrohydrodynamics, abbreviated as EHD, was first observed by Ferdinand Frederic Resus in 1808 while performing the electrophoresis of clay particles. EHD is also known as electro-fluid dynamics or as electrokinetics (the study is defined as the dynamics of electrically charged particles). A.K. Ghosh and Takashima [27] in 1979 investigated the electro-hydrodynamic instability of the viscoelastic fluid.

In the early 1920s, it has been seen that EHD sought the attention of researchers in the investigation of thermal instability of viscoelastic nanofluids in the field of engineering and science including chemical engineering, nuclear fusion, medicine, etc. Depending on the nature of the fluids and their flows, it was observed that magnetic field repercussions are superior if the fluids are highly electrically conducting. Because of this, electric forces had an important role in driving the motion if the fluid is having dielectric with low electrical conductivity. Hence, a proper study of theory is important to know the physics behind the complex flow etiquette of the nanofluid.

Many studies had been done to study the thermal convection in a dielectric fluid layer in the presence of a DC/or AC electric field. The study of this process of thermal convection in DC/or AC is called electro-thermal convection (ETC). El-Sayed [28] analyzed the problem of electro hydrodynamics (EHD) instability in the year 2008 in a horizontal layer of Oldroyd B-dielectric viscoelastic fluid through Brinkman porous medium. In the year

2009 Rudraiah [29] examined the repercussion of vertical electrical field on the electrothermal convection in a horizontal layer of dielectric fluid saturated by the porous medium. The repercussions of the AC electric field and the boundaries taken at the beginning of ETC in a scattered porous layer from below were examined by Shivkumara [30].

Hamad [31] studied the unsteady MHD-free convection flow over a vertical flat plate in a rotating frame of reference with a constant heat source in a nanofluid.

Previously the convective effect of viscoelastic nanofluid has been observed. A paper on the electrothermal instability of dielectric Oldroydnanoliquid saturating porous media has been examined by Poonam [32].

Awasthi [33] studied the nonlinear analysis of cylindrical flow and explains the heat and mass transfer. Uddin [34] studied the MHD flow with heat generation of nanofluids and explains the numerical study using PSO. Asthana [35] studied the heat and mass transfer flow of cylindrical flow. Awasthi [36] examined the Rayleigh Taylor instability and mass transfer of the annular layer. Awasthi [37] again studied the viscous potential flow of Rayleigh Taylor instability. Awasthi [38] studied Kelvin-Helmholtz instability for magnetohydrodynamic viscous corrections. Ismail [39] studied the instability of rivlin nanofluid with rotation saturating the porous medium.

Since with an application in many fields of viscoelastic non-Newtonian fluid, Oldroydnanofluid was applied to hemodynamics with the numerical simulation Hussain [33] also plays an important role because of its applications in the production of plastic sheet and removal/dismemberment of polymers from a slit die in polymer industry. In this paper, we investigate the compound effect of internal heat and electric field of Oldroyd-B-nanofluid using the Brinkman model under different types of gravity modulation. As far as we know, there is no observation of the compound effect of internal heating and AC electric field in a porous medium using the Brinkman model. We performed linear and nonlinear analysis on the system and obtained effect of different parameters on it. We also obtained the streamlines and discuss the results.

10.2 MATHEMATICAL STATEMENT OF THE PROBLEM

For the formulation of the governing equations, we consider an infinitely extended horizontal layer of incompressible dielectric Oldroyd-B nanofluid saturated by the porous layer between two parallel planes $z = 0$ and $z = d$. The temperature is taken on the upper and the lower boundary of the plane to be T_1 and T_0 such that $T_0 > T_1$. The boundaries are taken so that they are impermeable and thermally conducting. The lower surface of the parallel plane is taken to be grounded, while the upper surface is kept at an electric

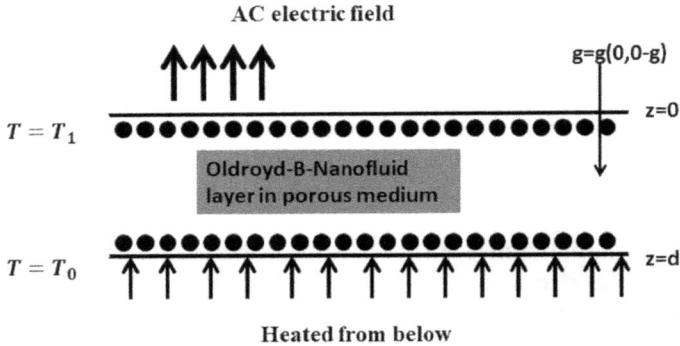

Figure 10.1 Schematic diagram.

potential whose root mean square is γ. Here the Cartesian form is taken such that the origin is taken along the horizontal plane and the z-axis is vertically upward. For the sake of simplicity, Darcy's law is considered and the Oberbeck-Boussinesque approximation technique is used. We make the following assumption using the Buongiorno [2] model as:

- fluid having incompressible flow
- there is no chemical reaction
- base fluid and nanoparticles are thermally equilibrium
- there is dilute mixture
- taken negligible viscous dissipation Biswas [34], Manna [35] and Biswas [36]
- taken negligible radiative heat transfer

The basic governing equations for a dielectric viscoelastic Oldroyd-B nanofluid in porous medium are Yadav [37], Gupta [38], Ramesh [39], and Rana [18]:

10.2.1 Equation of continuity

$$\nabla \cdot v_D = 0 \tag{10.2}$$

10.2.2 Equation of momentum-balance

$$\left(1 + \lambda_1 \frac{\partial}{\partial t}\right)\left[\frac{\rho}{\varepsilon}\frac{\partial v_D}{\partial t} + \frac{1}{\varepsilon^2}(v_D.\nabla)v_D\right] = \left(1 + \lambda_1 \frac{\partial}{\partial t}\right)[-\nabla p + \rho_p$$

$$+ (1 - \varphi)\rho_f\left(1 - \beta(T - T_1)\right)g - f_e] - \left(1 + \lambda_2 \frac{\partial}{\partial t}\right)\left(\bar{\mu}\nabla^2 - \frac{\mu}{k_1}\right)v_D \tag{10.3}$$

where ε denotes the porosity, also when φ is small in density of nanofluid is approximated by base fluid, λ_1 and λ_2 denote the relaxation and retardation time, μ is viscosity, $\bar{\mu}$ is effective viscosity, ρ is density, and p is pressure.

10.2.3 Equation of heat balance

$$(\rho c)_m \left[\frac{\partial}{\partial t} + \frac{1}{\varepsilon^2} (v_D . \nabla) \right] T = k_m \nabla^2 T + (\rho c)_p \left[D_B \nabla \phi . \nabla T + \frac{D_T}{T_1} \nabla T \nabla T \right]$$
$$+ Q(T - T_1) \tag{10.4}$$

10.2.4 Equation of nanoparticle volume fraction

$$\left[\frac{\partial}{\partial t} + \frac{1}{\varepsilon} (v_D . \nabla) \right] \phi = D_B \nabla^2 \phi + \frac{D_T}{T_1} \nabla^2 T \tag{10.5}$$

where ϕ is nanoparticle volume fraction, D_B is the brownian diffusion coefficient given by Einstein-Stokes-equation. $D_B = \dfrac{k_B T}{3\pi \mu_f d_p}$, $k_B k_f$, and k_s are defined as Boltzmann's constant, effective thermal conductivity of fluid, and solid-matrix phase, respectively, D_T is given as the coefficient of thermophoretic diffusion given by $D_B = \dfrac{\mu_f 0.26 k_f}{\rho_f (2k_f + k_p)} \phi$, where k_p is the effective thermal conductivity of the fluid phase.

The force of electric origin f_e for the incompressible nanofluid in eq. (10.3) can be expressed by Landau and Lifshitz (1960) as

$$f_e = \rho_e E - \frac{1}{2}(E \cdot E)\nabla \in + \frac{1}{2}\nabla \left[\rho \frac{\partial \in}{\partial \rho}(E \cdot E) \right] \tag{10.6}$$

where E stands for root mean square values of the electric field ρ_e is denoted as Charge density ϵ is the dielectic constant. In equation (10.6), the first term on the RHS is the coulomb force due to a free charge, and the second term depends on the gradient of ϵ. The third term, the electrostriction term, can be grouped with pressure p in eq. (10.2) and has no effect on the incompressible nanofluid. According to Maxwell's equations, as there is no free charge

$$\nabla \times \tilde{E} = 0 \qquad\qquad (10.7)$$

$$\nabla \cdot \left(\in \tilde{E} \right) = 0 \qquad\qquad (10.8)$$

In view of equation (10.6), $\tilde{E}$ can be expressed as

$$E = -\nabla \psi \qquad\qquad (10.9)$$

where ψ is defined as the root mean square value of electric potential such thate is assumed to be of the form

$$\in = \in_0 \left[1 - e \left(T - T_1 \right) \right] \qquad\qquad (10.10)$$

where e is defined as thermal expansion coefficient of dielectric constant and the value is assumed to be very small. Now further we suppose the temperature to be constant and nanoparticle flux including the effect of thermophoresis is assumed zero on the boundaries, and thus the boundary conditions are:

$$\left. \begin{aligned} w = 0, T = 1, D_B \frac{d\phi}{dz} + \frac{D_T}{T_1} \frac{dT}{dz} = 0, z = 0 \\ w = 0, T = 0, D_B \frac{d\phi}{dz} + \frac{D_T}{T_1} \frac{dT}{dz} = 0, z = d \end{aligned} \right\} \qquad (10.11)$$

and $g = (1 + \delta G(\Omega, t))$ is gravity acceleration.

10.3 BASIC SOLUTION

Taking the basic state to be inoperative, the quantities at the basic state are given as:

$$\begin{aligned} v = 0, T = T_b \left(z \right), p = p_b \left(z \right), \phi = \phi_b \left(z \right), \in = \in_b \left(z \right), \psi = \psi_b \left(z \right), \\ E = E_b \left(z \right) \end{aligned} \qquad (10.12)$$

The solution of the basic flow is obtained as follows:

$$T_b = T1 + \Delta T \frac{\sin\left[\sqrt{\frac{Q}{km}}(d-z)\right]}{\sin\left[\sqrt{\frac{Q}{km}}d\right]}, \phi_b = \left(\frac{-D_T}{D_B T1}\right) T_b + \phi_0,$$

$$E_b = \frac{E_0}{1 + e\Delta Tz/d},$$

$$\psi_b = \frac{-E_0 d}{e\Delta T} \log\left(1 + \frac{e\Delta T}{d}z\right)\hat{k}, \epsilon_b = \epsilon_0\left(1 + \frac{e\Delta T}{d}z\right)\hat{k}$$

(10.13)

10.4 PERTURBATION SOLUTION

Let the basic state as obtained in the above equations be a little bit perturbed so that we get the perturbed state as

$$v = v', p = p(z) + p', T = T_b(z) + T', \phi = \phi_b + \phi', \epsilon = \epsilon_b + \epsilon'$$

$$E = E_b + E', \psi = \psi_b + \psi', \lambda_1 = \frac{\alpha_m}{d^2}\lambda_1', \lambda_2 \frac{\alpha_m}{d^2}\lambda_2'$$

(10.14)

Here the prime denotes the perturbed quantities. Substitute the values of perturbed state in equation (10.2)–(10.9) and linearizing the term by neglecting the product of prime quantities. The pressure term is eliminated by momentum equation by taking the curl twice and retaining the vertical components only. Now the linearized perturbation equation in a non-dimensionalized form is written as:

$$\left(x^*, y^*, z^*\right) = \frac{(x, y, z)}{d}, \left(u^*, v^*, w^*\right) = \frac{(u, v, w)d}{\alpha_m}, t^* = \frac{t\alpha_m}{d^2}$$

$$p^* = \frac{pd^2}{\mu\alpha_m}, \phi^* = \frac{\phi - \phi_0}{\phi_1 - \phi_0}, T^* = \frac{T - T_0}{T_0 - T_1}, \psi^* = \frac{\psi}{eE_0\Delta Td}$$

where $\alpha_m = \dfrac{k_m}{(\rho c)_f}$

The linear stability equation (10.2)–(10.9) and (10.11) obtained after dropping the dashes in a non-dimensional form are as:

$$\nabla \cdot v_D = 0$$

(10.15)

$$\left(1+\lambda_1 \frac{\partial}{\partial t}\right)\frac{Da}{Pr\,\varepsilon}\left(\frac{\partial}{\partial t}\nabla^2 w\right) = \left(1+\lambda_1 \frac{\partial}{\partial t}\right)\left[\begin{array}{c} Ra\nabla^2 T - Rn\nabla^2 \phi \\ +R_{ae}\nabla^2\left(\theta - \dfrac{\partial \psi}{\partial z}\right) \end{array}\right]$$

$$+\left(1+\lambda_2 \frac{\partial}{\partial t}\right)\left(\nabla^2 D\tilde{a} - 1\right)\nabla^2 w \tag{10.16}$$

$$\frac{\partial T}{\partial t} + wf(z) = \nabla^2 T + \frac{Nb}{Le}f(z)\frac{\partial \phi}{\partial z} + \frac{NaNb}{Le}f(z)\frac{\partial T}{\partial z} \tag{10.17}$$

$$\frac{\partial \phi}{\partial t} + \frac{\partial \phi}{\partial z} = \frac{1}{Le}\nabla^2 \phi + \frac{Na}{Nb}\nabla^2 T \tag{10.18}$$

$$\nabla^2 \psi = \frac{\partial T}{\partial z} \tag{10.19}$$

$$f(z) = \frac{-\sqrt{Hs}\cos\left[\sqrt{Hs}(z-1)\right]}{\sin\sqrt{Hs}} \tag{10.20}$$

The non-dimensional parameters are defined as follows:

$Le = \dfrac{\alpha_m}{D_B}$, the Lewis number

$Ra = \dfrac{\rho g \beta (T - T_1) kd}{\mu \alpha_m}$, the thermal Rayleigh Darcy number

$Rm = \dfrac{\left[\rho_p \phi_0 + \rho(1 - \phi_0)\right]gd}{\mu \alpha_f}$, the basic density Rayleigh number

$Rn = \dfrac{(\rho_p - \rho_0)\phi_0 gkd}{\mu \alpha_m}$, the nanoparticle concentration Rayleigh number

$R_{ae} = \dfrac{e^2 \in E_0{}^2 (\nabla T)^2 k}{\mu \alpha_m}$, the AC electric Darcy number

$Hs = \dfrac{Qd^2}{km}$, internal heat source parameter

$Nb = \dfrac{(\rho c)_p \phi_0 \varepsilon}{(\rho c)}$, the modified particle density increment

$Na = \dfrac{D_T \nabla T}{D_B T_1 \phi_0}$, the modified diffusivity ratio

$\text{Dã} = \dfrac{\bar{\mu}k}{\mu d^2}$, the Brinkman Darcy number

The boundary condition in non-dimensional form becomes:

$$\left\{ w = \frac{\partial \psi}{\partial z} = T = 0, \frac{\partial \phi}{\partial z} + Na\frac{\partial T}{\partial z} = 0 \text{ at } z = 0,1 \right.$$

10.5 TYPES OF MODULATION

In this section, we will discuss the four types of gravity. Gravity modulation is defined as the vertical oscillation in a fluid layer in a constant gravitational field. Here we have imposed the gravity in four ways, i.e. sinusoidal, saw-tooth, day-night, and square wave forms. On applying the four types of modulation we infer about the case where the rate of convection takes place earlier, where $G(\Omega1,t)$ stands for gravity modulation.

Four types of modulation	Formula used
Trigonometric sin wave	$G(\Omega_1, t) = \sin(\Omega_i, t)$
Square wave	$G(\Omega_1, t) = \displaystyle\sum_{m=1}^{20} \frac{4\sin(mt\Omega_i)}{\pi m}$
Saw-tooth wave	$G(\Omega_1, t) = \displaystyle\sum_{m=1}^{10} \frac{2(-1)^{m-1}\sin(mt\Omega_i)}{\pi m}$
Day-night wave	$G(\Omega_1, t) = \displaystyle\sum_{m=1}^{10} \frac{2\left(\sin\left(\frac{\pi m}{2}\right) + \frac{3}{7}\sin\left(\frac{5\pi m}{6}\right)\right)\sin(mt\Omega_i)}{\pi^2 m^2}$ $+ \displaystyle\sum_{m=1}^{10} \frac{2\left(\cos\left(\frac{\pi m}{2}\right) + \frac{3}{7}\cos\left(\frac{5\pi m}{6}\right) - \frac{10}{7}\right)\cos(mt\Omega_i)}{\pi^2 m^2}$

10.6 STABILITY ANALYSIS

10.6.1 Linear stability

The differential (10.15)–(10.19) along with the boundary condition (10.20) constitutes the linear boundary value problem.

10.6.1.1 Normal mode technique

In the classical problem of thermal instability in Rayleigh–Benard convection, the basic state variables are assumed to be the function of z only, where z-direction is normal to x–y plane and therefore an arbitrary perturbation is considered in the form of a sinusoidal wave in 2-dimensional x–y plane in the following manner (Chandrasekhar 1961):

$$C(x,y,z,t) = \bar{C}(z)e^{\left[i\left(k_x x + k_y y\right)+t\right]} = \bar{C}(z)e^{i\left(k_x x + k_y y\right)}e^{\sigma_r t}e^{i\sigma_i t}$$

where C is an arbitrary perturbed quantity, k_1 and k_2 are the wave numbers linked with the perturbation along x and y directions respectively, σ ($= \sigma_r + i\sigma_i$) is the growth rate parameter, and $i = \sqrt{1}, a = \sqrt{k_x x^2 + k_y y^2}$ is the horizontal wave number. Applying normal mode technique as follows:

$$\begin{cases} w = W\exp\left(ik_x x + ik_y y + st\right) \\ T = \Theta\exp\left(ik_x x + ik_y y + st\right) \\ \phi = \Phi\exp\left(ik_x x + ik_y y + st\right) \\ \psi = \Psi\exp\left(ik_x x + ik_y y + st\right) \end{cases}$$

(10.21)

where s is taken to be the growth rate of disturbances, k_x and k_x are defined as the wave numbers in x and y directions, respectively. Applying the above expression (10.21) in equations (10.15)–(10.19) and making use of the basic state solutions (13) as follows:

$$-(1+\lambda_1 s)\frac{s}{Va}\left(D^2 - a^2\right)W - (1+\lambda_1 s)\begin{pmatrix} Rna^2\Phi - Raa^2\Theta - R_{ae}a^2\Theta \\ -R_{ae}a^2\Theta\dfrac{\partial\Psi}{\partial z} \end{pmatrix}$$

(10.22)

$$-(1+\lambda_2 s)D\tilde{a}\left(D^4 W - 2a^2 D^2 W + a^4 W\right) + (1+\lambda_2 s)\left(D^2 - a^2\right)W = 0$$

$$-Wf(z) + \left[D^2\Theta - a^2\Theta + Hs\Theta - s\Theta + \frac{\varepsilon NaNbf(z)}{Le}D\Theta\right]$$

$$+ \frac{\varepsilon Nbf(z)}{Le} + D\Phi = 0$$

(10.23)

$$\frac{Wf(z)}{\varepsilon} + \left[\frac{1}{Le}\left(D^2\Phi - a^2\Phi\right) - s\Phi\right] + \frac{1}{Le}\left(D^2 - a^2\right)\Theta = 0$$

(10.24)

$$\left(D^2 - a^2\right)\Psi - D\Theta = 0$$

(10.25)

where $D = \dfrac{\partial}{\partial z}$ and $a = \sqrt{k_x{}^2 + k_y{}^2}$ is defined as the resultant dimensionless wave number.

For the normal mode analysis, the boundary condition obtained are as follows:

at

$$\left.\begin{array}{l} W = 0,\ \Theta = 0,\ D\psi = 0 \\ D\Phi + N_a D\ \Theta = 0 \ \ at\ z = 0,1 \end{array}\right\} \tag{10.26}$$

From the above equations, s is the growth rate in a general complex quantity defined as $s = \omega_r + i\omega_i$, here when $\omega_r < 0$, the system is said to be always stable while for $\omega_r > 0$ the system is said to be unstable. For the neutral stability of the system, the real part of ω is zero. Hence here we $s = \omega_i$ where $s = \omega_i$ is the real dimensionless frequency. Now to obtain the analytical solution to the system of equations, the Galerkin-Weighted residuals are used. Accordingly for the process, the base function W, Θ, Φ, and φ are chosen in the following way:

$$W = A_m W_m,\ \ \Theta_m = B_m W_m,\ \ \Phi_m = C_m \Phi_m,\ \ \Psi_m = D_m \Psi_m \tag{10.27}$$

where

$$\begin{array}{l} W = A_1 \sin(\pi z),\ \ \Theta = B_1 \sin(\pi z),\ \ \Phi = (-Na)C_1 \sin(\pi z), \\ \Psi = D_1 \cos(\pi z) \end{array} \tag{10.28}$$

all of them satisfying boundary conditions.

A_m, B_m, C_m, D_m are the unknown coefficients, and m ranges from $m = 1$, 2, 3, ... N. Using the above base functions in equation (10.26) and multiplying the resultant equation first by W_m, second equation by Θ_m, third one by Φ_m and the fourth equation by the φ_m, integrating each one of them from limits zero to unity, we obtain a system of linear algebraic equations of 4N, in four unknowns $A_m, B_m, C_m,$ and D_m where $p = 1, 2, 3, ...$ N. The given system of algebraic equations has nontrivial solutions, when the determinant of the coefficient matrix vanish, following which gives the characteristic equations for the system. Further solving for Rayleigh number as the eigenvalue of the characteristic equation, we obtain the result for the first approximation as:

$$\begin{pmatrix} P_{11} & \dfrac{1}{2}a^2(\text{Ra}+\text{Ra}_e)(1+s\lambda_1) & P_{13} & P_{14} \\ P_{21} & P_{22} & P_{23} & 0 \\ P_{31} & P_{32} & P_{33} & 0 \\ 0 & P_{42} & 0 & P_{44} \end{pmatrix} \begin{pmatrix} A_1 \\ B_1 \\ C_1 \\ D_1 \end{pmatrix} = \begin{pmatrix} 0 \\ 0 \\ 0 \\ 0_1 \end{pmatrix} \tag{10.29}$$

For the nontrivial solution of (10.29), we have:
  where

$$P_{11} = \frac{-1}{2a}\left[\begin{array}{l} a^4 \tilde{\text{D}}a\text{Va}\left(1+s\lambda_2\right) + \pi^2\left(\text{Va}+s\left(-1-s\lambda_1 + \text{Va}\lambda_2 + \tilde{\text{D}}a\pi^2\text{Va}\lambda_2\right)\right) \\ +a^2\left(\text{Va}+2\tilde{\text{D}}a\pi^2\text{Va}+s\left(-1-s\lambda_1 + \text{Va}\lambda_2 + 2\tilde{\text{D}}a\pi^2\text{Va}\lambda_2\right)\right) \end{array}\right]$$

$$P_{12} = \frac{1}{2}a^2\left(\text{Ra}+R_{ae}\right)\left(1+s\lambda_1\right) \quad P_{13} = \frac{1}{2}a^2 Na\text{Rn}\left(1+s\lambda_1\right)$$

$$P_{14} = \frac{1}{2}a^2\pi R_{ae}\left(1+s\lambda_1\right) \quad P_{21} = -\frac{2\pi^2}{Hs-4\pi^2}$$

$$P_{22} = \frac{1}{Le\left(Hs-4\pi^2\right)}[Le\left(Hs-4\pi^2\right)$$
$$\times\left(a^2\_Hs + \pi^2 + s\right) + 2\sqrt{Hs}NaNb$$
$$\times\pi^2\varepsilon\tan\left[\frac{\sqrt{Hs}}{2}\right]]$$

$$P_{23} = \frac{\sqrt{Hs}NaNb\pi^2\varepsilon\tan\dfrac{\sqrt{Hs}}{2}}{Le\left(Hs-4\pi^2\right)} \quad P_{31}$$

$$= -\frac{2Na\pi^2}{Hs\varepsilon-4\pi^2\varepsilon}P_{32}$$

$$= \frac{\left(a^2+\pi^2\right)Na}{2Le}P_{42} = -\frac{\pi}{2}$$

$$P_{44} = \frac{1}{2}\left(-a^2-\pi^2\right)$$

10.6.2 Nonlinear analysis

In order to implement the nonlinear stability analysis, we take the Fourier series expression as follows Agarwal [40]:

$$\Psi = \sum_{i=1}^{\infty}\sum_{j=1}^{\infty}A_{ij}\left(t\right)\sin\left(j\pi z\right)\sin\left(i\pi ax\right) \tag{10.30}$$

$$\Theta = \sum_{i=1}^{\infty}\sum_{j=1}^{\infty} B_{ij}(t)\sin(j\pi z)\cos(i\pi ax) \tag{10.31}$$

$$\Phi = \sum_{i=1}^{\infty}\sum_{j=1}^{\infty} C_{ij}(t)\sin(j\pi z)\cos(i\pi ax) \tag{10.32}$$

$$\Upsilon = \sum_{i=1}^{\infty}\sum_{j=1}^{\infty} D_{ij}(t)\sin(j\pi z)\cos(i\pi ax) \tag{10.33}$$

Here for stream function we consider (1,1) mode and for nanoparticle concentration, fluid, and electric phase mode (1,1) and (0,2). We impose the following expressions for nonlinear stability analysis as Agarwal [40].

$$\Psi = A_{11}(t)\sin(\pi z)\sin(ax) \tag{10.34}$$

$$\Theta = B_{11}(t)\sin(\pi z)\cos(ax) + B_{02}(t)\sin(2\pi z) \tag{10.35}$$

$$\Phi = (-Na)C_{11}(t)\sin(\pi z)\cos(ax) + C_{02}(t)\sin(2\pi z) \tag{10.36}$$

$$\Upsilon = D_{11}(t)\cos(\pi z)\cos(ax) + D_{02}(t)\sin(2\pi z) \tag{10.37}$$

Substituting (10.34)–(10.37) into equations (10.22)–(10.25) and using the orthogonalization process of the Galerkin's technique we get the following:

$$A'_{11}(t) = \frac{1}{(a^2+\pi^2)(-1+Va\lambda_2)}$$
$$\begin{pmatrix} a^2 VaA_{11}(t) - \pi^2 VaA_{11}(t) + aRaVaB_{11}(t) \\ +aR_{ae}VaB_{11}(t) + aNaRnVaC_{11}(t) + a\pi R_{ae}VaD_{11}(t) \\ +aRaVa\lambda_1 B'_{11}(t) + aR_{ae}Va\lambda_1 B'_{11}(t) + aNaRn\lambda_1 VaC_{11}(t) \\ +a\pi R_{ae}Va\lambda_1 D_{11}(t) + a^2\lambda_1 A_{11}'(t) + \pi^2\lambda_1 A_{11}'(t) \end{pmatrix} \tag{10.38}$$

Replacing $A_{11}{'}(t) \to S_{11}(t)$ and $A_{11}{''}(t) \to S_{11}{'}(t)$ all other derivatives

$$S'_{11}(t) = \frac{1}{\mathrm{Le}\left(a^2 + \pi^2\right)\left(-Hs + 4\pi^2\right)}$$

$$\begin{bmatrix} \varepsilon\lambda_1(\varepsilon(a\mathrm{Va}(Hs^2\mathrm{LeRa}\lambda_1 + 4\pi^2(-\mathrm{LeR}_{ae} - Na(a^2 + \pi^2 Rn\lambda_1\mathrm{LeRa} \\ (-1 + a^2\lambda_1 + \pi^2\lambda_1))) + Hs(Na\left(a^2 Rn + \pi^2\left(Rn + 4Nb\mathrm{Ra}\varepsilon\right)\right)\lambda_1 + \\ \mathrm{Le}\left(Ra + R_{ae} - a^2\mathrm{Ra}\lambda_1 - 5\pi^2\mathrm{Ra}\lambda_1\right)))B_{11}(t) - aNa\mathrm{Va}(-4\pi^2 Rn \\ \left(-\mathrm{Le} + \left(a^2 + \pi^2\right)\lambda_1\right) + Hs(-\mathrm{LeRn} + a^2 Rn\lambda_1 + \pi^2\left(Rn + 2Nb\mathrm{Ra}\varepsilon\right) \\ \lambda_1))C_{11}(t) - \mathrm{Le}\left(Hs - 4\pi^2\right)(aR_{ae}B_{11}(t) + \left(a^2 + \pi^2\right)\left(-1 + \mathrm{Va}\lambda_2\right) \\ S_{11}(t)) + \mathrm{LeVa}A_{11}(t)(-a^2 Hs\varepsilon + 4a^2\pi^2\varepsilon - Hs\pi^2\varepsilon + 4\pi^4\varepsilon - \\ 4a^2 Na\pi^2 Rn\lambda_1 + a^2\pi\left(-Hs + 4\pi^2\right)\mathrm{Ra}\varepsilon\lambda_1 B_{02}(t) + a^2\pi\left(Hs - 4\pi^2\right) \\ Rn\varepsilon\lambda_1 C_{02}(t) + 4a^2\sqrt{Hs}\pi^2\mathrm{Ra}\varepsilon\lambda_1 \tan\dfrac{\sqrt{Hs}}{2} \end{bmatrix}$$

(10.39)

$$B'_{11}(t) = \frac{1}{\mathrm{Le}\left(Hs - 4\pi^2\right)}\begin{pmatrix} -Hs^2\mathrm{Le} - 4\mathrm{Le}\pi^4 + a^2\mathrm{Le}\left(Hs - 4\pi^2\right) \\ +Hs\pi^2\left(5\mathrm{Le} - 4NaNb\varepsilon\right)B_{11}(t) \\ +\pi(2HsNaNb\pi\varepsilon C_{11}(t) \\ +a\mathrm{Le}A_{11}(t)((Hs - 4\pi^2)B_{02}(t) - 4\sqrt{Hs}\tan\dfrac{\sqrt{Hs}}{2}))) \end{pmatrix}$$

(10.40)

$$B'_{02}(t) = \frac{\dfrac{1}{2\mathrm{Le}\left(Hs - 16\pi^2\right)}}{\times\begin{bmatrix} a\mathrm{Le}\pi\left(Hs - 16\pi^2\right)A_{11}(t)B_{11}(t) + 2(Hs^2\mathrm{Le} + 64\mathrm{Le}\pi^4 \\ +4Hs\pi^2\left(-5\mathrm{Le} + 4NaNb\varepsilon\right))B_{02}(t) + 16HsNb\pi^2\varepsilon C_{02}(t) \end{bmatrix}}$$

(10.41)

$$C'_{11}(t) = \frac{\left(a^2 + \pi^2\right)B_{11}(t)}{\mathrm{Le}} - \frac{\left(a^2 + \pi^2\right)C_{11}(t)}{\mathrm{Le}} \\ -\frac{a\pi A_{11}(t)\dfrac{4Na\pi}{Hs\varepsilon - 4\pi^2\varepsilon} - C_{02}(t)}{Na}$$

(10.42)

$$C'_{02}(t) = \frac{-\pi \left(8 Na\pi B_{02}(t) + aLeNaA_{11}(t)C_{11}(t) + 8\pi C_{02}(t) \right)}{2Le} \tag{10.43}$$

$$D'_{11}(t) = -\frac{1}{\pi} B'_{11}(t) \tag{10.44}$$

$$D'_{02}(t) = -\frac{1}{2\pi} B'_{02}(t) \tag{10.45}$$

$$E'_{11}(t) = -\frac{1}{\pi} B'_{11}(t) \tag{10.46}$$

$$E'_{02}(t) = -\frac{1}{2\pi} B'_{02}(t) \tag{10.47}$$

The introduced system of simultaneously ordinary differential equations above is answered with the help of a built-in tool NDSolve of Mathematica numerically.

10.7 HEAT TRANSPORT AND NANOPARTICLE CONCENTRATION

The fluid phase thermal Nusselt number Nu(t) is defined as Kiran (2016):

$$\text{Nu}(t) = \frac{\text{Heat transport by} \left(\text{conduction} + \text{convection} \right)}{\text{Heat transport by conduction}}$$

$$= 1 + \left[\frac{\int_0^{\frac{2\pi}{a}} \left(\frac{\partial \theta}{\partial z} \right) dx}{\int_0^{\frac{2\pi}{a}} \left(\frac{\partial \theta_b}{\partial z} \right) dx} \right]_{z=0} \tag{10.48}$$

$$\text{Nu}_T(t) = 1 - \frac{2\pi B_{02}(t) \tan \sqrt{Hs}}{\sqrt{Hs}} \tag{10.49}$$

In a similar way, we can find the thermal Nusselt numbers for the nanoparticle concentration, i.e. $\text{Nu}_\phi(t)$, as follows:

$$\mathrm{Nu}_\phi(t) = \frac{\text{Mass transport by} \left(\text{diffusion} + \text{Advection}\right)}{\text{Mass transport by molecular diffusion}}$$

$$= 1 + \frac{\left[\displaystyle\int_0^{\frac{2\pi}{a}} \left(\frac{1}{Le}\frac{\partial \phi}{\partial z} + \frac{Na}{Le}\frac{\partial \theta}{\partial z}\right) dx\right]}{\left[\displaystyle\int_0^{\frac{2\pi}{a}} \left(\frac{1}{Le}\frac{\partial \phi_b}{\partial z} + \frac{Na}{Le}\frac{\partial \theta_b}{\partial z}\right) dx\right]_{z=0}} \tag{10.50}$$

$$\mathrm{Nu}_\phi(t) = 1 + \frac{2\pi \left(NaB_{02}(t) + C_{02}(t)\right)\tan\sqrt{Hs}}{\sqrt{HsNa}} \tag{10.51}$$

10.8 RESULTS AND DISCUSSION

10.8.1 Linear stability analysis

Answering the linear stability for stationary convection, the Rayleigh number obtained is the same as obtained by [39], after ignoring all the external effect, i.e. Da = 0, Hs = 0 and R_{ae} = 0.

$$\mathrm{Ra} = \frac{\tilde{Da}\left(a^2 + \pi^2\right)^3}{a^2} + \frac{\left(\pi^2 + a^2\right)^2}{a^2} - \left(1 + \frac{Le}{\varepsilon}\right)NaRn \tag{10.52}$$

From Figure 10.2a, we obtained the marginal stability curves for Ra versus the wave number "a" for different values of the parameters. For stationary convection, the investigation is performed for different values of Hs, ε, Da, Na, Le, Rn. In the following figures, we depict the graph by taking values $Hs = 1, \tilde{Da} = 2, Na = 2, Rn = 4, Le = 100, Nb = 1, Va = 2, R_{ae} = 10$ and $\varepsilon = 0.4$ and changing the particular values to observe the repercussion. In all the observed curves, one common thing is observed that values of Ra starts from a higher point, falls speedily with increasing a, and steadily decreases with an increase in Ra again. In Figure 10.2a as the value of Hs increases the onset of convection gets faster, decreasing the thermal Rayleigh Darcy number. This is because as the value of Hs increases the width of the cell increases, thereby decreasing the thermal conductivity. Thermal conductivity is the amount of heat conducted to the whole body whereas internal heat is a material property depending on its specific and latent heat constant. Since decreasing km will surely decrease, the heat conduction through the body and the onset of convection takes place soon. Therefore the system will have a destabilizing effect. Figure 10.2b shows the repercussion of porosity on viscoelastic fluid. From the figure, it can be seen that as the parameter of porosity increases, the onset of convection is getting delayed with increases in thermal Rayleigh Darcy number. The repercussion observed is so because

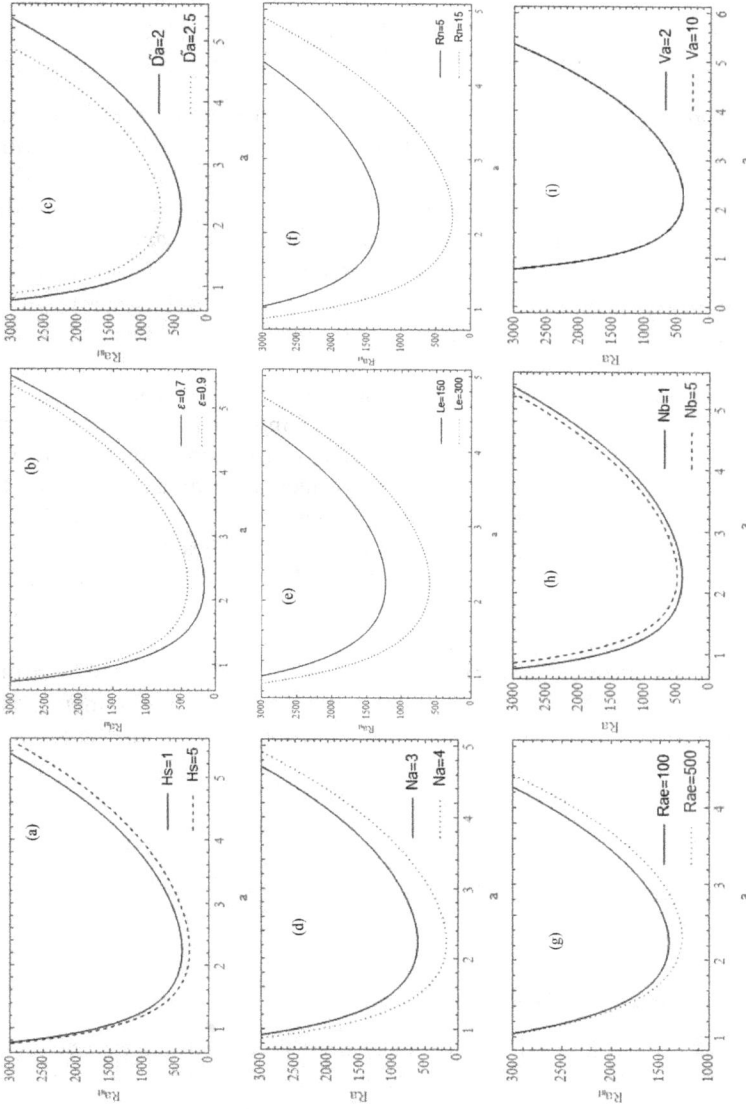

Figure 10.2 Marginal stability curves (a) effect of Hs, (b) effect of ε, (c) effect of Ďa, (d) effect of Na, (e) effect of Le, (f) effect of Rn, (g) effect of R_{ae}, (h) effect of Nb, (i) effect of Va.

with an increase in porosity, the surface area of the pores increases and thus heat transfer within the fluid decreases, i.e. the convection gets delayed. So the porosity will have a stabilizing effect on stationary convection. Similarly, on studying the repercussion of the Brinkman-Darcy number in Figure 10.2c, increase in permeability increases the ability of the porous material to allow fluid to pass through them. Hence the heat transfer through the fluid gets reduced, i.e. convection get delayed and the Brinkman-Darcy number has stabilizing repercussion on the stationary convection.

Now in Figure 10.2d, we observe that with an increase in the value of Na, the thermal Rayleigh Darcy number decreases, thereby making convection faster. This is so because on increasing Na the thermophoresis diffusion increases which results in the acceleration of particles from a hot region to the cold region and consequently heat moves from the hotter surface to the fluid rapidly. Hence Na has destabilizing repercussions on stationary convection. Further, the repercussion of Le on stationary convection can be seen in Figure 10.2e. With an increase in the parameter of Le, convection becomes faster and thus decrease in thermal Rayleigh Darcy number is observed. Therefore, Lewis number has a destabilizing repercussion on stationary convection.

From Figure 10.2f we observed the repercussion of Rn on the nanoparticle Rayleigh-Darcy number. With the increase in the concentration of Rn, the nanoparticle concentration has a great effect on heat transfer. Due to more number of particles, there will be more heat transfer in fluid and hence the convection becomes faster thereby making a system destabilize. In Figure 10.2g we observe that the increase in the value of R_{ae}, i.e. AC electric Rayleigh number the convection occurs more rapidly at a higher value of R_{ae} instead of a lower value of R_{ae}. Therefore, with an increase in the value of R_{ae}, the particles move so fast thereby causing more heat transfer within fluid particles. Hence Rayleigh Darcy's number decreases with an increase in R_{ae} and therefore R_{ae} shows the destabilizing repercussion on the system. Figure 10.2h depicts the repercussion of Nb, i.e. modified particle density. Here with an increase in the value of Nb the density of the particle increases, thereby decreasing in convection. Hence as the density increases, more heat is needed to start convection earlier. Therefore, Nb has a stabilizing effect on stationary convection. In Figure 10.2i no effect of Vadasz number is observed on stationary convection.

10.8.2 Nonlinear stability analysis

Proceeding nonlinear analysis, results have been discussed more precisely to know the basic physics of the problem, the equations obtained have time-dependent results. For observing the results of the nonlinear analysis, the basic equations (10.2)–(10.9) are checked and solved numerically using NDSolve of Mathematica where Nusselt number parameter (function of time T and concentration Φ) are obtained. Nusselt numbers for time and concentration are then plotted in Figures 10.3 and 10.4 under certain values of various parameters. After observing the graph, one common thing is

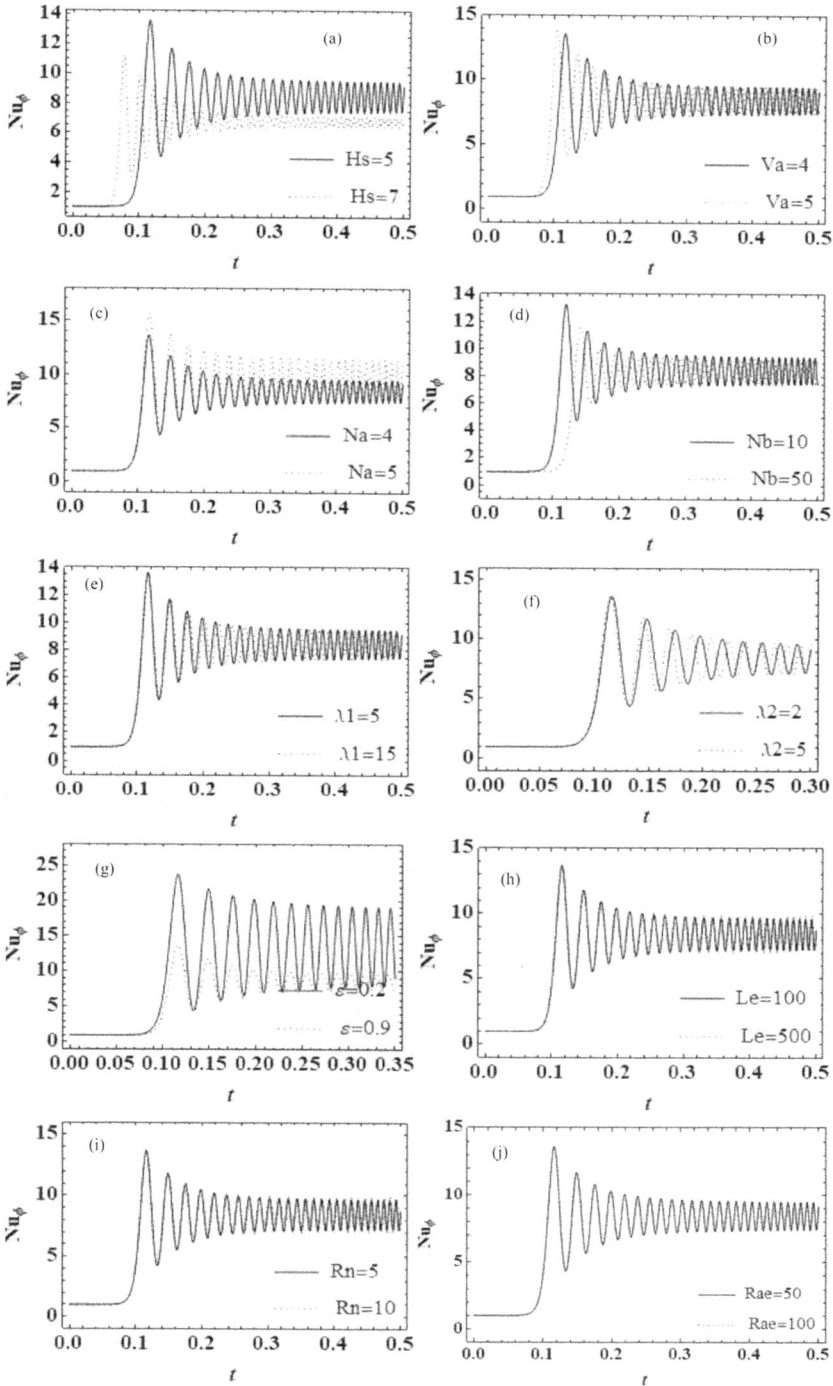

Figure 10.3 Graph of concentration Nusselt number (Nuφ) versus "t" with variation in different parameters.

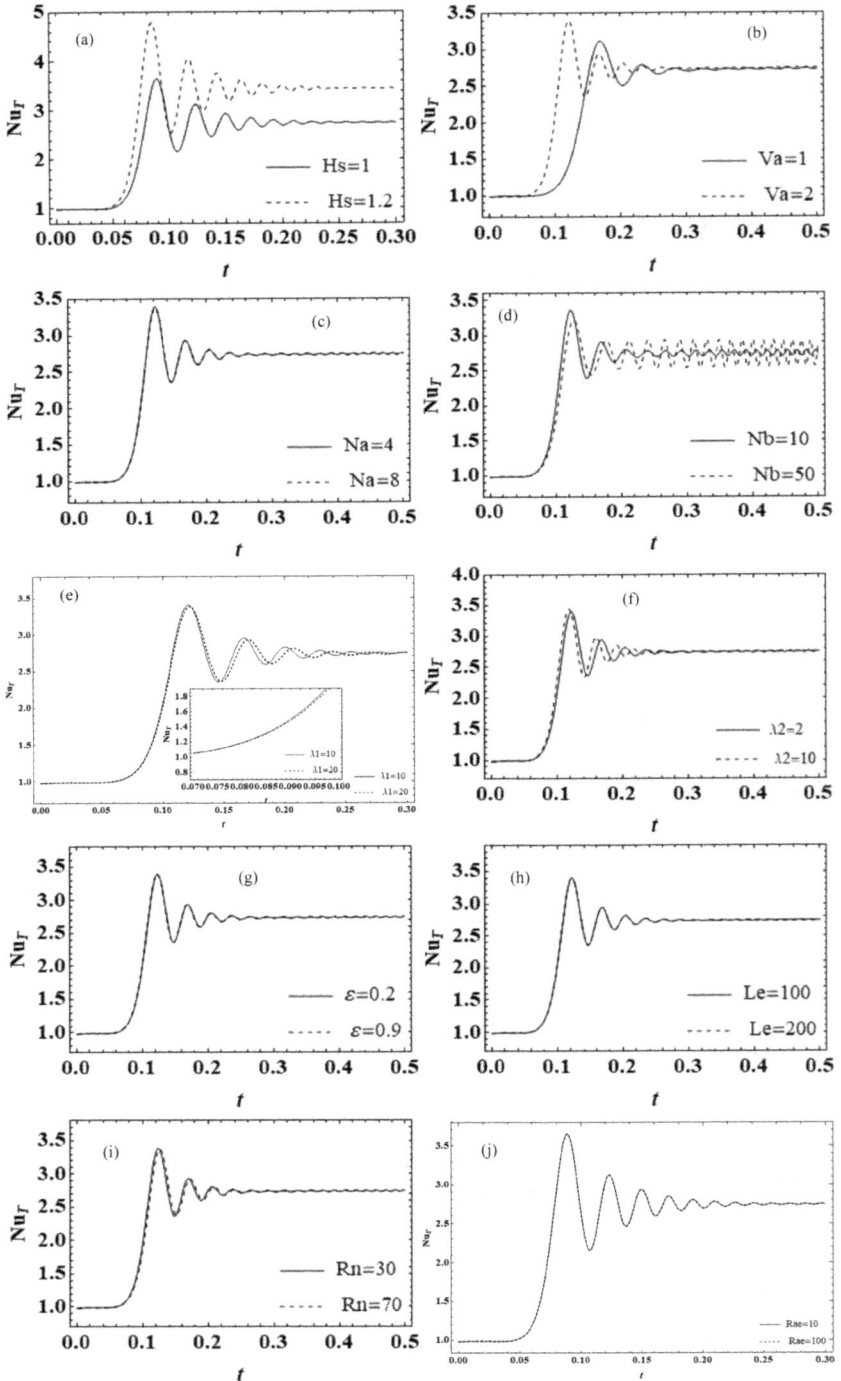

Figure 10.4 Graph of thermal Nusselt number (Nu_T) versus "t" with variation of different parameters.

observed in all the figures with an increase in time at the initial stage for a small interval of time, the heat and mass transfer is very significant, i.e. diffusion state is seen but as the time increases advection starts instantly and approaches to steady state.

From Figure 10.3a, we depict the repercussion of the internal heating over the Nusselt number for concentration. At the initial level on increasing the value of Hs, for a small interval of time a small mass transfer takes place, i.e. diffusion is observed but as time passes advection started rapidly and approaches to steady state. As the internal heat increases, the overall temperature of the system increases and hence mass transfer rate advances in the system. Hence Hs has stabilizing repercussions on the nonlinear stability of mass-transfer. Now from Figure 10.3b we observe the repercussion of the Vadasz number, where it can be seen that on increasing the value of the Vadasz number, convection get hastens as compared to the lower Vadasz number. This is observed so because on increasing Va the porosity also increases thereby increasing the cell size and hence the rate of diffusive mass transfer increases. Therefore Va has stabilizing repercussions on mass transfer. From Figure 10.3c we depict the repercussion of Na on the Nusselt number for mass transfer. It is very clearly observed through figures that on increasing the value of Na, there is an increase in thermophoresis diffusion (which is a mass-transfer process induced by a temperature gradient), as a result of which rapid mass-transfer takes place. Hence after a certain (small) time of interval of diffusion, advection\convection gets advanced and goes to a steady state.

From Figure 10.3d we observed the repercussion of Nb on the system. Since it can be seen from the figure that as the value of Nb gets higher, delay in convection is observed. At the initial level for a very small interval of time conduction is seen but as time increases convection is observed. More elaborately this is because, with an increase in value of Nb, there is a density particle increment in the fluid. Due to this, it takes time to get convected faster. Hence with an increase in value of Nb mass-transfer gets delayed. Relaxation time is the characteristic time in which a system relaxes under certain changes in external conditions. It plays a key parameter in characterizing the viscoelastic fluids. From Figure 10.3e we see that with an increase in the value of relaxation time, the time to get back to its position from the equilibrium state increases. Hence convection is delayed a little bit. As time passes it reaches to the steady state. From Figure 10.3f we observe the repercussion of retardation time. It can be seen that at a small interval of time mass transfer is through conduction but with time fluid gets convected rapidly at $t = 0.1$. On increasing the retardation time, the resistivity increases and due to this more mass transfer takes place. Therefore convection grows a little faster with an increase in retardation time.

From Figure 10.3g we express the effect of ε on mass transfer. At the initial level of time interval mass transfer is through advection but as time goes on convection starts and further goes to a steady state. So the porosity will have a stabilizing effect on stationary convection.

Moving forward to Figure 10.3h we depict the effect of Leon mass transfer which is observed to be slow. This is because with the increased value of Le, Brownian diffusion decreases which decreases the motion of particles and hence advection starts after a small interval of time and goes to a steady state. The effect of Rn is observed in Figure 10.3i where Rn does not affect the mass transfer. The effect observed is so because with the increased value of concentration, the mass of the nanoparticle increases and hence more time is needed. Hence negligible effect of Rn is observed. From Figure 10.3j the repercussion of R_{ae} is observed since R_{ae} shows a slightly small effect on mass transfer. This is because as the value of R_{ae} increases, thermal diffusivity decreases and hence mass transfer gets reduced. Hence negligible effect of R_{ae} is observed on mass transfer. We can compare results from Table 10.1.

Figure 10.4a depicts information about the heat transfer and its behavior on different parameters.

The repercussion of internal heat source parameter Hs over Nusselt number with time t is observed. With the increased value of Hs, the convection gets faster, and the heat-transfer rate increases. This is so because with an increase in internal heat parameter, particles tend to move fast, thereby increasing the convection rate and approaching a steady state as time flows.

Table 10.1 Concentration Nusselt number for different values of R_{ae} and t

t	$R_{ae} = 0$	$R_{ae} = 10$	$R_{ae} = 100$
	Nu ϕ	Nu ϕ	Nu ϕ
0.0	9.49853	9.48281	9.34266
0.1	12.6089	12.5964	12.4806
0.2	11.7034	11.7284	11.9444
0.3	12.3277	12.2955	11.9706
0.4	12.6214	12.6448	12.7085
0.5	10.4294	10.5744	11.7574
0.6	11.2278	11.0234	9.1022
0.7	12.2278	12.3217	10.2973
0.8	12.3493	11.7677	8.43047
0.9	12.0185	8.9544	11.273
1.0	9.45289	11.7952	8.65527
1.1	11.2455	11.2972	11.8933
1.2	11.8756	10.9085	11.84821
1.3	9.6118	11.648	9.48952
1.4	11.4077	9.00392	8.83795
1.5	10.7483	11.3512	9.14661

Figure 10.4b depicts the repercussion of the Vadasz number on the system. For a very small interval of time heat transfer is through conduction. On increasing Va the porosity also increases, thereby increasing the cell size and hence the rate of diffusive heat transfer increases. Due to this convection at higher value start more rapidly (with an increase in heat transfer). In Figure 10.4c we observe the repercussion of Na on heat transfer. On increasing the value of Na there is no effect observed and convection starts at the same interval of time. Now from Figure 10.4d on increasing the value of Nb particle concentration increases which decreases the rate of heat transfer. Hence convection gets delayed and reaches to steady state as time flows. Therefore Nb delayed the heat transfer in the system. The effect of the relaxation time can be observed in Figure 10.4e and it is observed that at a small interval of time heat transfer is through conduction which turns out to be convected just before $t = 0.1$ and no effect is observed on increasing time interval. Hence no effect is observed for λ_1 on heat transfer.

From Figure 10.4f we depict the repercussion of retardation time. It is observed that with an increase in the value of retardation time resistance to particles increases for the deformation and hence particle starts to convect more early with a longer time. Therefore λ_2 has a stabilizing effect on heat transfer. From Figure 10.4g the repercussion of porosity is observed, where it has been seen that heat transfer has not had much effect on heat transfer. Therefore, the convection rate for porosity remains the same and tends to steady state over time. Figure 10.4h shows the repercussion of Le where it is observed that with an increase in value of Le, the rate of heat transfer decreases. Figure 10.4i describes the repercussion of Rn which shows that an increase in particle density decreases the heat transfer rate. Hence convection gets delayed and approaches to steady state as time proceeds. Therefore Rn has destabilizing repercussions on heat transfer. From Figure 10.4j the effect of R_{ae} is observed. We see that the repercussion of R_{ae} on the thermal Nusselt number is nominal at time $t = 0.105$ to $t = 0.12$. On increasing the value of AC electric field nominal effect is observed in convection rate. This effect can be neglected for the sake of study. We can compare the results from Table 10.2.

The effect of streamlines, isotherms, and steady-state is observed in Figures 10.5–10.7. We observe the consequence of fluid flow to be very frail in magnitude at $t = 0$ to $t = 0.004$ for different values of R_{ae}. Convection cells start to develop as time goes on. The system has very nominal repercussions on the movement of fluid and therefore the heat transfer is taken through conduction only. As time increases, we investigate the change in fluid flow and observe the change (increase) in streamline. Here the conversion of conduction to a partial convection state is observed and heat transfer is slow. Further again on increasing time intervals, steady-state isotherms are observed repeatedly. Overall as time goes on conduction, partial convection, full convection, and finally a steady state of fluid flow are observed. Therefore heat transfer takes place as time flows.

Table 10.2 Value of thermal nusselt number for different values of R_{ae} with $R_{ae} = 0, 10, 100$ arcy lawd and shows that how oldroydnanofluid plays an important roledering them as a porous media.

t	$R_{ae} = 0$	$R_{ae} = 10$	$R_{ae} = 100$
	Nu_T	Nu_T	Nu_T
0.0	1.0606	1.06074	1.06206
0.1	3.0586	3.05798	3.05279
0.2	3.42262	3.42293	3 3.4256
0.3	3.43475	3.43453	3.4326
0.4	3.43861	3.43901	3.44232
0.5	3.4354	3.43608	3.44222
0.6	3.44792	3.44716	3.43839
0.7	3.43571	3.43458	3.43359
0.8	3.44266	3.44474	3.44979
0.9	3.44014	3.44288	3.44549
1.0	3.43821	3.44142	3.43859
1.1	3.44827	3.44378	3.44871
1.2	3.45117	3.44997	3.3.44871
1.3	3.45064	3.44921	3.44239
1.4	3.449	3.44038	3.43836
1.5	3.44355	3.45038	3.44314

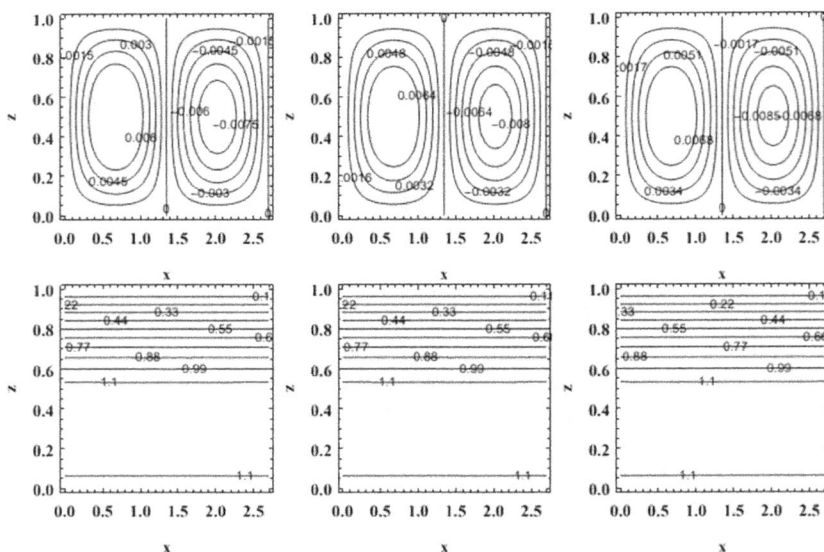

Figure 10.5 Variation of streamlines.

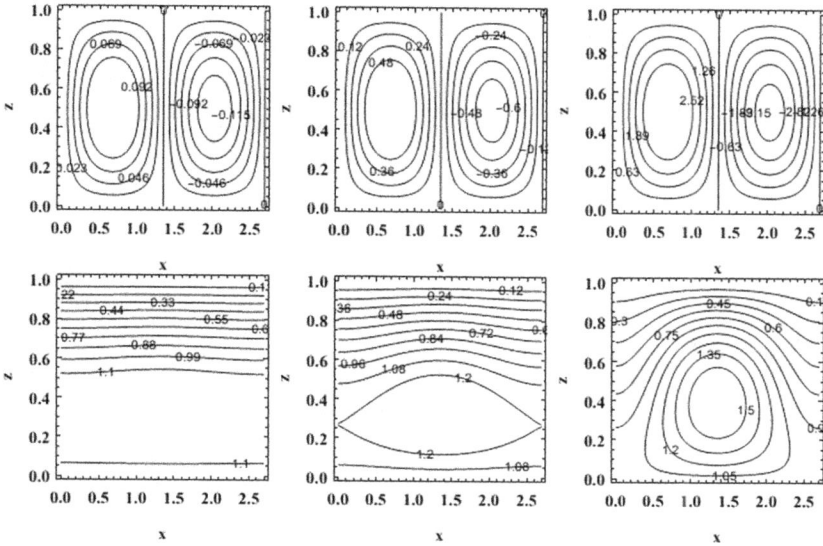

Figure 10.6 Variation of isotherm.

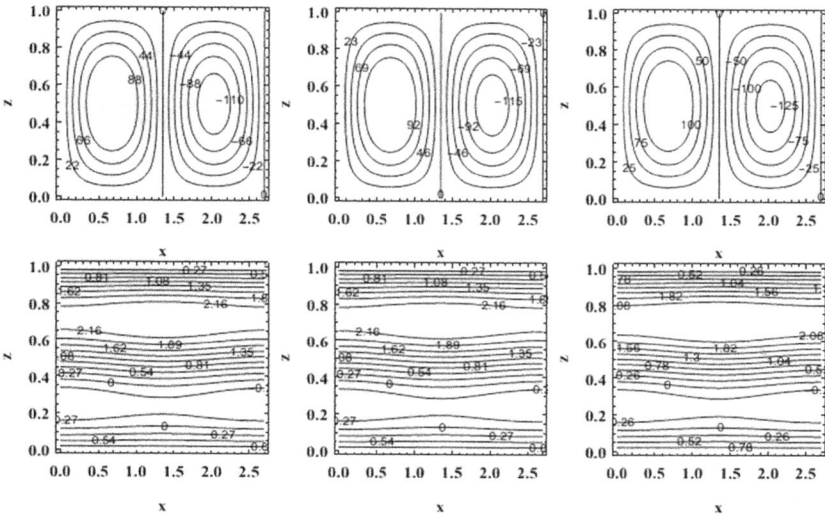

Figure 10.7 Steady state.

It is also observed that changing values of R_{ae} does not have much more repercussion at the same time interval. So as discussed in the Nusselt number for the heat transfer, the effect of R_{ae} is very nominal and can be neglected.

Further from Figure 10.8, we observe the effect of isohalines on mass transfer. It is clearly observed from figure that at the very initial time, the

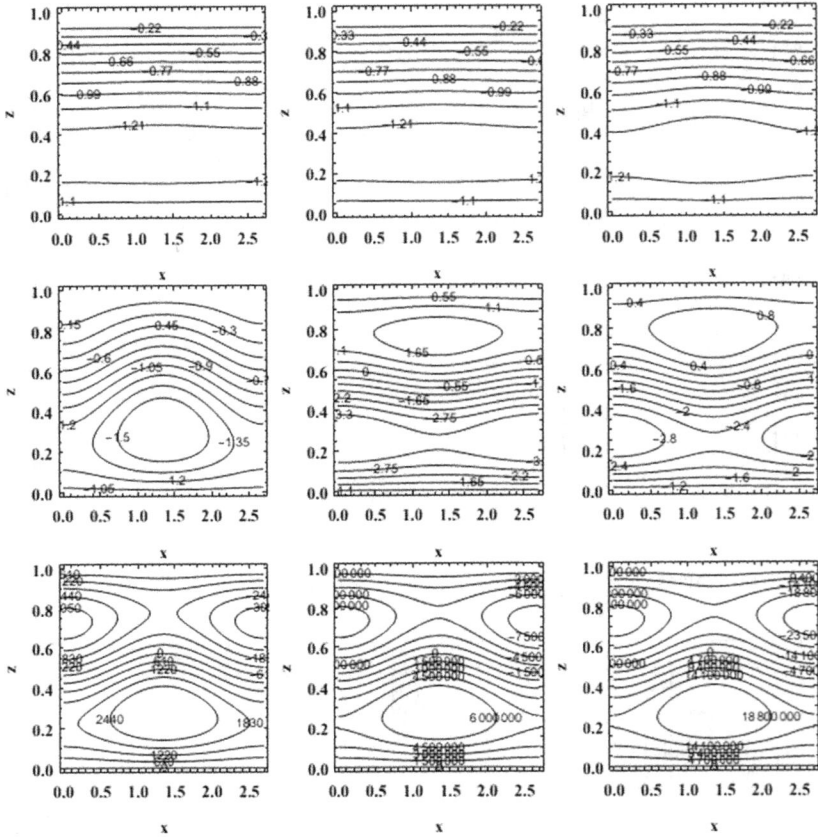

Figure 10.8 Variation of isohalines.

conduction state starts but as time goes on figures of partial convection and full convection start showing the effect of mass transfer. Hence mass-transfer rate increases with an increase in the convection rate.

The effect of steady-state on heat transfer can be observed in Figure 10.9, where Figure 10.9a shows the effect of porosity on the Nusselt number. With the increase in value of ε, the Nusselt number increases. This effect is so because the increase in the surface area increases with the convection rate. Figure 10.9b shows the effect of R_{ae} on Nu. With the increase in the value of R_{ae}, nominal effect on convection is observed which is neglected further. As the value of Nu increases, heat transfer goes to a constant state. Figure 10.9c shows that with an increase in the value of Na, diffusivity of particles increases and hence the heat-transfer rate increases. Figure 10.9d and 10.9e have a nominal effect on retardation and relaxation time. This is

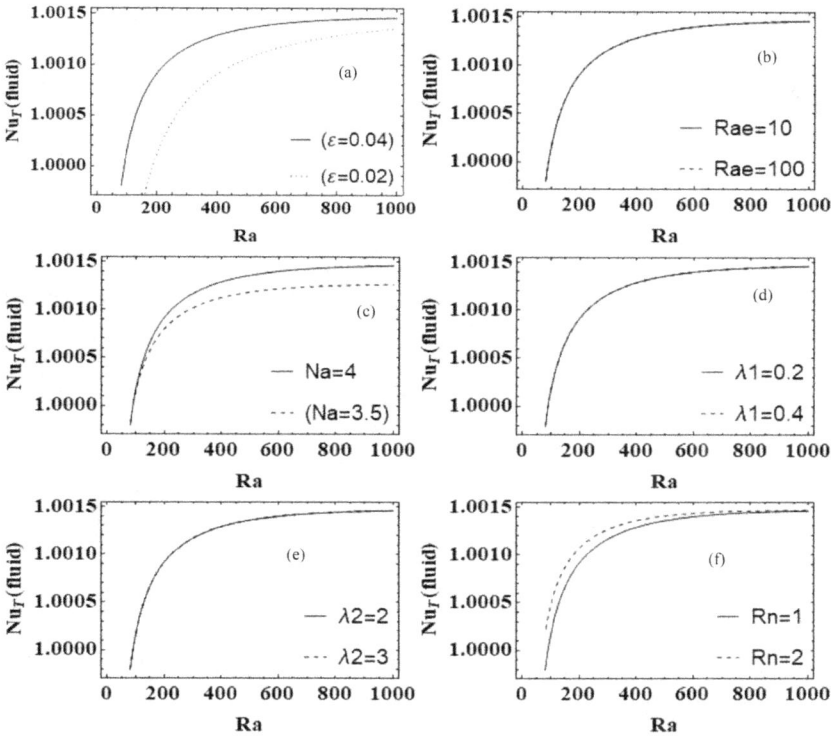

Figure 10.9 Graph of Nusselt number with thermal Rayleigh Darcy number with variation in different parameters.

because the steady state has no effect. From Figure 10.9f, the effect of Rn is shown. It is seen that with an increase in Rn, the concentration of the particles increases. Hence more heat is required and therefore convection takes place more rapidly.

10.9 EFFECT OF GRAVITY MODULATION

In Figures 10.10a and b, we have observed the effect of different types of gravity modulation, i.e. day-night, sinusoidal, square wave, and saw-tooth on the Nusselt number for the concentration and the Nusselt number for heat. It is clearly observed from the figures that day-night gravity modulation starts mass transfer and heat transfer earlier than a sinusoidal, square wave, and saw-tooth. Figure 10.10c depicts and compares the effect of gravity and electric field on heat transfer. At the initial stage, the heat transfer is through conduction, but as time increases, we observe that the effect of

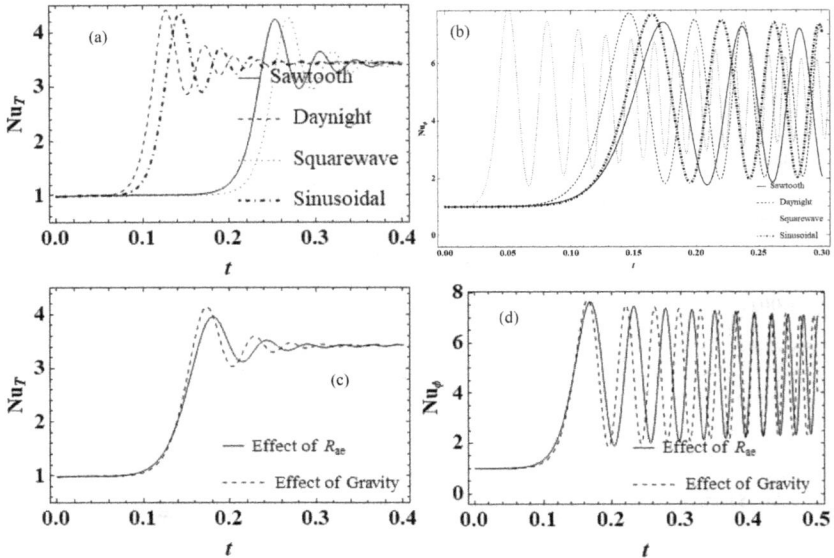

Figure 10.10 Graph showing the effect of (a) different type of gravity modulation on heat transfer, (b) different type of gravity modulation on mass transfer, (c) comparison of gravity modulation and AC electric effect on heat transfer, (d) comparison of gravity modulation and AC electric effect on mass transfer

electricity prevails over the effect of gravity. A similar result is obtained in the case of mass transfer in Figure 10.10d.

10.10 NUMERICAL UNCERTAINTY ANALYSIS

Numerical uncertainty results from the influence of discretization and iterative convergence error. This is the only uncertainty that cannot be eliminated but only can be minimized or bound in a solution. Uncertainty in heat and mass-transfer is analyzed. For this we solve equations (10.36–10.45) for different values of precision goal, accuracy goal, method, and other parameters (MaxStepsize, MaxStep, Real Exponent, Interpolation Error, Working Precision) as a default case from Mathematica and reported them in the form of tables (Tables 10.3 to 10.5). We compared the obtain solution with the default solution of the subroutine NDSolve and found that on increasing the values of different parameters, the solution is converging towards the default solution. In the same way, we can show the effect of different parameters which is also showing the converging effect.

Table 10.3 Effect of method on the sol of eq (10.38–10.47) by NDSolve of Mathematica

Method	Time	Nusselt number for heat transfer Nu_T	Error
Explicit RungeKutta method	0.1	1.06074	0.0000707303
	0.2	3.05798	−0.0000300052
	0.3	3.42293	2.79723×10^{-7}
	0.4	3.43454	3.13662×10^{-7}
	0.5	3.43901	1.17898×10^{-6}
Implicit RungeKutta method	0.1	1.06074	-8.4091×10^{8}
	0.2	3.05798	2.87119×10^{-6}
	0.3	3.42293	8.07464×10^{-7}
	0.4	3.43454	5.35619×10^{-8}
	0.5	2.94858	6.44203×10^{-8}
Adams method	0.1	1.06074	0.0000
	0.2	3.05798	0.0000
	0.3	3.42293	0.0000
	0.4	3.43454	0.0000
	0.5	3.43901	0.0000

Table 10.4 Effect of precision goal on the sol of eq (10.38–10.47) by NDSolve of Mathematica

Precision goal	Time	Nusselt number for mass transfer ($Nu\phi$)	Error
10	0.1	0.999992	-6.62015×10^{-12}
	0.2	1.00001	-4.8308×10^{-12}
	0.3	1.00001	-8.08753×10^{-12}
	0.4	1.00001	5.39124×10^{-13}
	0.5	1.00001	1.61753×10^{-11}
100	0.1	0.999992	-6.49758×10^{-11}
	0.2	1.00001	-4.62652×10^{-12}
	0.3	1.00001	-9.3805×10^{-12}
	0.4	1.00001	1.65423×10^{-13}
	0.5	1.00001	1.59037×10^{-11}
1000	0.1	0.999992	-6.49758×10^{-11}
	0.2	1.00001	-4.62652×10^{-12}
	0.3	1.00001	-9.3805×10^{-12}
	0.4	1.00001	1.65423×10^{-13}
	0.5	1.00001	1.59037×10^{-11}

Table 10.5 Effect of accuracy goal on the sol of eq (10.38–10.47) by NDSolve of Mathematica

Accuracy Goal	Time	Nusselt number for mass transfer (Nuϕ)	Error
10	0.1	0.999992	-9.81026×10^{-12}
	0.2	1.00001	-1.1622×10^{-12}
	0.3	1.00001	-1.9874×10^{-11}
	0.4	1.00001	-4.85612×10^{-12}
	0.5	1.00001	-2.46969×10^{-11}
100	0.1	0.999992	-1.80852×10^{-11}
	0.2	1.00001	-1.3899×10^{-11}
	0.3	1.00001	-1.807×10^{-11}
	0.4	1.00001	-2.4855×10^{-12}
	0.5	1.00001	2.68447×10^{-11}
1000	0.1	0.999992	-1.80852×10^{-11}
	0.2	1.00001	-1.3899×10^{-11}
	0.3	1.00001	-1.807×10^{-11}
	0.4	1.00001	-2.4855×10^{-12}
	0.5	1.00001	2.68447×10^{-11}

10.11 CONCLUSIONS

In this chapter we have looked into the combined repercussion of internal heating and AC electric field in a porous medium saturated by Oldroyd-B-nanofluid using Brinkman model for both linear and nonlinear stability analysis. Different parameters have different effects on linear (stationary) and nonlinear stability analysis. We notice the following effects:

- *Hs* the internal heat source parameter has destabilizing effect on stationary convection and advances earlier heat and mass-transfer in the system.
- The porosity has a stabilizing effect on the stationary convection.
- Delay in convection is observed due to the increase in Brinkman-Darcy number. Hence Da has a stabilizing effect on stationary convection.
- R_{ae} has a destabilizing effect on stationary convection with the increase in the value of the AC electric field.
- In nonlinear analysis of Oldroyd-B-fluid for mass transfer, with increase in the value of the Na rate of convection is very rapid at the same point of time, whereas just an opposite effect is observed in increasing porosity where the increase in the value of the porosity is less convected at the same point of time.

- Relaxation time has a destabilizing effect which is observed to be a little bit in mass-transfer, and nominal effect is observed in the case of heat transfer. While retardation time has a stabilizing effect on both mass-transfer and heat transfer.
- The effect of R_{ae} is seen very nominal in both heat and mass transfer. Hence the effect of R_{ae} can be neglected (refer Tables 10.1 and 10.2).
- Oscillatory convection is found to be debarred or rule out for any effect.
- Day-night profile show earlier rate of heat and mass transfer.
- On comparing the effect of gravity modulation and AC electric field, AC electric field prevail the heat and mass transfer.
- Numerical uncertainty results are obtained for method, accuracy goal, precision goal where the Adams method has zero error compared with the obtained results.
- No effect of relaxation time and retardation time is obtained in analysis of steady state graphs.

ACKNOWLEDGMENT

Funding: There is no funding for the current research work.
Conflicts of Interest: The authors have no conflicts of interest.

NOMENCLATURE

Latin Symbols

(x,y,z)	Cartesian or Euclidean coordinates (m)
Dã	the Brinkman-Darcy number
d	Dimensional layer depth (m)
Ra	Thermal Rayleigh-Darcy number
R_{ae}	AC Electric Rayleigh-Darcy number
Rm	Basic density Rayleigh number
Rn	Concentration Rayleigh number
Le	Lewis number
Nb	Modified particle-density increment
Na	Modified diffusivity ratio
Nu	Nusselt number
Km	effective thermal conductivity (W/m.k)
p	pressure(Pa)
g	Gravitational acceleration (m/s^2)
t	Time (s)
T	temperature (K)
T_1	temperature at the upper wall (K)

T_0 temperature at the lower wall (K)
v_D Darcy velocity (m/s)
D_B Brownian Diffusion coefficient (m²/s)
D_T thermophoretic diffusion coefficient
k_1 permeability
a wave number
Q Internal heat source
Hs   Internal Heat Source parameter

Greek Symbols

α_m Thermal diffusivity of the fluid (m²/s)
μ viscosity (kg/m.s)
$\bar{\mu}$ effective viscosity (kg/m.s)
ρ Density (kg/m³)
ϕ nanoparticle-volume fraction
$\in$ dielectric constant
ρ_e charge density
φ root mean square value of electric potential
Ψ Stream function

Subscripts and Superscripts

0   ref value
b basic value

Operators

$$\nabla^2 = \frac{\partial^2}{\partial x} + \frac{\partial^2}{\partial y} + \frac{\partial^2}{\partial z}$$

REFERENCES

[1] Choi, S. U., Eastman, J. A. (1995). Enhancing thermal conductivity of fluids with nanoparticles (No.ANL/MSD/CP-84938; CONF-951135-29). Argonne National Lab.(ANL), Argonne, IL (United States).

[2] Buongiorno, J., 2006, "Convective Transport in Nanofluids," *ASME Journal of Heat and Mass Transfer*, 128(3), 240–250

[3] Kuznetsov, A. V., Nield, D. (2010). Thermal instability in a porous medium layer saturated by a nanofluid: Brinkman model. *Transport in Porous Media*, 81(3), 409–422.

[4] Nield, D. A., Kuznetsov, A. V. (2009). Thermal instability in a porous medium layer saturated by a nanofluid. *International Journal of Heat and Mass Transfer*, 52(25–26), 5796–5801.

[5] Nield, D. A., Kuznetsov, A. V. (2011). The effect of vertical throughflow on thermal instability in a porous medium layer saturated by a nanofluid. *Transport in Porous Media*, 87(3), 765–775.

[6] Bhadauria, B. S., Agarwal, S. (2011). Convective transport in a nanofluid saturated porous layer with thermal non equilibrium model. *Transport in Porous Media*, 88(1), 107–131.

[7] Darcy, H. (1856). Les fontainespubliques de la ville de Dijon: exposition et application... Victor Dalmont.

[8] Sheikholeslami, M., Seyednezhad, M. (2018). Simulation of nanofluid flow and natural convection in a porous media under the influence of electric field using CVFEM. *International Journal of Heat and Mass Transfer*, 120, 772–781.

[9] Forchheimer, P. (1901). Wasserbewegungdurchboden. *Zeitschrift des Vereins Deutscher Ingenieure*, 45, 1782–1788.

[10] Brinkman, H. C. (1949). A calculation of the viscous force exerted by a flowing fluid on a dense swarm of particles. *Flow, Turbulence and Combustion*, 1(1), 27–34.

[11] Fetecau, C., Fetecau, C. (2003). The first problem of Stokes for an Oldroyd-B fluid. *International Journal of Non-Linear Mechanics*, 38(10), 1539–1544.

[12] Vieru, D., Fetecau, C., Fetecau, C. (2008). Flow of a generalized Oldroyd-B fluid due to a constantly accelerating plate. *Applied Mathematics and Computation*, 201(1–2), 834–842.

[13] Hayat, T., Shehzad, S. A., Mustafa, M., Hendi, A. (2012). MHD flow of an Oldroyd-B fluid through a porous channel. *International Journal of Chemical Reactor Engineering*, 10(1).

[14] Azeem Khan, W., Khan, M., Malik, R. (2014). Three-dimensional flow of an Oldroyd-B nanofluid towards stretching surface with heat generation/absorption. *PLoS One*, 9(8), e105107.

[15] Sheu, L. J. (2011). Thermal instability in a porous medium layer saturated with a viscoelastic nanofluid. *Transport in Porous Media*, 88(3), 461–477.

[16] Kumar, J. P., Umavathi, J. C., Murthy, C. (2017). Linear and Nonlinear analysis of thermal instability in a porous saturated by a nanofluid. *International Journal of Engineering, Science and Mathematics*, 6(4), 71–89.

[17] Srivastava, A., Bhadauria, B. S. (2016). Onset of convection in porous medium saturated by viscoelastic nanofluid: More realistic result. *Journal of Applied Fluid Mechanics*, 9(6), 3117–3125.

[18] Rana, G. C., Gautam, P. K., Saxena, H. Oscillatory Motions In An electro-thermal convection in shear thinning viscoelastic nanofluid layer in a porous medium OscilatornaKretanjaKodElectrotermalneKonvekcije u Pseudo-Plastic nomViskoelasticnomNanofluidnomSlojuPorozneSredine.

[19] Benjamin, T. B., Ursell, F. J. (1954). The stability of the plane free surface of a liquid in vertical periodic motion. *Proceedings of the Royal Society of London. Series A. Mathematical and Physical Sciences*, 225(1163), 505–515.

[20] Wadih, M., Roux, B. (1988). Natural convection in a long vertical cylinder under gravity modulation. *Journal of Fluid Mechanics*, 193, 391–415.

[21] Gresho, P. M., Sani, R. L. (1970). The effects of gravity modulation on the stability of a heated fluid layer. *Journal of Fluid Mechanics*, 40(4), 783–806.

[22] Murray, B. T., Coriell, S. R., McFadden, G. B. (1991). The effect of gravity modulation on solutal convection during directional solidification. *Journal of Crystal Growth*, 110(4), 713–723.

[23] Thomas, N. M., Maruthamanikandan, S. (2018, December). Gravity modulation effect on ferromagnetic convection in a Darcy-Brinkman layer of porous medium. *Journal of Physics: Conference Series*, 1139(1), 012022). IOP Publishing.

[24] Zhao, M., Wang, S., Li, S. C., Zhang, Q. Y., Mahabaleshwar, U. S. (2018). Chaotic Darcy-Brinkman convection in a fluid saturated porous layer subjected to gravity modulation. *Results in Physics*, 9, 1468-1480.

[25] Stergiou, Y., Hauser, M. J., Comolli, A., Brau, F., De Wit, A., Schuszter, G., ... Schwarzenberger, K. (2022). Effects of gravity modulation on the dynamics of a radial A+ B→ C reaction front. *Chemical Engineering Science*, 257, 117703.

[26] Smith, H. J. T., Blackburn, J. A. (1992). Experimental study of an inverted pendulum. *American Journal of Physics*, 60(10), 909–911.

[27] Takashima, M., Ghosh, A. K. (1979). Electrohydrodynamic instability in a viscoelastic liquid layer. *Journal of the Physical Society of Japan*, 47(5), 1717–1722.

[28] El-Sayed, M. F. (2008).Onset of electroconvective instability of Oldroydian viscoelastic liquid layer in Brinkman porous medium. *Archive of Applied Mechanics*, 78(3), 211–224.

[29] Rudraiah, N., Gayathri, M. S. (2009). Effect of thermal modulation on the onset of electrothermoconvection in a dielectric fluid saturated porous medium. *Journal of Heat Transfer*, 131(10).

[30] Shivakumara, I. S., Ng, C. O., Nagashree, M. S. (2011). The onset of electrothermoconvection in a rotating Brinkman porous layer. *International Journal of Engineering Science*, 49(7), 646–663.

[31] Hamad, M. A. A., Pop, I. (2011). Unsteady MHD free convection flow past a vertical permeable flat plate in a rotating frame of reference with constant heat source in a nanofluid. *Heat and Mass Transfer*, 47(12), 1517–1524.

[32] Gautam, P. K., Rana, G. C., and Saxena, H. Free electrothermo-convective instability in a dielectric Oldroydiannan ofluid layer in a porous medium. *Journal of Nanofluids* 12, 699–711 (2023).

[33] Awasthi, M. K. (2013). Nonlinear analysis of Rayleigh–Taylor instability of cylindrical flow with heat and mass transfer. *Journal of Fluids Engineering*, 135(6), 061205.

[34] Uddin, Z., Asthana, R., Kumar Awasthi, M., & Gupta, S. (2017). Steady MHD flow of nano-fluids over a rotating porous disk in the presence of heat generation/absorption: a numerical study using PSO. *Journal of Applied Fluid Mechanics*, 10(3), 871–879.

[35] Asthana, R., Awasthi, M. K., &Agrawal, G. S. (2014).Viscous potential flow analysis of Kelvin–Helmholtz instability of a cylindrical flow with heat and mass transfer. *Heat Transfer—Asian Research*, 43(6), 489–503.

[36] Awasthi, M. K. (2019). Rayleigh–Taylor Instability of swirling annular layer with mass transfer. *Journal of Fluids Engineering*, 141(7), 071202.

[37] Awasthi, M. K. (2013). Viscous corrections for the viscous potential flow analysis of Rayleigh–Taylor instability with heat and mass transfer. *Journal of Heat Transfer*, 135(7), 071701.

[38] Awasthi, M. K., Asthana, R., &Agrawal, G. S. (2012).Viscous corrections for the viscous potential flow analysis of magnetohydrodynamic Kelvin-Helmholtz instability with heat and mass transfer. *The European Physical Journal A*, 48, 1–10.

[39] Ismail, & Bhadauria, B. S. (2022, August). Thermal instability of Rivlin-Ericksen elastico-viscous nanofluid saturated by a porous medium with rotation. In *International workshop of Mathematical Modelling, Applied Analysis and Computation* (pp. 436–455). Cham: Springer Nature Switzerland.

[40] Agarwal, S., Sacheti, N. C., Chandran, P., Bhadauria, B. S., Singh, A. K. (2012). Non-linear convective transport in a binary nanofluid saturated porous layer. *Transport in Porous Media*, 93, 29–49.

Chapter 11

Thermal instability in Hele-Shaw Cell with variable forces for Rivlin-Ericksen Nanofluid

S.N. Rai, B.S. Bhadauria, Anurag Srivastava, and Anish Kumar

Babasaheb Bhimrao Ambedkar University, Lucknow, India

11.1 INTRODUCTION: BACKGROUND AND DRIVING FORCES

The acceptance of nanoliquids is no longer hidden from the community of scientists and researchers. In the primary stage, the awareness behind nanoliquids was to make a new thing through the inclusion of base liquid to the nanoparticles, which would make like – a liquid that has the thermal-conductivity of metal as well as nanoparticles. The name "nanofluids/nanoliquids" was first adapted by Choi (1995). The mathematical equations for convective-transport in the nanoliquids majorly having the properties of Brownian-diffusion/thermophoresis were projected by Buongiorno (2006). Buongiorno et al. (2009) studied the benchmark investigation on thermal-conductivity of nanoliquids. Kuznetsov and Nield (2010) used a Brinkman model to study convective instability in the porous materials filled with nanoliquids. Thermal convection is a vital phenomenon in industrial organizations due to its extensive application in heat-exchangers, thermal systems, electronic freezing, and so forth. In illumination, this, thermal-convection in a nanoliquid layer has been an area of intense current attention. Numerous authors utilizing these governing-equations studied the problem of thermal-convection in nanoliquids Neild and Bejan (2017) have done most of their studies on thermal instability in nanoliquids and porous materials that are saturated with nanoliquids. Under conditions of local thermal nonequilibrium Bhadauria and Srivastava (2022) studied the joint impacts of through-flow as well as internal heating in nanoliquids saturated porous materials.

Rivlin-Ericksen (R-E) liquid, one category of viscoelastic liquid, is theoretically familiarized by Rivlin-Ericksen (1955). The unsteady flow of the R-E liquid by a uniform circulation of dust particles over channels of unlike cross-segments in the existence of time-periodic pressure-gradients is examined by Srivastava (1971). Rana and Kumar (2010) examined the thermal-instability (T-I) of the R-E viscoelastic rotating liquid filled by suspended particles as well as a variable-gravity-field in the porous medium (P-M). Rana and Thakur (2012) calculated the impact of suspended particles on T-I in the R-E viscoelastic liquid in a Brinkman P-M. Chand and Rana (2012)

DOI: 10.1201/9781003465171-11

investigated the T-I of the R-E viscoelastic nanoliquid saturated through a P-M. The revised-model of T-I in the R-E viscoelastic nanoliquid in the P-M is studied by Chand et al. (2015). R-E elastic-viscous nanoliquid double-diffusive convection stability exploration in a P-M has been reported by Rana and Chand (2015). Rana et al. (2016) proposed a more reasonable model of T-I of the R-E nanoliquid saturated through the Darcy-Brinkman P-M. Saini and Sharma (2017) examined the influence of upright through flow in the R-E viscoelastic nanoliquid in the non-Darcy P-M. Ismail and Bhadauria (2022) examined the T-I of R-E elastic-viscous nanoliquid saturated by a P-M under rotation.

In order to feel as well as visualize the hydrodynamical motion entrance in the Earth, the laboratory equipment was set-up. Mathematical depiction of this laboratory-equipment was initially specified by Hele-Shaw (1898) in 1898 as well as examined a slow 2-D (two dimensional) laminar movement in a uniform P-M in the very thin-slot squeezed among equidistant walls and mentioned to this complete arrangement as Hele-Shaw (H-S) cell. There are numerous engineering as well as natural science uses for the Hele-Shaw cell. Hele-Shaw cells are extensively utilized in a variety of fields, including hydrogeology, soil mechanics, petroleum engineering, and energy storage. Thermal instability within the H-S-cell using a viscous-liquid having inconstant-density was observed by Wooding (1960). Hartline and Lister (1977) examined the T-I procedure within the H-S-cell and deduces that for marginal-stability a system of equations is alike to that explaining movement through a P-M. Aniss et al. (1995) studied the GBL equation from the mathematical equations for H-S-cell in the boundless interval. Later this formative work, a huge number of works have been finished in numerous articles in the literature (Bhadauria et al., 2005; Boulal et al., 2008; Souhar & Aniss, 2012; Wakif et al., 2016). Yadav (2019) examined the combined influence of magnetic field as well as throughflow over T-I of nanoliquid within H-S-cell. Linear/nonlinear stability exploration on T-I within H-S-cell saturated through-nanoliquid under the impact of throughflow as well as gravity-modulation applying Brinkman-model has been examined by Bhadauria and Kumar (2021). The joint impact of g-jitter and thermal-difference on R-E nanoliquids within the HS-cell was investigated by Bhadauria et al. (2022).

A state when a container containing a weighty liquid oscillates vertically by a definite frequency as well as amplitude is known as g-jitter (gravity modulation). The impact of g-jitter on the presence of upright waves on a free surface of liquids in the container has been studied by Benjamin and Ursell (1954). They applied Mathieu's equation to discuss the stability norms Gresho and Sani (1970) investigated a stability of flat layer of liquids heated from below or above for the situation of the time-periodic buoyancy-force which is caused by shaking a liquid layer, consequently producing the sinusoidal modulation of gravity-field. They established that g-jitter can meaningfully impress the stability restrictions of the system. Numerous works (Clever et al., 1993; Bhadauria et al., 2012, 2013; Bhadauria & Kiran, 2014; Kumar et al., 2023) have been dedicated to analyzing linear/nonlinear stability exploration of the beginning of convection through the impact of g-jitter. Out of several, some of the nearby associated works are informed here.

The primary investigation describing the effect of magnetic fields in a flat liquid layer was specified by Thompson (1951) as well as by Chandrasekhar (1961). At high magnetic field strengths, Nakagawa (1957) claims that the Chandrasekhar number influences constant critical Rayleigh number. Nakagawa (1959) continued his investigation in the rotating liquid layer and his test achieves that Chandrasekhar's hypothetical evaluations regarding wavenumber of interventions at marginal constancy over magnetic rotation effect. Taking into account both the buoyant and magnetic forces acting in a flat liquid layer, Finlayson (1970) calculated their joint conclusions. Obtaining a critical Rayleigh number was achieved by implementing normal-mode technique. T-I in a flat layer of magnetic liquids has been identified by Gotoh and Yamada (1982). Ozoe and Maruo (1987) studied the influence of magnetic fields on the creation of convection through the numeric-evaluation to attain heat-transportation. Siddheshwar and Pranesh (2000) examined the influence of inner angular momentum under magneto-convection within electrically conducting liquids over a gravitational-field and a time dependent-temperature. Bhadauria (2006) have examined the influence of temperature modulation under magneto convection applying the Fluquet idea and rigid-rigid boundaries. Afterward, Bhadauria (2008) continued his investigation to study the simultaneous effect of a magnetic field/temperature-profile under convection in the electrically conducting fluid applying rigid-rigid/free-free restrictions. Siddheshwar et al. (2012) studied the heat-transportation for stationary magnetic-convection in Newtonian-liquids over gravity/thermal modulation by achievement of weak nonlinear instability investigation and implementing a cubic GBL technique.

The effect of the upright magnetic fields under creation of T-I within nanoliquid layer has been investigated by Yadav (2013; Gupta, 2013). The properties of heat transportation of nanoliquids are improved by implementation of external magnetic fields. However, this improvement is unique to upright magnetic fields, as described by Sheikholeslami et al. (2014). In an upright lid-driven cavity with spinning circular tube that conducts heat, Bansal and Chatterjee (2015) investigated the consequence of the magnetic fields on convective transfer of nanoliquids. Numerous works (Awasthi et al., 2012; Awasthi, 2013, 2019; Asthana et al., 2014; Uddin et al., 2017) have been dedicated to studying heat and mass transfer. Out of several, some of the nearby associated works are informed here.

The impact of time varying sinusoidal-magnetic fields upon ferro-liquids heated from overhead by applying Fluquet technique is reported by Aniss et al. (2001). They have estimated that the suitable substitute of the ratio of gravitational and magnetic forces produce a connection between a (sub)harmonic convection way. It has been reported by Bhadauria and Kiran (2014) that T-I based on magnetic number can occur through an upright magnetic field that is time-dependent. The effect of the inside-heat source on a stationary magnetic-instability of pair stress liquids by magnetic-modulation has been studied by Keshri et al. (2019). In Hele-Shaw cell, Rai et al. (2023) explored the T-I of nanoliquids through four different categories of magnetic-modulation Rai and Bhadauria (2023) investigated T-I in nanoliquids filled in a H-S cell through three different categories of rotational-modulation with the effects of magnetic-field/through-flow. Recent research (Rai et al., 2023) investigated T-I in Walter-B nanoliquid filled in the H-S cell under three different categories of magnetic-modulation with through-flow.

The novelty of this manuscript is to perform linear/nonlinear stability exploration and also a comparative study between magnetic-field and gravity modulation in an electrically conducting R-E nanoliquid within the Hele-Shaw cell. To the best of our information, no analogous study is done in the literature, discussing these significant features simultaneously.

11.2 PROBLEM FORMULATIONS

Consider a Rivlin-Ericksen nanoliquid layer, which is electrically conducting and is upright and constrained by two parallel free-free permeable confinement at $z = 0$ and $z = h$. We consider a system that is stretched along the x-axis while being constrained along the y-axis. This system is constrained by perpendicular impermeable borders (phase-walls) at $y = 0$ and $y = h (< < b)$. The system will function as endlessly extensible parallel horizontally plates on both the x and y directions if the domain has no restrictions in the y-direction. The outcomes in this situation will look like those of a traditional Rayleigh-Bénard convection problem. The presentation is given in Figure 11.1.

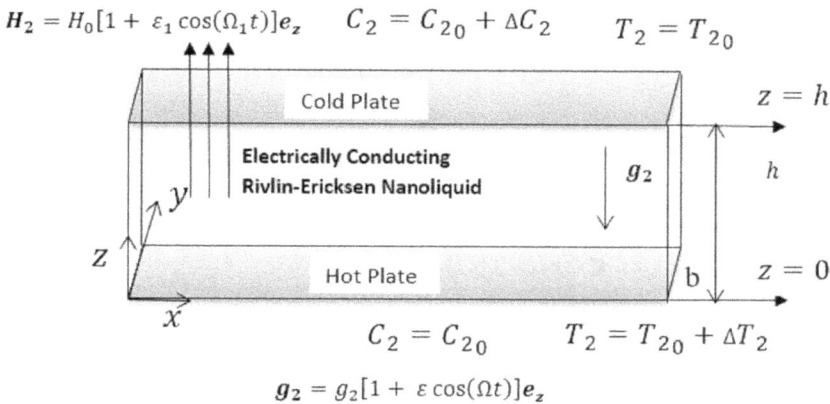

$$H_2 = H_0[1 + \varepsilon_1 \cos(\Omega_1 t)]e_z \qquad C_2 = C_{20} + \Delta C_2 \qquad T_2 = T_{20}$$

$$C_2 = C_{20} \qquad T_2 = T_{20} + \Delta T_2$$

$$g_2 = g_2[1 + \varepsilon \cos(\Omega t)]e_z$$

Figure 11.1 Schematic of Hele-Shaw cell convective problem.

11.3 GOVERNING EQUATIONS

The governing equations are occupied from Bhadauria and Kiran (2014) and Bhadauria et al. (2022):

$$\nabla \cdot V_2 = 0, \tag{11.1}$$

$$\nabla \cdot H_2 = 0, \tag{11.2}$$

$$\rho_{2_0}\left[\frac{\partial}{\partial t} + V_2 \cdot \nabla_2\right]V_2 = -\nabla_2 p_2 - \frac{1}{K_2}\left(\mu_2 + \mu\frac{\partial}{\partial t}\right)$$
$$V_2 + \mu_2\nabla_2{}^2V_2 + \rho_2 g_2 + \mu_e\left(H_2.\nabla_2\right)H_2, \tag{11.3}$$

$$\left[\frac{\partial}{\partial t} + V_2 \cdot \nabla_2\right]T_2 = \alpha_2\nabla_2{}^2T_2 + \frac{(\rho c)_p}{(\rho c)_f}$$
$$\left[D_B\nabla_2 C_2.\nabla_2 T_2 + \left(\frac{D_{T_2}}{T_{T_0}}\right)\nabla_2 T_2.\nabla_2 T_2\right], \tag{11.4}$$

$$\left[\frac{\partial}{\partial t} + V_2 \cdot \nabla_2\right]C_2 = D_B\nabla_2{}^2C_2 + \left(\frac{D_{T_2}}{T_{T_0}}\right)\nabla_2{}^2T_2, \tag{11.5}$$

$$\left[\frac{\partial}{\partial t} + V_2 \cdot \nabla_2\right]H_2 = \left(H_2.\nabla_2\right)V_2 + \eta\nabla_2{}^2H_2, \tag{11.6}$$

$$\rho_2 = C_2\rho_{2_p} + \rho_{2_0}\left(1 - C_2\right)\left\{1 - \beta_{T_2}\left(T_2 - T_{2_0}\right)\right\}, \tag{11.7}$$

$$H_2 = H_0\left\{1 + \varepsilon_1\cos\left(\Omega_1 t\right)\right\}e_z, \tag{11.8}$$

$$g_2 = g_0\left\{1 + \varepsilon\cos\left(\Omega t\right)\right\}e_z. \tag{11.9}$$

11.4 BOUNDARY CONDITIONS

The boundary conditions (BCs) are considered from Bhadauria and Kiran (2014):

$$V_2 = 0, \quad T = T_{2_0} + \Delta T_2, \qquad C_2 = C_{2_0}, \qquad \text{at } z = 0,$$
$$V_2 = 0, \quad T = T_{2_0}, \qquad\qquad C_2 = C_{2_0} + \Delta C_2, \quad \text{at } z = h. \Big\} \qquad (11.10)$$

11.5 PRIMARY STATE

$$V_2 = V_{2_b} = 0, \, T_2 = T_{2_b}(z,t), \, C_2 = C_{2_b}, \, p_2 = p_{2_b}(z,t),$$
$$\rho_2 = \rho_2(z,t), \, H_2 = H_0 e_z. \tag{11.11}$$

$$\rho_2 \cdot g_2 + \nabla_2 p_{2_b} = 0, \tag{11.12}$$

$$\frac{d^2 T_{2_b}}{dz^2} = 0, \tag{11.13}$$

$$\frac{d^2 C_{2_b}}{dz^2} = 0. \tag{11.14}$$

Utilizing the BC's eqn. (11.10) into eqns. (11.13) as well as (11.14), the primary-state solutions are reduced as:

$$T_{2_b}(z) = T_{2_0} + \Delta T_2 \left(1 - \frac{z}{h} \right), \tag{11.15}$$

$$C_{2_b}(z) = C_{2_0} + \Delta C_2 \left(\frac{z}{h} \right). \tag{11.16}$$

11.6 PERTURBATION STATE

Now, we overlaid infinitesimal perturb on the primary state:

$$V_2 = V_{2_b} + \chi V'_2, T_2 = T_{2_b} + \chi T'_2, C_2 = C_{2_b} + \chi C'_2, p_2 \tag{11.17}$$
$$= p_{2_b} + \chi p'_2, \rho_2 = \rho_{2_b} \times \rho'_2, H_2 = H_0 + \chi H'_2,$$

where $p'_2, V'_2, H'_2, T'_2, C'_2$, and ρ'_2 represent the perturbed quantities.
Put on eqn. (17) into eqns. (11.1–11.8), we have

$$\nabla \cdot V'_2 = 0, \tag{11.18}$$

$$\nabla \cdot H'_2 = 0, \tag{11.19}$$

$$\rho_{2_0}\left[\frac{\partial}{\partial t} + V'_2 \cdot \nabla_2\right]V'_2 = -\nabla_2 p'_2 - \frac{1}{K_2}\left(\mu_2 + \mu\frac{\partial}{\partial t}\right)V'_2$$
$$+\mu_2\nabla_2^2 V'_2 + \rho'_2 g_2 e_z + \mu_e\left(H'_2.\nabla_2\right)H'_2 + \mu_e H_0\frac{\partial H'_2}{\partial z}, \tag{11.20}$$

$$\left[\frac{\partial}{\partial t} + V'_2 \cdot \nabla_2\right]T'_2 + w'_2\frac{\partial T_{2_b}}{\partial z} = \alpha_2\nabla_2^2 T'_2 + \frac{(\rho c)_p}{(\rho c)_f} \tag{11.21}$$
$$[D_B\nabla_2 C'_2 \cdot \nabla_2 T'_2 + \left(\frac{D_{T2}}{T_{2_0}}\right)\nabla_2 T'_2 \cdot \nabla_2 T'_2],$$

$$\left[\frac{\partial}{\partial t} + V'_2 \cdot \nabla_2\right]C'_2 + w'_2\frac{\partial C_{2_b}}{\partial z} = D_B\nabla^2_2 C'_2 + \left(\frac{D_{T2}}{T_{2_0}}\right)\nabla_2^2 T'_2, \tag{11.22}$$

$$\left[\frac{\partial}{\partial t} + V'_2 \cdot \nabla_2\right]H'_2 - \left(H'_2 \cdot \nabla'_2\right)V'_2 - H_0\frac{\partial V'_2}{\partial z} = \eta\nabla^2_2 H'_2, \tag{11.23}$$

$$\rho'_2 = \left(\rho_{2_p} - \rho_{2_0} - \beta_{T2}\rho_{2_0}T_{2_0}\right)C'_2 - \beta_{T2}\rho_{2_0}T'_2, \tag{11.24}$$

$$H_{2_b} = H_0\left\{1 + \varepsilon_1 \cos\left(\Omega_1 t\right)\right\}, \tag{11.25}$$

$$g_{2_b} = g_0\left\{1 + \varepsilon \cos\left(\Omega t\right)\right\}. \tag{11.26}$$

Since only two-dimensional disturbances are considered in the present investigation, so a stream-function ψ and magnetic potential ϕ are familiarized as $(u'_2, 0, w'_2) = (\psi_z, 0, -\psi_x)$ and $(H'_{2x}, 0, H'_{2z}) = (\phi_z, 0, -\phi_x)$. On elimination of density and pressure-terms as of eqns. (11.20) and (11.24), after introduction of nondimensional quantities, delivered as:

$$t = \frac{b^2}{\alpha_2} t^*, (x,y,z) = \left(x^*, y^*, z^*\right)b, \ V'_2 = \frac{\alpha_2}{b} V_2^*,$$

$$C'_2 = \Delta C_2 \cdot C_2^*, T'_2 = \Delta T_2 \cdot T_2^*, \nabla'_2 = \frac{\nabla_2^*}{b},$$

$$\Omega_1 = \frac{\alpha_2}{b^2} \Omega_1^*, \ \Omega = \frac{\alpha_2}{b^2} \Omega^*, \ \psi = \alpha_2 \psi^*, \phi = b H_0 \phi^*,$$

where $\alpha_2 = \dfrac{\kappa}{(\rho c)_f}$.

$$(11.27)$$

$$\frac{Hs}{P_r} \frac{\partial}{\partial t}\left(\nabla_2^2 \psi\right) - Hs\nabla_2^4 \psi + \nabla_2^2 \psi + g_n R_a \frac{\partial T_2}{\partial x}$$

$$- g_n R_n \frac{\partial C_2}{\partial x} - Hs Q H_n \times \frac{P_r}{P_{rm}} \frac{\partial}{\partial z}\left(\nabla_2^2 \phi\right) - F \frac{\partial\left(\nabla_2^2 \psi\right)}{\partial t}$$

$$+ \frac{Hs}{P_r} = \times \frac{\partial\left(\psi, \nabla_2^2 \psi\right)}{\partial(x,z)} - Hs Q \times \frac{P_r}{P_{rm}} \frac{\partial\left(\phi, \nabla_2^2 \phi\right)}{\partial(x,z)},$$

$$(11.28)$$

$$\left(\frac{\partial}{\partial t} - \nabla_2^2\right) T_2 - \frac{\partial\psi}{\partial x} \frac{\partial T_{2b}}{\partial z}\left(\frac{N_B}{L_e}\right) \nabla_2 C_2 \cdot \nabla_2 T_2$$

$$+ \left(\frac{N_A N_B}{L_e}\right) \nabla_2 T_2 \cdot \nabla_2 T_2 + \frac{\partial\left(\psi, T_2\right)}{\partial(x,z)},$$

$$(11.29)$$

$$\left(\frac{\partial}{\partial t} - \frac{1}{L_e}\nabla_2^2\right) C_2 - \frac{\partial\psi}{\partial x} \frac{\partial C_{2b}}{\partial z} = \left(\frac{N_A}{L_e}\right) \nabla_2^2 T_2 + \frac{\partial\left(\psi, C_2\right)}{\partial(x,z)},$$

$$(11.30)$$

$$\left(\frac{\partial}{\partial t} - \frac{P_r}{P_{rm}}\nabla_2\right)\phi - g_m \frac{\partial\psi}{\partial z} = \frac{\partial\left(\psi, \phi\right)}{\partial(x,z)},$$

$$(11.31)$$

$$T_{2b} = 1 - z,$$

$$(11.32)$$

$$C_{2b} = z,$$

$$(11.33)$$

where

$$H_n = \left[1 + \varepsilon_1 \cos\left(\Omega_1 t\right)\right], \; g_n = \left[1 + \varepsilon \cos\left(\Omega t\right)\right],$$

$$R_a = \frac{\rho_{2_0} g_0 \beta_{T_2} K_2 h \Delta T_2}{\mu_2 \alpha_2}, R_n = \frac{g_0 \beta_{T_2} K_2 h \left(\rho_{2_p} - \rho_{2_0} - \beta_{T_2} \rho_{2_0} T_{2_0}\right) \Delta C_2}{\mu_2 \alpha_2},$$

$$P_r = \frac{\mu_2}{\alpha_2 \rho_{2_0}}, \; P_{rm} = \frac{\mu_2}{\eta \rho_{2_0}}, \; Hs = \frac{K_2}{h^2}, \; L_e = \frac{\alpha_2}{D_B}, \; N_A = \frac{D_2 \Delta T_2}{D_B T_{2_0} \Delta C_2},$$

$$N_B = \frac{\left(\rho c\right)_p \Delta C_2}{\left(\rho c\right)_f}, \; F = \frac{\mu \alpha_2}{\mu_2 h^2}, \; Q = \frac{\mu_e H_0^2 h^2}{\mu_2 \eta}.$$

Under a supposition of isothermal/stress-free BC's, a desired system of eqns. (11.28)–(11.31) is solved:

$$\left. \begin{aligned} \psi = 0 = \Delta_2^2 \psi, \; T_2 = 0, \; C_2 = 0, \frac{\partial \phi}{\partial z} = 0, \quad \text{at } z = 0, \\ \psi = 0 = \Delta_2^2 \psi, \; T_2 = 0, \; C_2 = 0, \frac{\partial \phi}{\partial z} = 0, \quad \text{at } z = 1. \end{aligned} \right\} \tag{11.34}$$

11.7 LINEAR STABILITY EXPLORATION

We investigate for the solutions to eqns. (11.28)–(11.31) which have been implemented from Bhadauria and Srivastava (2022) for a stationary phase of convection in the following manner:

$$\psi = A \sin\left(ax\right)\sin\left(\pi z\right), \tag{11.35}$$

$$T_2 = B \cos\left(ax\right)\sin\left(\pi z\right), \tag{11.36}$$

$$C_2 = C \cos\left(ax\right)\sin\left(\pi z\right), \tag{11.37}$$

$$\phi = D \sin\left(ax\right)\cos\left(\pi z\right), \tag{11.38}$$

where a is the wave-number and A, B, C and D are scalars. Using the orthogonality criterion and eqns. (11.35)–(11.38) in the linearized form of eqns. (11.28)–(11.31) yields the following:

$$
\begin{bmatrix}
-\dfrac{1}{2}\left(a^{2}+\pi^{2}\right)\left(\left(a^{2}+\pi^{2}\right)Hs+1\right) & -\dfrac{aR_{n}}{2} & \dfrac{aR_{n}}{2} & -\dfrac{\pi\left(a^{2}+\pi^{2}\right)HsP_{r}Q}{2P_{rm}} \\[2ex]
\dfrac{a}{2} & \dfrac{\left(a^{2}+\pi^{2}\right)}{2} & 0 & 0 \\[2ex]
-\dfrac{a}{2} & \dfrac{\left(a^{2}+\pi^{2}\right)N_{A}}{2L_{e}} & \dfrac{\left(a^{2}+\pi^{2}\right)}{2L_{e}} & 0 \\[2ex]
-\dfrac{\pi}{2} & 0 & 0 & \dfrac{\left(a^{2}+\pi^{2}\right)P_{r}}{2P_{rm}}
\end{bmatrix}
$$

$$
\times
\begin{bmatrix} A \\ B \\ C \\ D \end{bmatrix}
=
\begin{bmatrix} 0 \\ 0 \\ 0 \\ 0 \end{bmatrix}
\tag{11.39}
$$

For the nontrivial solution of eqn. (11.39), we have

$$
R_{a} = \frac{\left(a^{2}+\pi^{2}\right)^{2}+Hs\left(\left(a^{2}+\pi^{2}\right)^{3}+\pi^{2}\left(a^{2}+\pi^{2}\right)Q\right)}{a^{2}} - \left(L_{e}+N_{A}\right)R_{n}.
\tag{11.40}
$$

11.8 WEAKLY NONLINEAR STABILITY EXPLORATION

We use the subsequent manners in Fourier-series expansions for weakly nonlinear stability assessment because it is supposed that all physical variables are independent of y:

$$
\psi = A_{11}(t)\sin(ax)\sin(\pi z),
\tag{11.41}
$$

$$
T_{2} = B_{11}(t)\cos(ax)\sin(\pi z) + B_{02}(t)\sin(2\pi z),
\tag{11.42}
$$

$$
C_{2} = C_{11}(t)\cos(ax)\sin(\pi z) + C_{02}(t)\sin(2\pi z),
\tag{11.43}
$$

$$
\phi = D_{11}(t)\sin(ax)\cos(\pi z) + D_{02}(t)\cos(2\pi z).
\tag{11.44}
$$

We obtain by substituting expressions (11.41–11.44) in eqns. (11.28)–(11.31), ignoring the second/third terms of the RHS of eqn. (11.28) because they are small by Bhadauria et al. (2022), and using orthogonalization method of Galerkin's approach.

$$A'_{11}(t) = \frac{P_r\left(-\left(a^2 + \pi^2\right)P_{rm}\left(\left(a^2 + \pi^2\right)Hs + 1\right)A_{11}(t)\right)}{\left(a^2 + \pi^2\right)P_{rm}\left(FP_r + Hs\right)}$$

$$+ \frac{P_r\left(\pi\left(a^2 + \pi^2\right)HsP_rQD_{11}(t)H_n - aP_{rm}g_n\left(R_aB_{11}(t) - R_nC_{11}(t)\right)\right)}{\left(a^2 + \pi^2\right)P_{rm}\left(FP_r + Hs\right)}, \tag{11.45}$$

$$B'_{11}(t) = -\left(a^2 + \pi^2\right)B_{11}(t) - A_{11}(t)a\left(\pi B_{02}(t) + 1\right), \tag{11.46}$$

$$B'_{02}(t) = \frac{1}{2}\pi\left(aA_{11}(t)B_{11}(t) - 8\pi B_{02}(t)\right), \tag{11.47}$$

$$C'_{11}(t) = aA_{11}(t)\left(1 - \pi C_{02}(t)\right) - \frac{\left(a^2 + \pi^2\right)\left(N_AB_{11}(t) + C_{11}(t)\right)}{L_e}, \tag{11.48}$$

$$C'_{02}(t) = \frac{\pi\left(aL_eA_{11}(t)C_{11}(t) - 8\pi\left(N_AB_{02}(t) + C_{02}(t)\right)\right)}{2L_e}, \tag{11.49}$$

$$D'_{11}(t) = \pi A_{11}(t)H_n - \left(a^2 + \pi^2\right)\frac{P_r}{P_{rm}}D_{11}(t), \tag{11.50}$$

$$D'_{02}(t) = -4\pi^2\frac{P_r}{P_{rm}}D_{02}(t). \tag{11.51}$$

A numerical solution is found for the autonomous simultaneous ODE system stated above.

11.9 HEAT AND MASS TRANSFER

The nanoliquid Nusselt-number is taken from Bhadauria et al. (2022).

$$Nu_T(t) = 1 + \left[\frac{\int_0^{2\pi/a}\left(\frac{\partial T_2}{\partial z}\right)dx}{\int_0^{2\pi/a}\left(\frac{dT_{2b}}{dz}\right)dx}\right]_{z=0}. \tag{11.52}$$

Put on eqns. (11.32) and (11.42) into eqn. (11.52), we have

$$Nu_T(t) = 1 - 2\pi B_{02}(t).$$ (11.53)

$$Nu_C(t) = 1 + \left[\frac{\int_0^{2\pi/a} \left(\frac{\partial C_2}{\partial z} + N_A \frac{\partial T_2}{\partial z} \right) dx}{\int_0^{2\pi/a} \left(\frac{dC_{2b}}{dz} \right) dx} \right]_{z=0}.$$ (11.54)

Put on eqns. (11.33, 11.42 and 11.43) into eqn. (11.54), we have

$$Nu_C(t) = 1 + 2\pi \left(N_A B_{02}(t) + C_{02}(t) \right).$$ (11.55)

11.10 RESULTS AND DISCUSSION

11.10.1 Stationary convection

In Figure (11.2), we depict fluctuation of critical Hele-Shaw Rayleigh (HSR) number R_{a0} against H-S-number Hs for several dimension-free parameters Q, L_e, N_A and R_n. Figure 11.2a, depicts that on rising the amount of Q then rise the magnitude of R_{a0} and hence Q has the stabilization nature to form the system more stable. Further, R_{a0} increases on rising the value of Hs then also the H-S number has the stabilization nature. Figure 11.2b shows that

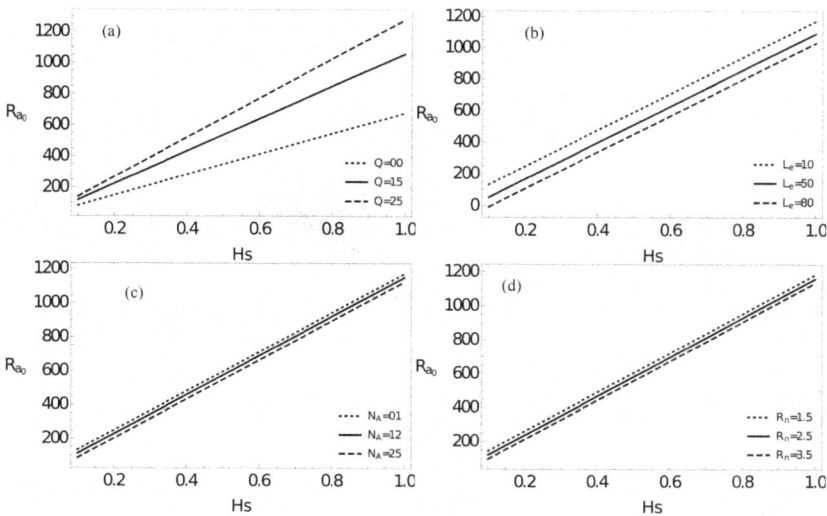

Figure 11.2 Fluctuation of critical HSR-number against HS-number with $Q = 20$, $L_e = 10$, $N_A = 2$, $R_n = 2$.

on rising the magnitude of L_e then decrease the amount of R_{a0} and therefore L_e has the destabilization nature on the system. Figures 11.2c and d also demonstrate similar nature like Figure 11.2b.

11.10.2 Heat and mass transfer

This text reports the impact of time-dependent magnetic-field (MF)/gravity modulation by a weakly nonlinear stability exploration to examine the behavior of the Nusselt-number in the H-S-cell to study heat/mass transportation of the R-E nanoliquid. For the heat/mass transportation the following parameters are considered: F, Hs, P_r, P_{rm}, Q, L_e, N_A, ε_1, ε, Ω_1, Ω, and R_n. For examining the influence of the magnetic-field/gravity modulation with small amplitude on heat/mass transportation are taken. The amount of amplitude of modulation is taken between 0.1 and 0.5. The findings are also depicted graphically. Fixed parameter values are opted from Bhadauria et al. (2022) and Rai et al. (2023) as $F = 0.5$, $Hs = 0.8$, $P_r = 5$, $P_{rm} = 1.6$, $Q = 20$, $L_e = 10$, $N_A = 0.2$, $\varepsilon_1 = \varepsilon = 0.3$, $\Omega_1 = \Omega = 10$, and $R_n = 2$.

In Figures 11.3–11.8, we show the graph of heat-transportation Nusselt-number Nu_T as well as mass-transportation Nusselt-number Nu_C with time t for several parameters applying trigonometrical cosine-waveform of magnetic-field/gravity modulation.

Figures 11.3a and 11.6a display that on increasing the amount of kinematic-viscoelasticity permeability F then both Nu_T and Nu_C decrease; that is, heat as well as mass transportation in a system diminishes. This happens

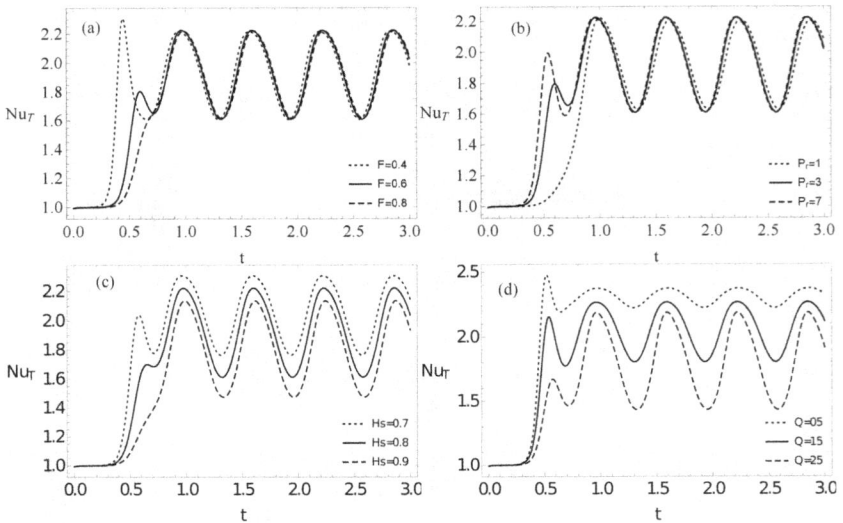

Figure 11.3 Transportation of heat for $F = 0.5$, $H_s = 0.8$, $P_r = 5$, $P_{rm} = 1.6$, $Q = 20$, $L_e = 10$, $N_A = 0.2$, $\varepsilon_1 = 0.3$, $\Omega_1 = 10$, $\varepsilon = 0.3$, $\Omega = 10R_n = 2$.

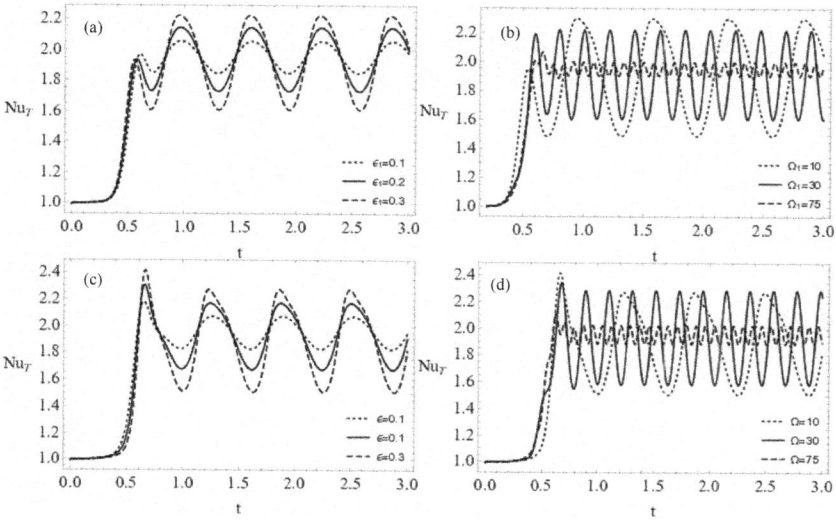

Figure 11.4 Transportation of heat for $F = 0.5$, $H_s = 0.8$, $P_r = 5$, $P_{rm} = 1.6$, $Q = 20$, $L_e = 10$, $N_A = 0.2$, $\epsilon_1 = 0.3$, $\Omega_1 = 10$, $\epsilon = 0.3$, $\Omega = 10$ $R_n = 2$.

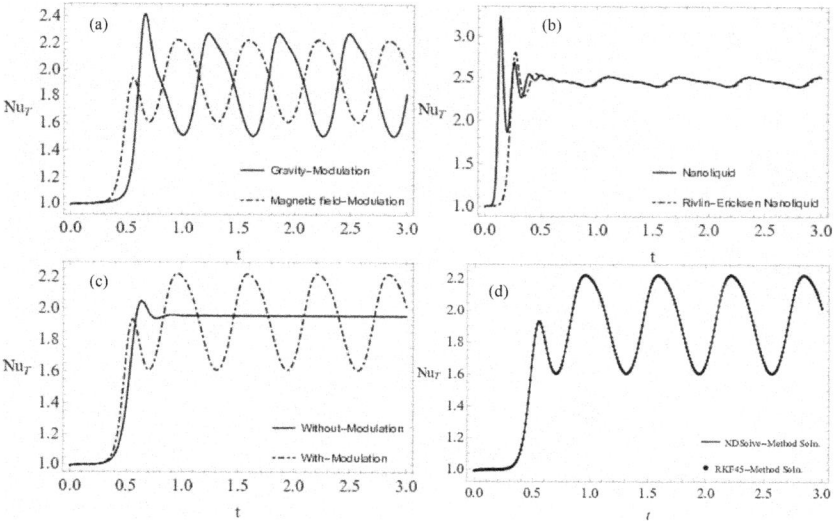

Figure 11.5 Heat transportation for $F = 0.5$, $H_s = 0.8$, $P_r = 5$, $P_{rm} = 1.6$, $Q = 20$, $L_e = 10$, $N_A = 0.2$, $\epsilon_1 = 0.3$, $\Omega_1 = 10$, $\epsilon = 0.3$, $\Omega = 10$ $R_n = 2$.

as on rising F, permeability will rise. As a result, it takes additional time for the onset of convection. Figures 11.3b and 11.6b depict that the influence of the R-E nanoliquid Prandtl-number (P_r) increases (when a kinematic viscosity increases or a thermal diffusivity reduces) heat as well as mass transportation increases. Figures 11.3c and 11.6c demonstrate that on rising the

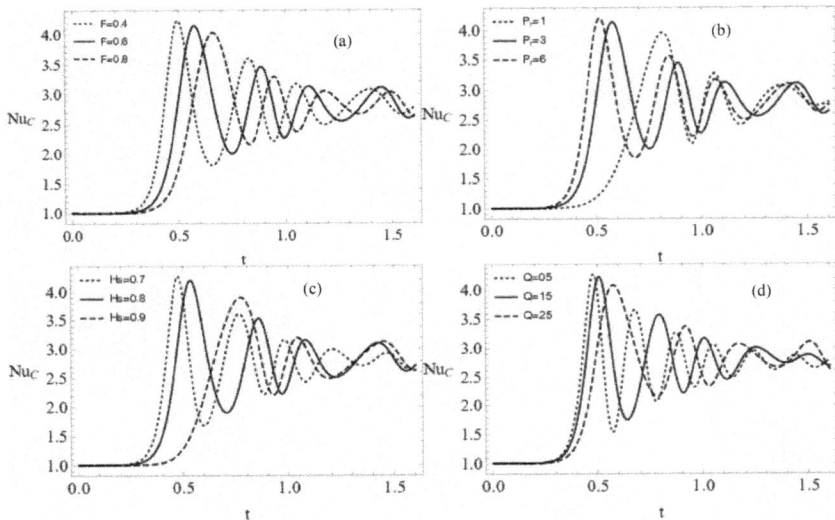

Figure 11.6 Transportation of mass for $F = 0.5$, $H_s = 0.8$, $P_r = 5$, $P_{rm} = 1.6$, $Q = 20$, $L_e = 10$, $N_A = 0.2$, $\epsilon_1 = 0.3$, $\Omega_1 = 10$, $\epsilon = 0.3$, $\Omega = 10$ $R_n = 2$.

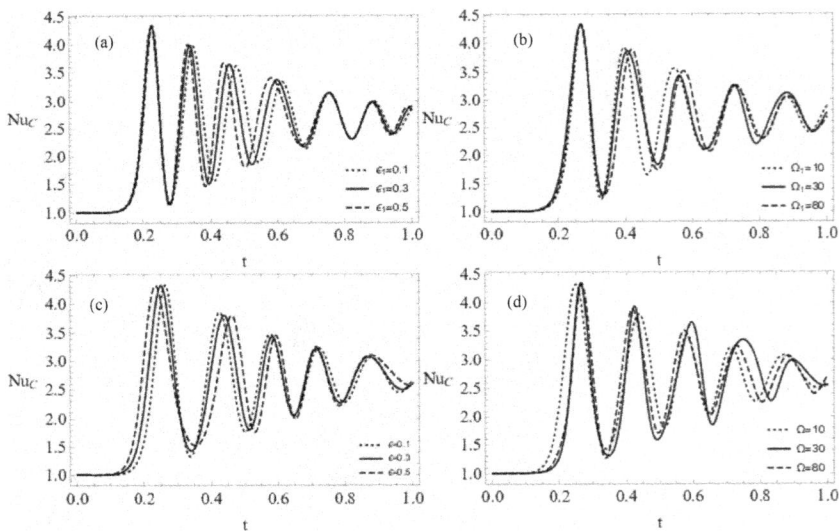

Figure 11.7 Transportation of mass for $F = 0.5$, $H_s = 0.8$, $P_r = 5$, $P_{rm} = 1.6$, $Q = 20$, $L_e = 10$, $N_A = 0.2$, $\epsilon_1 = 0.3$, $\Omega_1 = 10$, $\epsilon = 0.3$, $\Omega = 10$ $R_n = 2$.

value of Hs, Nu_T, and Nu_C reduce, that is, heat as well as mass transportation in a system decreases. This happens because on rising the H-S number the permeability increases. As an outcome, its takings more time to initiate the convection. Since eqn. (11.39), we realize that Hs is a product of the magnitude of the MF-modulation, which means that a magnitude of the MF

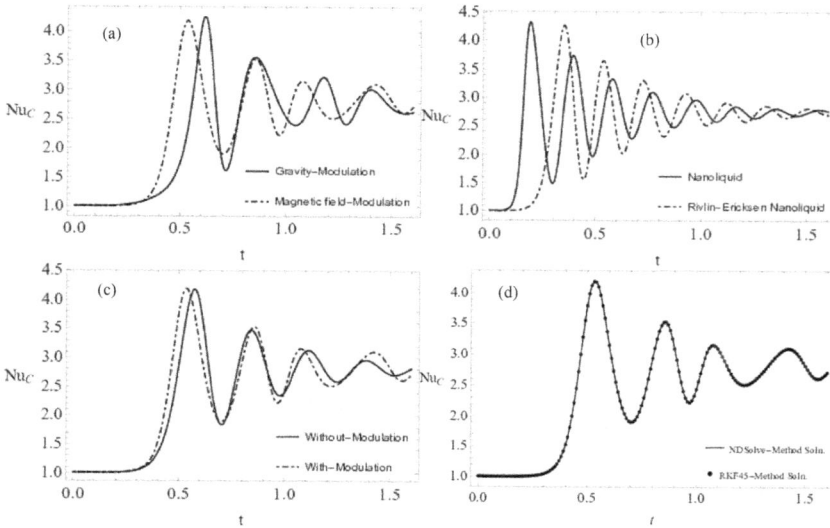

Figure 11.8 Mass transportation for $F = 0.5, H_s = 0.8, P_r = 5, P_{rm} = 1.6, Q = 20, L_e = 10,$
$N_A = 0.2, \epsilon_1 = 0.3, \Omega_1 = 10, \epsilon = 0.3, \Omega = 10$ $R_n = 2$.

modulation is affected by Hs. Figures 11.3d and 11.6d describe on increasing the amount of Q then rate of heat/mass transportation reduces in a system and consequently Q stabilizes a system.

Figures 11.4a and 11.7a depict that the increase in amplitude of MF-modulation (ϵ_1) rises the rate of Nu_T/Nu_C in a system and consequently enhances heat as well as mass transportation and therefore ϵ_1 destabilizes the system, which are similar outcomes investigated by Keshri et al. (2019). Figures 11.4c and 11.7c demonstrate that with the rises in amplitude of gravity modulation (ϵ) then rate of Nu_T/Nu_C rises in a system and consequently enhance the heat as well as mass transportation and therefore ϵ destabilize the system. In Figures 11.4b and 11.7b, we describe the outcomes of the frequency of the MF-modulation on the Nu_T/Nu_C. It is examined that for minor Ω_1 an extra heat as well as mass transport is formed. As Ω_1 increases, Nu_T/Nu_C diminishes. Whenever Ω_1 rises from 10 to 80, Nu_T/Nu_C diminishes significantly, and accordingly, the magnetic-modulating frequency diminishes heat/mass transport. Additionally, the modulation over magneto-convection absolutely vanishes with a rise in the amplitude of the modulated frequencies. From now the influence of modulating frequency (Ω_1) is to stabilize a system. In Figures 11.4d and 11.7d, we found similar result for gravity modulation like in Figures 11.4b and 11.7b

In Figures 11.5a and 11.8a, we describe the comparison of MF/gravity modulation on the Nu_T/Nu_C rate with time (t). In comparison with MF-modulation, gravity-modulation has the highest rate of heat/mass transfer.

$$\left(Nu_T \,/\, Nu_C\right)^{\text{Gravity-Modulation}} > \left(Nu_T \,/\, Nu_C\right)^{\text{MF-Modulation}}$$

But near the beginning of advection/convection, heat as well as mass transport is initiated earlier in the case of MF-modulation as compared to gravity-modulation. In Figures 11.5b and 11.8b, we demonstrate the comparison of nanoliquid/R-E nanoliquid on the Nu_T/Nu_C rate with time (t). In comparison with R-E nanoliquid, we notice that the amount of heat as well as mass transport is maximal in the case of nanoliquids.

$$\left(Nu_T \,/\, Nu_C\right)^{\text{Nanoliquid}} > \left(Nu_T \,/\, Nu_C\right)^{\text{R-E Nanoliquid}}$$

Moreover, onset convection/advection is initiated earlier in the case of nanoliquid as compared to R-E nanoliquid. In Figures 11.5c and 11.8c, we depict the comparison of with/without modulation on the Nu_T/Nu_C amount with time (t). We notice that amount of heat as well as mass transport is maximal in the case of modulation as equated to without-modulation.

$$\left(Nu_T \,/\, Nu_C\right)^{\text{With-Modulation}} > \left(Nu_T \,/\, Nu_C\right)^{\text{Without-Modulation}}$$

Moreover, onset convection/advection is initiated earlier in the case of modulation as compared to without-modulation.

In Figures 11.5d and 11.8d, we compare the solutions for the Nu_T/Nu_C rate with time (t). By using the RKF45-Technique and the Mathematica NDSolve-Technique, we are working to confirm the accuracy of our results. We draw the conclusion that convergence of all of Mathematica NDSolve-Technique solutions for heat as well as mass transport has been confirmed using RKF45-Technique because we observe that the results are nearly the same in both approaches, that is:

$$\left(Nu_T \,/\, Nu_C\right)^{\text{NDSolve-Technique}} = \left(Nu_T \,/\, Nu_C\right)^{\text{RKF45-Technique}}$$

Bhadauria and Kiran (2014) report the same outcome as well.

11.11 CONCLUSIONS

The study of beginning of instability and heat as well as mass transport of Rivlin-Ericksen nanoliquid within H-S cell under magnetic-field/gravity modulation has been carried out. Following is a concise summary of the implications of Section 11.10:

- Kinematic-viscoelastic permeability (F), nanoliquid magnetic number (Q), and H-S-number (Hs) have a stabilizing character on a system.

- A system is unstable due to effects of concentration Rayleigh number R_n, modified diffusivity ratio N_A, and nanoliquid Lewis number L_e.
- Increasing F stabilizes the system by causing a reduction in mass as well as heat transport.
- Increasing Hs stabilizes the system by causing a reduction in mass as well as heat transport.
- As per the value of P_r increases, the amount of heat/mass transport rises whereas Q diminishes.
- Even if P_r destabilizes the system, Q makes it more stable.
- The impact of rising $\varepsilon_1/\varepsilon$ is to enhance the heat/mass transportation due to convection/advection.
- The influence of modulation initiates to vanish at appropriately huge values of Ω_1/Ω.
- As amplitude of Ω_1/Ω increases, heat as well as mass transportation in a system diminishes.
- The amount of heat and mass transported by the system decreases as Ω_1/Ω's magnitude increases.
- Rivlin-Ericksen nanoliquid is more stabilizing as compared to a normal nanoliquid.
- Gravity-modulation is more effective as compared to magnetic-field modulation as far as heat/mass transport is concerned.
- The heat and mass transfer happens in the conduction state at short time scales (t), while convection is used to evaluate it at longer time scales.

NOMENCLATURE

t	time
V_2	velocity of Rivlin-Ericksen (R-E) nanoliquids
C_2	volume fraction of the nanoparticles
$K_2\left(=\dfrac{b^2}{12}\right)$	permeability of R-E nanoliquids-flow
D_B	coefficient of Brownian diffusion
p_2	pressure
h	dimensional layer separation distance
D_T	coefficient of thermophoresis diffusion
$(\rho c)_f/(\rho c)_p$	heat-capacities
β_{T2}	coefficient of thermal expansion
μ_2	viscosity of R-E nanoliquids
μ	kinematic-viscoelasticity of R-E nanoliquids
ρ_{20}	density of base liquids
ρ_2	density of R-E nanoliquids
ρ_{2p}	density of nanoparticles

T_2	temperature of R-E nanoliquids
H_2	induced magnetic-field
g_2	gravity
g_0	mean-gravity
η	electrical resistivity
η	electrical resistivity
H_0	magnetic-field imposed
μ_e	magnetic permeability of R-E nanoliquids
e_z	unit vector in upward direction
ε	amplitude of gravity modulation
ε_1	amplitude of magnetic field (MF) modulation
Ω	frequency of gravity modulation
Ω_1	frequency of MF modulation
$\alpha_2\left(=\dfrac{\kappa}{(\rho c)_f}\right)$	thermal diffusivity of R-E nanoliquids
κ	effective thermal conductivity of R-E nanoliquids

Operator

$$\nabla_2^2 \equiv \frac{\partial^2}{\partial x^2} + \frac{\partial^2}{\partial y^2} + \frac{\partial^2}{\partial z^2}$$

Nondimensional parameters

R_a	Hele-Shaw(H-S) thermal Rayleigh number
R_{a0}	critical Hele-Shaw Rayleigh number
R_n	concentration Rayleigh number
P_r	Prandtl number
P_{rm}	magnetic Prandtl number
Hs	H-S number
L_e	Lewis number
N_A	modified diffusivity ratio
N_B	modified particle density increment
F	kinematic-viscoelasticity permeability
Q	magnetic number (Chandrasekhar number)

REFERENCES

Aniss, S., Souhar, M., & Brancher, J. P. (1995). Asymptotic study and weakly nonlinear analysis at the onset of Rayleigh–Bénard convection in Hele–Shaw cell. *Physics of Fluids*, 7(5), 926–934.

Aniss, S. D., Belhaq, M., & Souhar, M. (2001). Effects of a magnetic modulation on the stability of a magnetic liquid layer heated from above. *Journal of Heat Transfer*, 123(3), 428–433.

Awasthi, M. K., Asthana, R., & Agrawal, G. S. (2012). Viscous corrections for the viscouspotential flow analysis of magnetohydrodynamic Kelvin-Helmholtz instability with heatand mass transfer. *The European Physical Journal A*, 48, 1–10.

Awasthi, M. K. (2013a). Viscous corrections for the viscous potential flow analysis of Rayleigh–Taylor instability with heat and mass transfer. *Journal of Heat Transfer*, 135(7), 071701.

Awasthi, M. K. (2013b). Nonlinear analysis of Rayleigh–Taylor instability of cylindrical flow with heat and mass transfer. *Journal of Fluids Engineering*, 135(6), 061205.

Asthana, R., Awasthi, M. K., & Agrawal, G. S. (2014). Viscous potential flow analysis of kelvin–Helmholtz instability of a cylindrical flow with heat and mass transfer. *Heat Transfer—Asian Research*, 43(6), 489–503.

Awasthi, M. K. (2019). Rayleigh–Taylor Instability of swirling annular layer with masstransfer. *Journal of Fluids Engineering*, 141(7), 071202.

Buongiorno, J. (2006). Convective transport in nanofluids. *The ASME Journal of Heat Transfer*, 128, 240–250.

Buongiorno, J., Venerus, D. C., Prabhat, N., McKrell, T., Townsend, J., Christianson, R., & Zhou, S. Q. (2009). A benchmark study on the thermal conductivity of nanofluids. *Journal of Applied Physics*, 106(9), 094312–094314.

Bhadauria, B. S., Bhatia, P. K., & Debnath, L. (2005). Convection in Hele–Shaw cell with parametric excitation. *International Journal of Non-Linear Mechanics*, 40(4), 475–484.

Boulal, T., Aniss, S., Belhaq, M., & Azouani, A. (2008). Effect of quasi-periodic gravitational modulation on the convective instability in Hele-Shaw cell. *International Journal of Non-Linear Mechanics*, 43(9), 852–857.

Bhadauria, B. S., & Kumar, A. (2021). Throughflow and gravity modulation effect on thermal instability in a Hele-Shaw cell saturated by nanofluid. *Journal of Porous Media*, 24(6), 31–51.

Bhadauria, B. S., Rai, S. N., & Srivastava, A. K. (2022). Weakly nonlinear analysis of combined effect of g-jitter and thermal difference on a Rivlin-Ericksen nanofluid in Hele-Shaw cell. *In Conference Proceedings of Science and Technology*, 5, 106–114.

Benjamin, T. B., & Ursell, F. J. (1954). The stability of the plane free surface of a liquid in vertical periodic motion. *Proceedings of the Royal Society of London. Series A. Mathematical and Physical Sciences*, 225(1163), 505–515.

Bhadauria, B. S., Siddheshwar, P. G., & Suthar, O. P. (2012). Nonlinear thermal instability in a rotating viscous fluid layer under temperature/gravity modulation. *ASME Journal of Heat Transfer*, 134(10), 094312–094314.

Bhadauria, B. S., Hashim, I., & Siddheshwar, P. G. (2013). Study of heat transport in a porous medium under G-jitter and internal heating effects. *Transport in Porous Media*, 96, 21–37.

Bhadauria, B. S., & Kiran, P. (2014a). Weak nonlinear oscillatory convection in a viscoelastic fluid-saturated porous medium under gravity modulation. *Transport in Porous Media*, 104(3), 451–467.

Bhadauria, B. S. (2006). Time-periodic heating of Rayleigh–Benard convection in a vertical magnetic field. *Physica Scripta*, 73(3), 296.

Bhadauria, B. S. (2008). Combined effect of temperature modulation and magnetic field on the onset of convection in an electrically conducting-fluid-saturated porous medium. *Journal of Heat Transfer*, 130(5), 052601.

Bansal, S., & Chatterjee, D. (2015). Magneto-convective transport of nanofluid in a vertical lid-driven cavity including a heat-conducting rotating circular cylinder. *Numerical Heat Transfer, Part A: Applications*, 68(4), 411–431.

Bhadauria, B. S., & Kiran, P. (2014b). Weak nonlinear analysis of magneto-convection under magnetic field modulation. *Physica Scripta*, 89(9), 095209–095210.

Bhadauria, B. S., & Srivastava, A. (2022). Combined effect of internal heating and through-flow in a nanofluid saturated porous medium under local thermal non-equilibrium. *Journal of Porous Media*, 25(2), 75–95.

Choi, S. (1995). Enhancing Thermal Conductivity of Fluids with Nanoparticles, in *Development and Applications of Non-Newtonian Flows*, D.A. Signier and H.P. Wang, Eds., New York: ASME, 99–105.

Chand, R., & Rana, G. C. (2012). Thermal Instability of Rivlin–Ericksen elastico-viscous nanofluid saturated by a porous medium. *Journal of Fluids Engineering*, 134(12), 121203.

Chand, R., Rana, G. C., & Singh, K. (2015). Thermal instability in a Rivlin-Ericksen elastico-viscous nanofluid in a porous medium: A revised model. *International Journal of Nanoscience and Nanoengineering*, 2(1), 1–5.

Clever, R., Schubert, G., & Busse, F. H. (1993). Two-dimensional oscillatory convection in a gravitationally modulated fluid layer. *Journal of Fluid Mechanics*, 253, 663–680.

Chandrasekhar, S. (1961). *Hydrodynamic and Hydromagnetic Stability*. Oxford University Press, London.

Finlayson, B. (1970). Convective instability of ferromagnetic fluids. *Journal of Fluid Mechanics*, 40(4), 753–767.

Gresho, P. M., & Sani, R. L. (1970). The effects of gravity modulation on the stability of a heated fluid layer. *Journal of Fluid Mechanics*, 40(4), 783–806.

Gotoh, K., & Yamada, M. (1982). Thermal convection in a horizontal layer of magnetic fluids. *Journal of the Physical Society of Japan*, 51(9), 3042–3048.

Gupta, U., Ahuja, J., & Wanchoo, R. K. (2013). Magneto convection in a nanofluid layer. *International Journal of Heat and Mass Transfer*, 64, 1163–1171.

Hele-Shaw, H.S.J. (1898). Experiments on the nature of surface resistance of water and streamline motion under certain experimental conditions. *Transactions of the Royal Institution of Naval Architects*, 40, 21–46.

Hartline, B. K., & Lister, C. R. B. (1977). Thermal convection in a Hele-Shaw cell. *Journal of Fluid Mechanics*, 79(2), 379–389.

Ismail, & Bhadauria, B. S. (2022). Thermal Instability of Rivlin-Ericksen Elastico-Viscous Nanofluid Saturated by a Porous Medium with Rotation. *In International workshop of Mathematical Modelling, Applied Analysis and Computation*, Cham: Springer Nature Switzerland, 436–455.

Kuznetsov, A. V., & Nield, D. (2010). Thermal instability in a porous medium layer saturated by a nanofluid: Brinkman model. *Transport in Porous Media*, 81, 409–422.

Keshri, O. P., Kumar, A., & Gupta, V. K. (2019). Effect of internal heat source on magneto-stationary convection of couple stress fluid under magnetic field modulation. *Chinese Journal of Physics*, 57, 105–115.

Kumar, A., Bhadauria, B. S., Kumar, A., & Rai, S. N. (2023). Effect of different types of gravity modulation on the instability of micro-polar nano-fluid of finite extent in horizontal directions. *Chinese Journal of Physics*, https://doi.org/10.1016/j.cjph. 2023.04.015

Neild, D.A., & Bejan, A. (2017). *Convection in Porous Media*. Springer-Verlag, New York.

Nakagawa, Y. (1957). Experiments on the inhibition of thermal convection by a magnetic field. *Proceedings of the Royal Society of London. Series A. Mathematical and Physical Sciences*, 240(1220), 108–113.

Nakagawa, Y. (1959). Experiments on the instability of a layer of mercury heated from below and subject to the simultaneous action of a magnetic field and rotation. II. *Proceedings of the Royal Society of London. Series A. Mathematical and Physical Sciences*, 249(1256), 138–145.

Ozoe, H., & Maruo, E. (1987). Magnetic and gravitational natural convection of melted silicon-two-dimensional numerical computations for the rate of heat transfer: Heat transfer, combustion, power, thermophysical properties. *JSME International Journal*, 30(263), 774–784.

Rivlin, R. S., & Ericksen, J. L. (1955). Stress-deformation relaxations for isotropic materials. *Journal of Rational Mechanics and Analysis*, 4, 323–425.

Rana, G. C., & Kumar, S. (2010). Thermal instability of Rivlin-Ericksen elastico-viscous rotating fluid permeating with suspended particles under variable gravity field in porous medium. *Studia Geotechnica et Mechanica*, 32(4), 39–54.

Rana, G. C., & Thakur, R. C. (2012). Effect of suspended particles on thermal convection in Rivlin-Ericksen fluid in a Darcy-Brinkman porous medium. *Journal of Mechanical Engineering and Sciences*, 2, 162–171.

Rana, G. C., & Chand, R. (2015). Stability analysis of double-diffusive convection of Rivlin-Ericksen elastico-viscous nanofluid saturating a porous medium: A revised model. *Forschung im Ingenieurwesen*, 79(1–2), 87–95.

Rana, G. C., Chand, R., & Sharma, V. (2016). Thermal instability of a Rivlin-Ericksen nanofluid saturated by a Darcy-Brinkman porous medium: A more realistic model. *Engineering Transactions*, 64(3), 271–286.

Rai, S. N., Bhadauria, B. S., Kumar, A., & Singh, B. K. (2023a). Thermal instability in nanoliquid under four types of magnetic-field modulation within Hele-Shaw cell. *ASME Journal of Heat and Mass Transfer*, 145(7), 072501.

Rai, S. N., & Bhadauria, B. S. (2023). Thermal instability in electrically conducting nanoliquid filled in Hele–Shaw cell under 3-types of rotational-speed modulation with impact of through-flow and magnetic-field. *Chinese Journal of Physics*, https://doi.org/10.1016/j.cjph.2023.05.018

Rai, S. N., Bhadauria, B. S., Srivastava, A., & Kumar, A. (2023b). Thermal instability in Walter-B nanoliquid filled in Hele-Shaw cell under 3-types of magnetic-field modulation with through-flow. *Special Topics & Reviews in Porous Media: An International Journal*. https://doi.org/10.1615/SpecialTopicsRevPorousMedia.2023047492

Srivastava, L. P. (1971). Unsteady flow of Rivlin–Ericksen fluid with uniform distribution of dust particles through channels of different cross sections in the presence of time dependent pressure gradient. *Bulletin of the Technical University of Istanbul*, 194, 19.

Saini, S., & Sharma, Y. D. (2017). The effect of vertical throughflow in Rivlin-Ericksen elastico-viscous nanofluid in a non-Darcy porous medium. *Nanosystems: Physics, Chemistry, Mathematics*, 8(5), 606–612.

Souhar, K., & Aniss, S. (2012). Effect of Coriolis force on the thermosolutal convection threshold in a rotating annular Hele-Shaw cell. *Heat and Mass Transfer*, 48(1), 175–182.

Siddheshwar, P. G., & Pranesh, S. (2000). Effect of temperature/gravity modulation on the onset of magneto-convection in electrically conducting fluids with internal angular momentum. *Journal of Magnetism and Magnetic Materials*, 219(2), 153–162.

Siddheshwar, P. G., Bhadauria, B. S., Mishra, P., & Srivastava, A. K. (2012). Study of heat transport by stationary magneto-convection in a Newtonian liquid under temperature or gravity modulation using Ginzburg–Landau model. *International Journal of Non-Linear Mechanics*, 47(5), 418–425.

Sheikholeslami, M., Bani Sheykholeslami, F., Khoshhal, S., Mola-Abasia, H., Ganji, D. D., & Rokni, H. B. (2014). Effect of magnetic field on Cu–water nanofluid heat transfer using GMDH-type neural network. *Neural Computing and Applications*, 25, 171–178.

Thompson, W. B. (1951). CXLIII. Thermal convection in a magnetic field. *The London, Edinburgh, and Dublin Philosophical Magazine and Journal of Science*, 42(335), 1417–1432.

Uddin, Z., Asthana, R., Kumar Awasthi, M., & Gupta, S. (2017). Steady MHD flow of nano-fluids over a rotating porous disk in the presence of heat generation/absorption: anumerical study using PSO. *Journal of Applied Fluid Mechanics*, 10(3), 871–879.

Wooding, R. A. (1960). Instability of a viscous liquid of variable density in a vertical Hele-Shaw cell. *Journal of Fluid Mechanics*, 7(4), 501–515.

Wakif, A., Boulahia, Z., & Sehaqui, R. (2016). The effect of the rotation on the onset of convection in a Hele–Shaw cell saturated by a Newtonian nanofluid: A revised model. *Elixir Engineering is an Engineering*, 92, 38976–38985.

Yadav, D. (2019). The effect of pulsating throughflow on the onset of magneto convection in a layer of nanofluid confined within a Hele-Shaw cell. *Proceedings of the Institution of Mechanical Engineers, Part E: Journal of Process Mechanical Engineering*, 233(5), 1074–1085.

Yadav, D., Bhargava, R., & Agrawal, G. (2013). Thermal instability in a nanofluid layer with a vertical magnetic field. *Journal of Engineering Mathematics*, 80, 147–164.

Chapter 12

MHD flow of a micropolar hybrid nanofluid between two parallel disks

Reshu Gupta

Applied Science Cluster (Mathematics), UPES, Dehradun, India

12.1 INTRODUCTION

The fluids which have a pairing between the spin of every particle and the apparent velocity are known as micropolar fluids. These fluids contain microscopic-sized particles. Eringen (1964) was the founder of the theory of such fluids. This theory is modified by himself (ERINGEN, 1966). According to modified theory, micropolar fluids are a new sub-class of microfluids. Agarwal (2020, 2021) discussed the flow and heat and mass transfer of micropolar fluid in different models of the disks. A nanometer-sized particle is defined as a nanoparticle and a fluid which contains such nanoparticles is recognized as a nanofluid. These fluids are formed by the suspension of the nanometer-sized particles in the base fluid. Such fluids contribute as a catalyst to escalate the rate of heat transfer in fluids.

First, Choi (1995) used the term "nanofluid" in his article. A hybrid nanofluid is an updated form of nanofluids in which multiple nanoparticles are required to rectify the flaws of a single nanofluid. The theory of hybrid nanofluids was first carried out by Turuc et al. (R. TURCU, 8 C.E.). Waini et al. (2020) and Nabwey and Mahdy (2021) discussed hybrid nanofluid flow in different models of sheets. Faizal et al. (2020) examined the properties of heat transfer and the flow of micropolar hybrid nanofluid over a sheet. Bilal et al. (2021), Chu et al. (2023), and Gumber et al. (2022) studied the same fluid in the model of plates. The model of disks is very interesting among researchers. Agarwal (2022a, 2021) used the same model for different conditions. Gupta and Agrawal (2023) and Gupta et al. (2023) analyzed flow manner of the micropolar nanofluid and the Casson nanofluid between two parallel disks. Gupta (2022a, 2022b, 2023) applied semi analytical methods in their studies. Agarwal (2022b) used DTM to find heat transfer profile between two parallel plates. Gul et al. (2021), Waqas et al. (2021), and Reddy et al. (2021) used the nanofluid and hybrid nanofluid in the model of disks in their studies.

The chapter aims to examine the MHD flow of a micropolar hybrid nanofluid between two parallel disks. In this study, we take titanium dioxide (TiO_2) and copper (Cu) as hybrid nanoparticles, which are stayed in water as a base fluid. This combination forms a TiO$_2$-Cu/water hybrid nanofluid. DTM is applied to solve the obtained nonlinear ODEs. The influence of several physical parameters is discussed and pictured by a graph.

12.2 MATHEMATICAL FORMULATION

A cylindrical coordinate system is used by considering the MHD flow of a hybrid nanofluid with micropolar theory through two parallel and porous disks. Titanium dioxide (TiO$_2$) and copper (Cu) are considered as hybrid nanoparticles with water as a base fluid. The origin of the cylindrical system coincides with $r = 0$. Both disks are at a distance of 2ℓ apart. The lower and the upper disks rest at the plane $z = -\ell$ and $z = \ell$, respectively. The injection velocity of both disks is v . u and w are considered the velocity components along the r and z directions. A uniform magnetic field of intensity B_0 is induced as shown in Figure 12.1.

The governing equations of the flow in cylindrical form, suggested by Eringen (Eringen, 1964; Eringen, 1966) are

$$\frac{\partial u}{\partial r} + \frac{u}{r} + \frac{\partial w}{\partial z} = 0 \tag{12.1}$$

$$\left(\mu_{bnf} + \kappa\right)\left(\frac{\partial^2 u}{\partial r^2} + \frac{1}{r}\frac{\partial u}{\partial r} - \frac{u}{r^2} + \frac{\partial^2 u}{\partial z^2}\right) - \kappa\frac{\partial N}{\partial z} - \frac{\partial p}{\partial r}$$

$$- \sigma_{bnf}B_0^2 u = \rho_{bnf}\left(u\frac{\partial u}{\partial r} + w\frac{\partial w}{\partial z}\right) \tag{12.2}$$

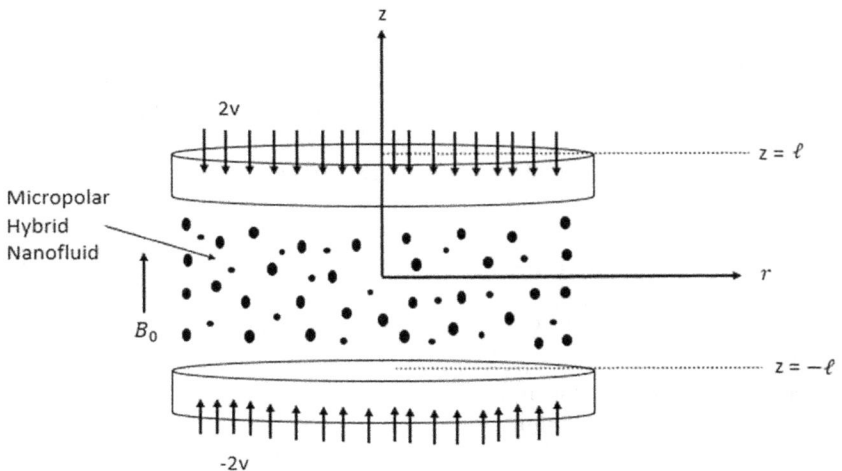

Figure 12.1 Flow model.

$$\left(\mu_{bnf}+\kappa\right)\left(\frac{\partial^2 w}{\partial r^2}+\frac{1}{r}\frac{\partial w}{\partial r}-\frac{u}{r^2}+\frac{\partial^2 w}{\partial z^2}\right)+\kappa\left(\frac{\partial N}{\partial r}+\frac{N}{r}\right)$$

$$-\frac{\partial p}{\partial z}=\rho_{bnf}\left(u\frac{\partial w}{\partial r}+w\frac{\partial w}{\partial z}\right) \tag{12.3}$$

$$\gamma_{bnf}\left(\frac{\partial^2 N}{\partial r^2}+\frac{1}{r}\frac{\partial N}{\partial r}-\frac{N}{r^2}+\frac{\partial^2 N}{\partial z^2}\right)+\kappa\left(\frac{\partial u}{\partial z}-\frac{\partial w}{\partial r}\right)$$

$$-2\kappa N=\rho_{bnf}j\left(u\frac{\partial w}{\partial r}+w\frac{\partial N}{\partial z}\right) \tag{12.4}$$

where μ_{bnf}, ρ_{bnf}, and σ_{bnf} are the effective viscosity, effective density, and electrical conductivity of the hybrid nanofluid, κ is the vertex viscosity, $\gamma_{bnf}\left[=\left(\mu_{bnf}+\dfrac{\kappa}{2}\right)j\right]$ is the spin gradient viscosity depending upon the hybrid nanofluid dynamic viscosity and j is the micro inertia.

The boundary conditions are

At $z=\ell : u=0, w=2v, N=0,$

At $z=-\ell : u=0, w=-2v, N=0.$ \tag{12.5}

According to Von Kármán (1921), profiles are

$$u=-rF'\left(z\right), w=2F\left(z\right), N=-rG\left(z\right) \tag{12.6}$$

On solving Equations (12.2), (12.3), and (12.4) and by using Equation (12.6), we have

$$\left(\mu_{bnf}+\kappa\right)F^{iv}-\kappa G''-2\rho_{bnf}FF'''-\sigma_{bnf}B_0^2 F''=0 \tag{12.7}$$

$$\gamma_{bnf}G''+\kappa F''-2\kappa G-\rho_{bnf}j\left(F'G-2FG'\right)=0 \tag{12.8}$$

Use the following parameters proposed by Valipour et al. (2015)

$$f\left(\eta\right)=\frac{F\left(z\right)}{v}, g\left(\eta\right)=\frac{\ell^2 G\left(z\right)}{v}, \eta=\frac{z}{\ell} \tag{12.9}$$

$$\left(\mu_{hnf} + \kappa\right) f^{iv} - \kappa g'' - 2\rho_{hnf}\ell^2 v f f'' - \sigma_{hnf} B_0^2 \ell^2 f'' = 0 \tag{12.10}$$

$$\gamma_{hnf} g'' + \kappa\ell^2\left(f'' - 2g\right) - \rho_{hnf} j v \ell\left(f'g - 2fg'\right) = 0 \tag{12.11}$$

μ_{hnf}, ρ_{hnf} (Waqas et al., 2021) and $\dot{A}_1, \dot{A}_2, \dot{A}_{31}$ and $\dot{A}_3$ are introduced as follows

$$\rho_{hnf} = \rho_f\left(1-\tilde{\phi}_c\right)\left(\left(1-\tilde{\phi}_t\right) + \tilde{\phi}_t\frac{\rho_t}{\rho_f}\right) + \tilde{\phi}_c\rho_{cu}$$

$$\mu_{hnf} = \frac{\mu_f}{\left(1-\tilde{\phi}_t\right)^{2.5}\left(1-\tilde{\phi}_c\right)^{2.5}}$$

$$\dot{A}_1 = \left(1-\tilde{\phi}_t\right)^{2.5}\left(1-\tilde{\phi}_c\right)^{2.5}, \dot{A}_2 = \left(1-\tilde{\phi}_c\right)\left(\left(1-\tilde{\phi}_t\right) + \tilde{\phi}_t\frac{\rho_t}{\rho_f}\right) + \tilde{\phi}_c\frac{\rho_{cu}}{\rho_f},$$

$$\dot{A}_{31} = \frac{\sigma_{t+}2\sigma_f - 2\tilde{\phi}_t\left(\sigma_f - \sigma_t\right)}{\sigma_{t+}2\sigma_f + \tilde{\phi}_t\left(\sigma_f - \sigma_t\right)}, \dot{A}_3 = \frac{\sigma_{cu+}2\dot{A}_{31}\sigma_f - 2\tilde{\phi}_c\left(\dot{A}_{31}\sigma_f - \sigma_{cu}\right)}{\sigma_{cu+}2\dot{A}_{31}\sigma_f + \tilde{\phi}_c\left(\dot{A}_{31}\sigma_f - \sigma_{cu}\right)} \tag{12.12}$$

Substituting the above values from Equation (12.12) in Equations (12.10) and (12.11), we get

$$\left(1 + \dot{A}_1 K_x\right) f^{iv} - \dot{A}_1 K_x g'' - 2R_a\dot{A}_1\dot{A}_2 f f''' - M_o\dot{A}_1\dot{A}_3\dot{A}_{31} f'' = 0 \tag{12.13}$$

$$\left(1 + \dot{A}_1\frac{K_x}{2}\right) g'' + \dot{A}_1 K_x\left(f'' - 2g\right) - \dot{A}_1\dot{A}_2 R_a\left(f'g - 2fg'\right) = 0 \tag{12.14}$$

where $K_x\left[= \frac{\kappa}{\mu_{hnf}}\right]$ is the material parameter, $M_o\left[= \frac{\sigma_f\ell^2 B^2}{\mu_f}\right]$ is the magnetic parameter, $R_a\left[= \frac{\rho_f v\ell}{\mu_f}\right]$ is the Reynolds number, ρ_t and ρ_{cu} are the density of nanoparticles and ρ_f is the density of base fluid. In the present study, TiO$_2$ and Cu are taken as hybrid nanoparticles and water as the base fluid, i.e., $\rho_t = 4250$ kg/m^3, $\rho_{cu} = 8933$ kg/m^3, and $\rho_f = 997.1$ kg/m^3.

The boundary conditions are

$$f(1) = 1, f'(1) = 0, g(1) = 0,$$

$$f(-1) = -1, f'(-1) = 0, g(-1) = 0. \tag{12.15}$$

From equation (12.13) and equation (12.14), it can be observed that $f'(\eta)$ is an even function while $g(\eta)$ is an odd function. The axial velocity is symmetric about $z = 0$. Therefore, we consider this model in the region $0 \le \eta \le 1$.
So transformed boundary conditions are

$$f(0) = 0, \; f''(0) = 0, \; g(0) = 0,$$

$$f(1) = 1, \; f'(1) = 0, \; g(1) = 0. \tag{12.16}$$

Shear stress on the upper and the lower disks are given by

$$\tau_w = -\left(\mu_{hnf} + \kappa\right)\left[\frac{\partial u}{\partial z}\right]_{z = \pm \ell} = \frac{rv}{\ell^2} \frac{\mu_f}{\left(1 - \tilde{\phi}_t\right)^{2.5}\left(1 - \tilde{\phi}_c\right)^{2.5}}\left[1 + K_x\right]f''(\pm 1).$$

12.3 METHODOLOGY OF DTM

k times differentiation of $f(x)$ in this method is given by

$$\hat{F}(k) = \frac{1}{k!}\left(\frac{d^k f(x)}{dx^k}\right)_{x = x_0} \tag{12.17}$$

The inverse transformation of $\hat{F}(k)$ is defined by

$$f(x) = \sum_{k=0}^{\infty} \hat{F}(k)(x - x_0)^k \tag{12.18}$$

$f(x)$ can be shown in the form of finite series and hence Equation (12.18) can be stated as

$$f(x) = \sum_{k=0}^{n} \hat{F}(k)(x - x_0)^k \tag{12.19}$$

From Equation (12.18) and Equation (12.19), we get

$$f(x) = \sum_{k=0}^{\infty}(x - x_0)^k \frac{1}{k!}\left(\frac{d^k f(x)}{dx^k}\right)_{x=x_0}$$

which represents the Taylor series form of $f(x)$ at $x = x_0$. Following theorems T_i ($i \le 10$) can be concluded from Equation (12.18) and Equation (12.19):

T_1: If $f(x) = g(x) \pm h(x)$ then $\hat{F}(k) = \hat{G}(k) \pm \hat{H}(k)$.

T_2: If $f(x) = cg(x)$ then $\hat{F}(k) = c\hat{G}(k)$, where c is a constant.

T_3: If $f(x) = \dfrac{d^m g(x)}{dx^m}$ then $\hat{F}(k) = \dfrac{(k+m)!}{k!}\hat{G}(k+m)$.

T_4: If $f(x) = g(x)h(x)$ then $\hat{F}(k) = \sum_{k_1=0}^{k}\hat{G}(k_1)\hat{H}(k-k_1)$

T_5: If $f(x) = e^{ix}$ then $\hat{F}(k) = \dfrac{x^k}{k!}$

T_6: If $f(x) = x^n$ then $\hat{F}(k) = \delta(k-n)$ where $\delta(k-n) = \begin{cases} 1, k = n \\ 0, k \ne n \end{cases}$

T_7: If $f(x) = g_1(x)g_2(x).........g_n(x)$ then

$$\hat{F}(k) = \sum_{k_{n-1}=0}^{k}\sum_{k_{n-2}=0}^{k_{n-1}}\sum_{k_1=0}^{k_2}\hat{G}_1(k_1)\hat{G}_2(k_2 - k_1).............\hat{G}_n(k - k_{n-1})$$

T_8: If $f(z) = (1 + z)^q$ then $\hat{F}(k) = \dfrac{q(q-1).........(q-k+1)}{k!}$

T_9: If $f(z) = \sin(\omega z + \alpha)$ then $\hat{F}(k) = \dfrac{\omega^k}{k!}\sin\left(\dfrac{\pi k}{2} + \alpha\right)$

T_{10}: If $f(z) = \cos(\omega z + \alpha)$ then $\hat{F}(k) = \dfrac{\omega^k}{k!}\cos\left(\dfrac{\pi k}{2} + \alpha\right)$

12.4 APPLICATION OF DTM

First, convert Equations (12.13) and (12.14) in their iterative form by using DTM theorems.

$$\left(1 + \dot{A}_1 K_x\right)(k+4)(k+3)(k+2)(k+1)\hat{F}(k+4) - \dot{A}_1 K_x(k+2)$$

$$(k+1)\hat{G}(k+2) - 2\dot{A}_1\dot{A}_2 R_a\sum_{n=0}^{k}(n+3)(n+2)(n+1)\hat{F}(n+3)$$

$$\hat{F}(k-n) - M_o\dot{A}_1\dot{A}_3\dot{A}_{31}(k+2)(k+1)\hat{F}(k+2) = 0 \qquad (12.20)$$

$$\left[1 + \frac{\dot{A}_1}{2} K_x\right](k+2)(k+1)\hat{G}(k+2) + \dot{A}_1 K_x((k+2)(k+1)\hat{F}(k+2) - 2$$

$$\hat{G}(k)) - \dot{A}_1 \dot{A}_2 R_a \sum_{n=0}^{k} (n+1)\hat{F}(n+1)\hat{G}(k-n) - 2\sum_{n=0}^{k} (n+1)$$

$$\hat{G}(n+1)\hat{F}(k-n) = 0$$

$$(12.21)$$

Solve the above equations under the following boundary conditions

$$\hat{F}(0) = 0, \hat{F}(1) = a, \hat{F}(2) = 0, \hat{F}(3) = b, \hat{G}(0) = 0, \hat{G}(1) = c.$$

The values of a, b, and c can be calculated with the help of Equation (12.15) and then we get the approximate solution of $f(\eta)$ and $g(\eta)$.

12.5 RESULTS AND DISCUSSION

In this section, flow and microrotation profile of TiO_2-Cu with water micropolar hybrid nanofluid between two parallel disks are investigated by using DTM. The effects of the different parameters like Reynolds number, magnetic field and material parameters, volume fraction parameter for TiO_2 and Cu on axial velocity $f(\eta)$, radial velocity $f'(\eta)$, and microrotation $g(\eta)$ are discussed. The densities of titanium, copper, and water are taken as $\rho_t = 4250$ kg/m^3, $\rho_{cu} = 8933$ kg/m^3, and $\rho_f = 997.1$ kg/m^3, respectively. It is worth noting that the axial velocity increases from the lower to the upper disk for all different parameters, and there is very little difference in numeric values for the different values of all parameters. The nature of the radial velocity and microrotation is presented in Figures 12.2–12.9. Figures 12.2 and 12.3 present the effect of the Reynolds number on the radial velocity and the microrotation for the fixed values $\phi_t = \phi_c = 0.01$, $K_x = 1$, $M_o = 0.5$ respectively. Figure 12.2 shows the decrease nature of velocity in the entire region. It can also be depicted that the velocity reduces up to the mid of both disks and increases thereafter by increasing the Reynolds number. Figure 12.3 has negative microrotation in the entire region. It has approximately symmetrical behavior. $g(\eta)$ declines near the lower part and raises near the upper part. Also, microrotation increases with the increased values of the Reynolds number. Figures 12.4 and 12.5 depict the nature of the profile for various values of material parameter while keeping other variables fixed like $\phi_t = \phi_c = 0.01$, $R_a = 0.5$, $M_o = 0.5$. Figure 12.4 shows that the velocity has decreased behavior from the lower to the upper disk. It can also be depicted that the velocity increased with rise in the K_x near the lower disk while decreased with rise in the K_x near the upper disk. Figure 12.5 shows

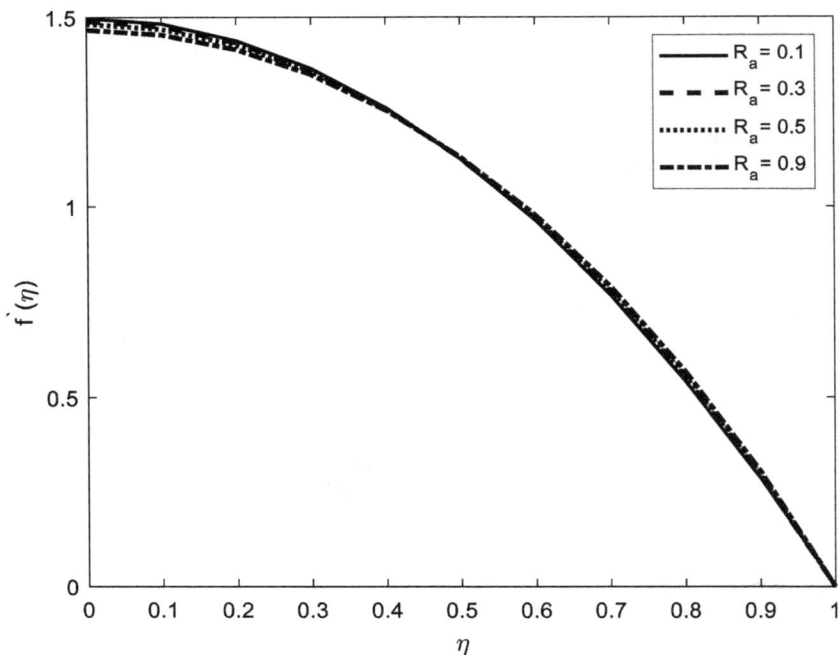

Figure 12.2 Impact of Reynolds number on $f'(\eta)$.

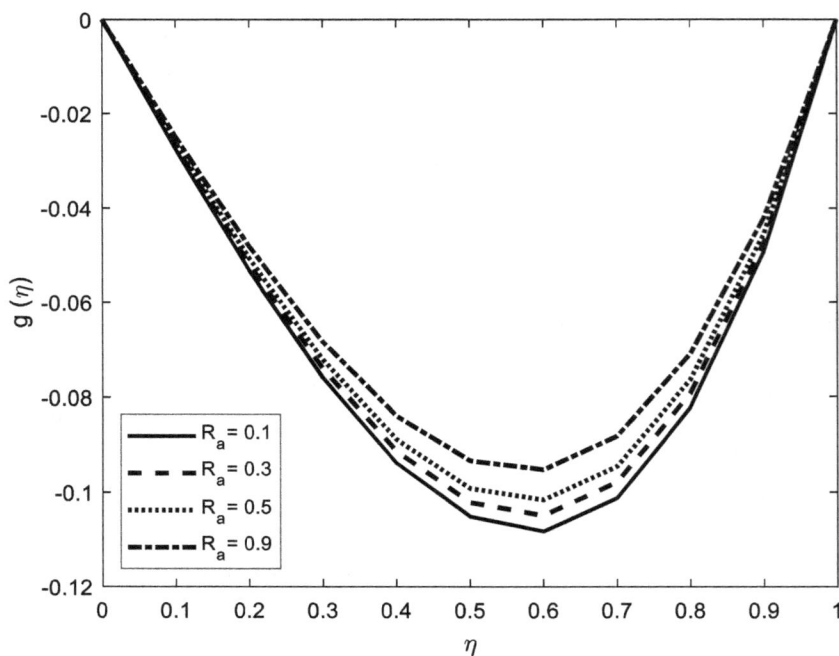

Figure 12.3 Impact of Reynolds number on $g(\eta)$.

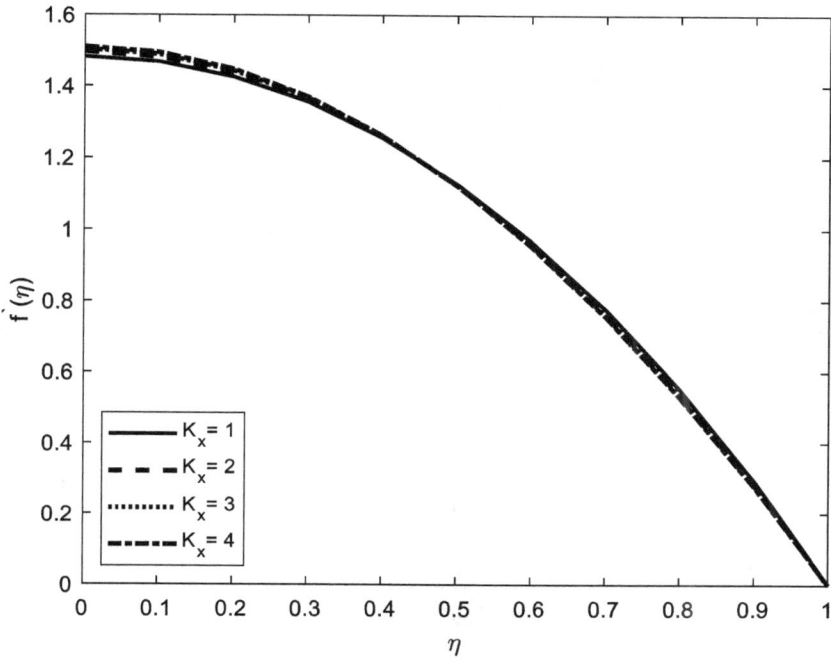

Figure 12.4 Impact of material parameter on $f'(\eta)$.

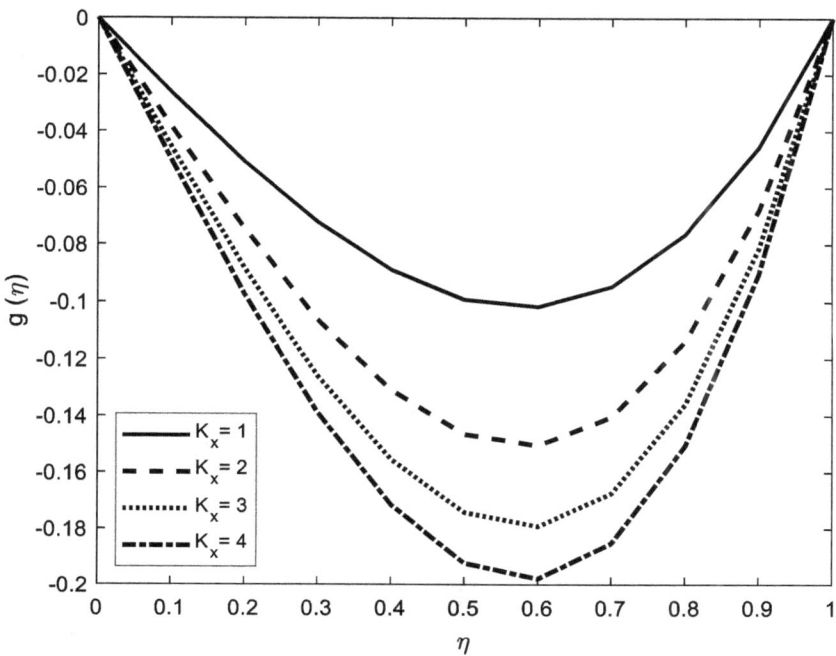

Figure 12.5 Impact of material parameter on $g(\eta)$.

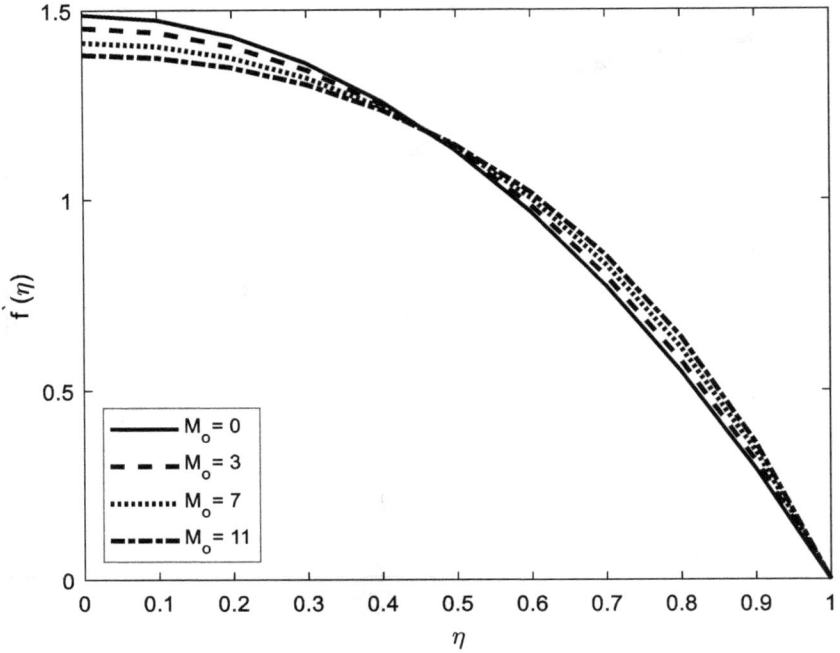

Figure 12.6 Impact of magnetic field parameter on $f'(\eta)$.

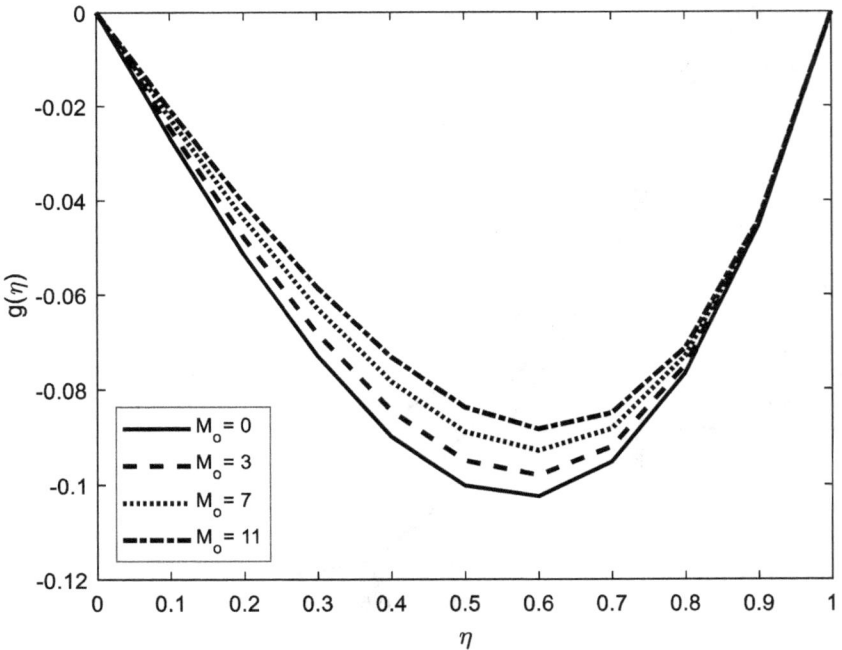

Figure 12.7 Impact of magnetic field parameter on $g(\eta)$.

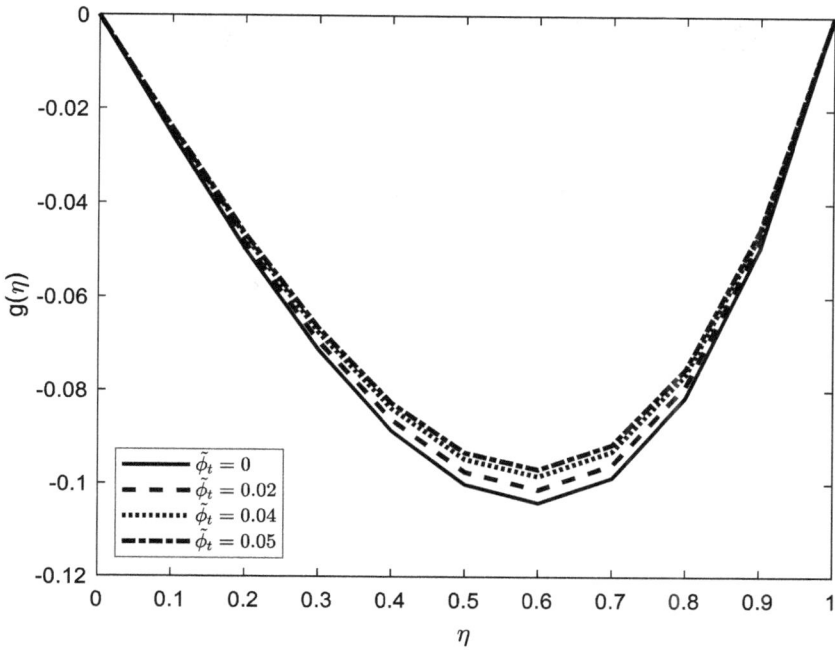

Figure 12.8 Impact of $\tilde{\phi}_t$ on $g(\eta)$.

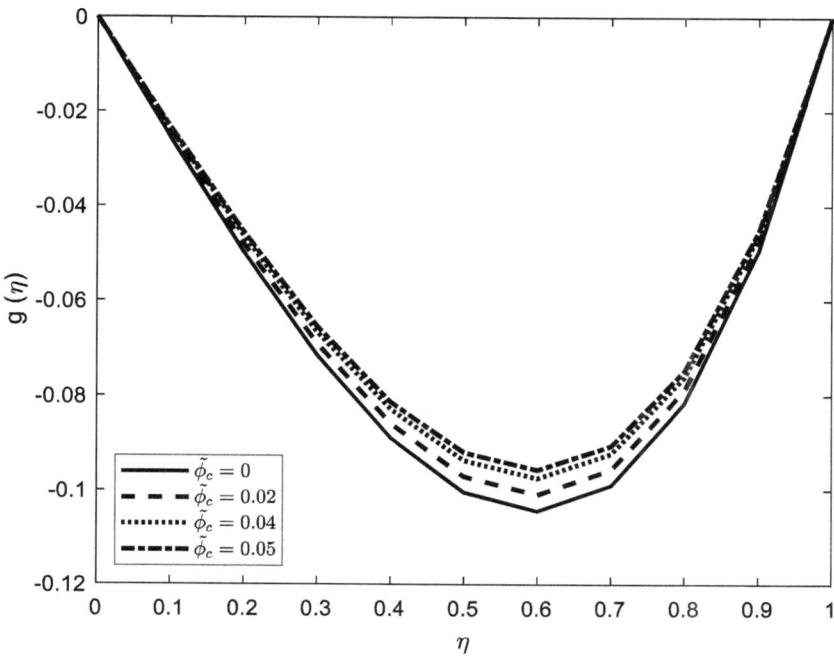

Figure 12.9 Impact of $\tilde{\phi}_c$ on $g(\eta)$.

Table 12.1 Values of $f'(\eta)$, $f(\eta)$ and $g(\eta)$ when $K_x = 0.5$, $R_a = 0.5$, $M_o = 0.5$, $\tilde{\phi}_t = \tilde{\phi}_c = 0.01$

η	$f'(\eta)$		$f(\eta)$		$g(\eta)$	
	DTM	NM	DTM	NM	DTM	NM
0.1	1.450898	1.456657	0.145950	0.146548	0.014108	0.015981
0.3	1.346481	1.349726	0.427461	0.428997	0.038687	0.043986
0.5	1.131300	1.130407	0.677174	0.678959	0.052546	0.060238
0.7	0.791988	0.787390	0.871724	0.872928	0.049174	0.057190
0.9	0.306827	0.302847	0.984269	0.984496	0.023011	0.027378

Table 12.2 Shear stress at the upper disk

R_a	K_x	M_o	$\tilde{\phi}_t$	$\tilde{\phi}_c$	$-f''(1)$	
					DTM	NM
0.1	0.5	0.5	0.01	0.01	3.081353	3.080474
0.5	0.5	0.5	0.01	0.01	3.254612	3.248212
0.1	1	0.5	0.01	0.01	3.022912	3.022352
0.1	5	0.5	0.01	0.01	2.839481	2.838901
0.1	0.5	1	0.01	0.01	3.146368	3.145265
0.1	0.5	3	0.01	0.01	3.391914	3.393168
0.1	0.5	0.5	0.03	0.01	3.079691	3.078811
0.1	0.5	0.5	0.05	0.01	3.077968	3.077090
0.1	0.5	0.5	0.01	0.03	3.088732	3.087723
0.1	0.5	0.5	0.01	0.05	3.095550	3.094418

the negative behavior of the microrotation. Also, it is going to more negative on increasing the values of the material parameter.

Figures 12.6 and 12.7 describe the character of the profiles influenced by the magnetic field when the other parameters values are $\phi_t = \phi_c = 0.01$, $R_a = 0.5$, $K_x = 1$. According to the Figure 12.6, the radial velocity decreases near the lower disk and increases near the upper disk on increasing the intensity of the magnetic field. Figure 12.7 describes the less negative nature of the microrotation on rising the values of the magnetic field. Its nature is not specific near the upper disk. The nature of the microrotation profile is discussed in Figures 12.8 and 12.9 for the various values of the volume fraction of nanoparticles for constant values of $M_o = 4$, $R_a = 0.1$, $K_x = 1$. The same behavior can be seen by both figures.

12.6 CONCLUSIONS

The MHD flow of a micropolar hybrid nanofluid between two porous disks is discussed. The obtained higher-order nonlinear ODEs are evaluated by using DTM. The obtained outcomes are compared with the numerical method results which verify the exactness and validity of the DTM. The author makes the following conclusions:

- The nature of the radial velocity is decreasing in the entire gap length for all parameters.
- The nature of the microrotation profile is less negative on increasing volume fraction of TiO_2 and Cu.
- The comparison of numeric values with literature shows that DTM is a better way for evaluating nonlinear ODEs.

REFERENCES

Agarwal, R. (2020). *Analytical Study of Micropolar Fluid Flow between Two Porous Disks PJAEE, 17 (12) (2020) Analytical Study of Micropolar Fluid Flow between Two Porous Disks.*

Agarwal, R. (2021). Heat and mass transfer in electrically conducting micropolar fluid flow between two stretchable disks. *Materials Today: Proceedings, 46,* 10227–10238. https://doi.org/10.1016/j.matpr.2020.11.614

Agarwal, R. (2022a). An analytical study of non-Newtonian visco-inelastic fluid flow between two stretchable rotating disks. In *Palestine Journal of Mathematics* (Vol. 11).

Agarwal, R. (2022b). Squeezing MHD flow along with heat transfer between parallel plates by using the differential transform method. *Journal of Nano- and Electronic Physics, 14*(6). https://doi.org/10.21272/jnep.14(6).06010

Agarwal, R., & Kumar Mishra, P. (2021). Analytical solution of the MHD forced flow and heat transfer of a non-Newtonian visco-inelastic fluid between two infinite rotating disks. *Materials Today: Proceedings, 46,* 10153–10163. https://doi.org/10.1016/j.matpr.2020.10.632

Ahmad Faizal, N. F., Md Ariffin, N., Rahim, Y. F., Hafidzuddin, M. E. H., & Wahi, N. (2020). MHD and slip effect in micropolar hybrid nanofluid and heat transfer over a stretching sheet with thermal radiation and non-uniform heat source/sink. *CFD Letters, 12*(11), 121–130. https://doi.org/10.37934/cfdl.12.11.121130

Bilal, M., Gul, T., Alsubie, A., & Ali, I. (2021). Axisymmetric hybrid nanofluid flow with heat and mass transfer amongst the two gyrating plates. *ZAMM - Journal of Applied Mathematics and Mechanics / Zeitschrift Für Angewandte Mathematik Und Mechanik, 101*(11). https://doi.org/10.1002/zamm.202000146

Choi, S. U. S. (1995). Enhancing thermal conductivity of fluids with nanoparticles. *1995 International Mechanical Engineering Congress and Exhibition, San Francisco, CA (United States), 12-17 Nov 1995; Other Information: PBD: Oct 1995.*

Chu, Y., Bashir, S., Ramzan, M., & Malik, M. Y. (2023). Model-based comparative study of magnetohydrodynamics unsteady hybrid nanofluid flow between two infinite parallel plates with particle shape effects. *Mathematical Methods in the Applied Sciences, 46*(10), 11568–11582. https://doi.org/10.1002/mma.8234

Eringen, A. C. (1964). Simple microfluids. *International Journal of Engineering Science, 2*(2), 205–217. https://doi.org/10.1016/0020-7225(64)90005-9

Eringen, A. C. (1966). Theory of micropolar fluids. *Journal of Mathematics and Mechanics, 16*(1), 1–18.

Gul, T., Kashifullah Bilal, M., Alghamdi, W., Asjad, M. I., & Abdeljawad, T. (2021). Hybrid nanofluid flow within the conical gap between the cone and the surface of a rotating disk. *Scientific Reports, 11*(1), 1180. https://doi.org/10.1038/s41598-020-80750-y

Gumber, P., Yaseen, M., Rawat, S. K., & Kumar, M. (2022). Heat transfer in micropolar hybrid nanofluid flow past a vertical plate in the presence of thermal radiation and suction/injection effects. *Partial Differential Equations in Applied Mathematics, 5*, 100240. https://doi.org/10.1016/j.padiff.2021.100240

Gupta, R. (2022a). Comparative study of micropolar fluid flow between two disks. *Journal of Harbin Institute of Technology, 54*. https://doi.org/10.11720/JHIT.54112022.02

Gupta, R. (2022b). Homotopy perturbation method for the MHD second-order fluid flow through a channel with permeable sides. *Journal of Harbin Institute of Technology, 54*. https://doi.org/10.11720/JHIT.54112022.06

Gupta, R., & Agrawal, D. (2023). Flow analysis of a micropolar nanofluid between two parallel disks in the presence of a magnetic field. *Journal of Nanofluids, 12*(5), 1320–1326. https://doi.org/10.1166/jon.2023.2021

Gupta, R., Selvam, J., Vajravelu, A., & Nagapan, S. (2023). Analysis of a squeezing flow of a Casson nanofluid between two parallel disks in the presence of a variable magnetic field. *Symmetry, 15*(1). https://doi.org/10.3390/sym15010120

Kármán, T. V. (1921). Über laminare und turbulente Reibung. *ZAMM - Journal of Applied Mathematics and Mechanics / Zeitschrift Für Angewandte Mathematik Und Mechanik, 1*(4), 233–252. https://doi.org/10.1002/zamm.19210010401

Nabwey, H. A., & Mahdy, A. (2021). Transient flow of micropolar dusty hybrid nanofluid loaded with Fe3O4-Ag nanoparticles through a porous stretching sheet. *Results in Physics, 21*, 103777. https://doi.org/10.1016/j.rinp.2020.103777

Turcu, R., Darabont, A. L., Nan, A. N., Macovei, D., Bica, D., & Biro, L. P. (n.d.). New polypyrrole-multiwall carbon nanotubes hybrid materials. *Journal of Optoelectronics and Advanced Materials, 8*(2), 643–647.

Reddy, M. G., Kumar, N., Prasannakumara, B. C., Rudraswamy, N. G., & Ganesh Kumar, K. (2021). Magnetohydrodynamic flow and heat transfer of a hybrid nanofluid over a rotating disk by considering Arrhenius energy. *Communications in Theoretical Physics, 73*(4), 045002. https://doi.org/10.1088/1572-9494/abdaa5

Gupta, R. (2023). Flow of a second-order fluid due to disk rotation. In CRC Press (Ed.), *Advances in Mathematical and Computational Modeling of Engineering Systems* (pp. 315–333).

Valipour, P., Ghasemi, S. E., & Vatani, M. (2015). Theoretical investigation of micropolar fluid flow between two porous disks. *Journal of Central South University, 22*(7), 2825–2832. https://doi.org/10.1007/s11771-015-2814-1

Waini, I., Ishak, A., & Pop, I. (2020). Hybrid nanofluid flow induced by an exponentially shrinking sheet. *Chinese Journal of Physics, 68*, 468–482. https://doi.org/10.1016/j.cjph.2019.12.015

Waqas, H., Farooq, U., Naseem, R., Hussain, S., & Alghamdi, M. (2021). Impact of MHD radiative flow of hybrid nanofluid over a rotating disk. *Case Studies in Thermal Engineering, 26*, 101015. https://doi.org/10.1016/j.csite.2021.101015

Chapter 13

Modeling Rayleigh-Taylor instability in nanofluid layers

Mukesh Kumar Awasthi

Babasaheb Bhimarao Ambedkar University, Lucknow, India

Ashwani Kumar

Technical Education Department Uttar Pradesh Kanpur, India

Nitesh Dutt

COER University, Roorkee, Uttarakhand, India

13.1 INTRODUCTION

RT instability is a well-known hydrodynamic phenomenon that occurs at the boundary between two fluids of diverse densities when subjected to a gravitational field. This instability has been extensively studied in various natural and engineering systems, from astrophysical environments like supernovae and accretion disks to terrestrial applications such as oceanography and inertial confinement fusion.

The phenomenon of RT instability has been extensively studied by various researchers. Sharp [1] provided a comprehensive and straightforward explanation for its occurrence. Jacobs and Catton [2] tackled the three-dimensional problem of RT instability, considering a nonviscous liquid overlying a less dense gas. Meanwhile, Kull [3] developed the fundamental theory for Rayleigh-Taylor instability in accelerated fluids and also explored bubble dynamics and nonlinear perturbations in both two and three dimensions. In a related study, Milkaelian [4] focused on the RM and RT instabilities, providing hybrid, approximate, and exact solutions.

When dealing with viscous fluids, the irrotational flow theory can also be useful to investigate RT instability. Joseph et al. [5] utilized this theory to examine the interface between two viscous fluids with different viscosities. On the other hand, Asthana et al. [6] formulated the nonlinear perturbations and instability problem in cylindrical configuration, considering the irrotational flow theory of viscous fluids and demonstrating the stabilizing influence of viscosity. Awasthi et al. [7] analyzed RT instability using nonlinear theory. The irrotational theory of viscous fluids has been applied by Awasthi [8] to study the rotational impact on the RT instability in cylindrical geometry. The RT instability in cylindrical geometry was examined by Awasthi [9] using nonlinear analysis.

DOI: 10.1201/9781003465171-13

249

In recent years, the investigation of RT instability has expanded to include the effects of nanoparticles dispersed within the fluids, giving rise to a new and intriguing area of research known as "Rayleigh-Taylor instability in nanofluids."

Nanofluids, which consist of a base fluid (liquid or gas) with suspended nanoparticles at the nanoscale, have emerged as a promising class of advanced materials with exceptional thermal and transport properties. The incorporation of nanoparticles into the fluid alters its thermophysical characteristics, leading to improved heat transfer capabilities and potential applications in energy, electronics cooling, and nanomedicine, among others. The behavior of nanofluids at fluid interfaces and under external perturbations has drawn considerable interest, as it introduces novel dynamics and can significantly influence their overall performance. Awasthi and his coauthors [10–12] used potential flow theory to analyze the interface between viscous and nanofluid. Moatimid and Hassan [13] delved into the convective linear stability of Walter's type viscoelastic nanofluid within a vertical layer. Additionally, Moatimid and Gaber [14] conducted a linear stability of two nanofluid layers when fluid is conducting. Exploring the temporal instability, Moatimid et al. [15] incorporated Marangoni effects into the analysis, utilizing the potential flow theory of viscous fluids.

Several research studies have already explored the behavior of nanofluids subjected to Rayleigh-Taylor instability [17], investigating both nonlinear and linear regimes of the instability growth. Theoretical models, numerical simulations, and experimental studies [16] have been explored to analyze the impact of nanoparticle properties (e.g., size, shape, concentration) and fluid characteristics (e.g., density, viscosity) on the stability and dynamics of the nanofluid interface.

This chapter aims to present a comprehensive review of the existing literature on RT instability in nanofluids, synthesizing the current state of knowledge and identifying the key factors influencing the instability growth and evolution. Additionally, we will discuss the potential applications and practical implications of the findings, shedding light on the opportunities and challenges in utilizing nanofluids in buoyancy-driven systems.

In this context, the study of RT instability in nanofluids becomes paramount due to its relevance in various practical scenarios and fundamental implications in fluid mechanics. Understanding the interplay between buoyancy-driven instabilities and nanoparticle dispersion can provide insights into designing advanced nanofluid-based systems with optimized thermal and transport properties.

This chapter studies the RT instability occurring at the interface between nanofluid and a viscous fluid. The analysis employs the irrotational flow theory of viscous fluids to examine the linear perturbed equations, where the perturbed velocity is assumed to be irrotational, and both fluids have nonzero viscosity. The consideration of viscosity is also incorporated into the normal stress balance equation. By deriving the dispersion relationship, the authors obtain a two-degree polynomial dependent on density, viscosity, surface tension, and other relevant factors. The study establishes the neutral stability criterion, and the stability of the interface is further discussed through a variety of plots.

13.2 MODELING

Consider a stratified configuration involving a layer of nanofluid atop a viscous incompressible fluid. The two fluids have distinct properties, with the nanofluid characterized by density ρ_n, and viscosity μ_n. The boundary between the nanofluid and the viscous fluid is denoted by $z = 0$, as depicted in Figure 13.1. The nanofluid layer is confined between $z = h_n$, while the viscous fluid is confined between $z = -h_f$.

The governing equations for the nanofluid phase are expressed as follows:

$$\left.\begin{aligned} \nabla \cdot q_n &= 0 \\ \rho_n\left(\frac{\partial q_n}{\partial t} + (q_n \cdot \nabla)q_n\right) &= -\nabla p_n + \rho_n g + \mu_n \nabla^2 q_n \end{aligned}\right\} \tag{13.1}$$

Here, q_n represents the velocity in the nanofluid phase, and p_n corresponds to the pressure in nanofluid layer.

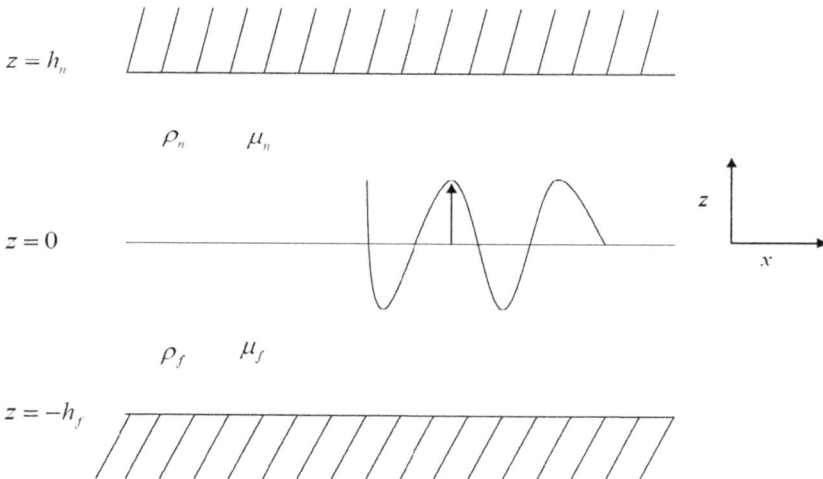

Figure 13.1 Schematic diagram of the problem.

The viscosity of nanofluids is dependent on various factors such as the volume fraction of nanoparticles, viscosity of the base fluid, the nanolayer of fluid particles surrounding the nanoparticles, and the shape of the nano-sized metal particles. The viscosity of the nanofluid can be represented as:

$$\mu_n = \mu_{eff}\left(1 - \frac{\phi_{ag}}{\phi_m}\right)^{-[\delta]\phi_m} \tag{13.2}$$

Here, μ_{eff} is referred to as apparent viscosity. The shape parameter δ accounts for the rotating behavior of nanoparticles about their axes, considering them as prolate spheroids. The maximum possible volume fraction for prolate spheroids is represented by ϕ_m, which falls between 0.68 and 0.75 based on experimental and simulation data. In this study, a value of $\phi_m = 0.74$ is considered. The aggregate volume fraction, ϕ_{ag}, is given by:

$$\phi_{ag} = \phi_{mod}\left(\frac{r_a}{r}\right)^{3-d} \tag{13.3}$$

where ϕ_{mod} is the equivalent volume fraction computed using the formula:

$$\phi_{mod} = \phi\left(1 + \frac{\gamma}{a}\right)\left(1 + \frac{\gamma}{b}\right)^2 \tag{13.4}$$

Here, γ denotes the thickness of the interfacial layer, and a and b are the lengths of the semi-major and semi-minor axes of the nanoparticles, respectively. The fractal index d indicates the extent of changes in the packing fraction from the center to the edge of aggregates.

The density of the viscoelastic nanofluid, ρ_n, can be expressed as a combination of the densities of the metal particle (ρ_p) and the base fluid (ρ_b):

$$\rho_n = \phi\rho_p + (1 - \phi)\rho_b \tag{13.5}$$

Moving on to the governing equations for the viscous incompressible fluid, denoted by q_f, they are given by:

$$\begin{rcases} \nabla \cdot q_f = 0 \\ \rho_f\left(\frac{\partial q_f}{\partial t} + (q_f \cdot \nabla)q_f\right) = -\nabla p_f + \rho_f g + \mu_f \nabla^2 q_f \end{rcases} \tag{13.6}$$

13.3 STABILITY ANALYSIS

13.3.1 Basic state

In the symmetry state, both fluids are not moving, resulting in $q_n = 0$ and $q_f = 0$. Additionally, the boundary is flat in this state, leading to constant pressures as $p_n = 0 = p_f$.

13.3.2 Perturbed state

An extremely small perturbation is applied to the symmetry state, causing the interface to be displaced to $z = \Gamma(x, t)$. In the perturbed state, the velocities and pressures become $q_n = 0 + q'_n$, $q_f = 0 + q'_f$ and $p_n + p'_n$, $p_f + p'_f$, respectively.

The linearized equations can be written as follows:

$$\left. \begin{array}{l} \nabla \cdot q'_n = 0 \\[2mm] \rho_n \dfrac{\partial q'_n}{\partial t} = -\nabla p'_n + \rho_n g + \mu_n \nabla^2 q'_n \end{array} \right\} \tag{13.7}$$

$$\left. \begin{array}{l} \nabla \cdot q'_f = 0 \\[2mm] \rho_f \dfrac{\partial q'_f}{\partial t} = -\nabla p'_f + \rho_f g + \mu_v \nabla^2 q'_f \end{array} \right\} \tag{13.8}$$

Assuming the perturbed flow is irrotational, we have $q'_n = \nabla \varphi_n$; $q'_f = \nabla \varphi_f$, leading to continuity equations from (13.7) and (13.8):

$$\nabla^2 \varphi_n = 0; \nabla^2 \varphi_f = 0 \tag{13.9}$$

13.3.3 Interfacial and boundary conditions

The condition at the interface is given as

$$\left. \begin{array}{l} \dfrac{\partial \phi_n}{\partial z} = \dfrac{\partial \Gamma}{\partial t} \\[3mm] \dfrac{\partial \phi_f}{\partial z} = \dfrac{\partial \Gamma}{\partial t} \end{array} \right\} \quad \text{at} \quad z = 0 \tag{13.10}$$

Furthermore, there is no flow across the boundary, resulting in:

$$\left. \begin{array}{l} \dfrac{\partial \phi_n}{\partial z} = 0 \quad \text{at} \quad z = h_n \\[3mm] \dfrac{\partial \phi_f}{\partial z} = 0 \quad \text{at} \quad z = -h_f \end{array} \right\} \tag{13.11}$$

The interfacial stress balance at the interface can be mathematically represented as

$$p_n - p_f - 2\mu_n \frac{\partial^2 \varphi_n}{\partial z^2} + 2\mu_f \frac{\partial^2 \varphi_f}{\partial z^2} = \sigma \frac{\partial^2 \Gamma}{\partial x^2} \tag{13.12}$$

13.4 DISPERSION RELATIONSHIP

To examine the stability, we utilize the well-established normal mode process. The interface elevation has the form $\Gamma(x, t) = \Gamma_0 Exp(-i\varsigma t + ikx)$, and remaining parameters $T(x, z, t)$ are represented as $T(x, z, t) = T(z)Exp(-i\varsigma t + ikx)$.

The functions φ_n, φ_f that satisfy conditions (13.10) and (13.11) are as follows:

$$\phi_n = \frac{i\varsigma}{k} \frac{\cosh\left(k\left(z - h_n\right)\right)}{\sinh\left(kh_n\right)} \Gamma_0 \exp\left(-i\varsigma t + ikx\right) \tag{13.13}$$

$$\phi_f = -\frac{i\varsigma}{k} \frac{\cosh\left(k\left(z + h_f\right)\right)}{\sinh\left(kh_f\right)} \Gamma_0 \exp\left(-i\varsigma t + ikx\right) \tag{13.14}$$

The linear form of equation (13.12) is given as

$$\left[\rho_n \left(\frac{\partial \varphi_n}{\partial t} + g\Gamma\right) + 2\mu_{eff}(1 - \frac{\phi_{ag}}{\phi_m})^{-[\delta]\phi_m} \frac{\partial^2 \varphi_n}{\partial z^2}\right]$$
$$-\left[\rho_f \left(\frac{\partial \varphi_f}{\partial t} + g\Gamma\right) + 2\mu_f \frac{\partial^2 \varphi_f}{\partial z^2}\right] = -\sigma \frac{\partial^2 \Gamma}{\partial x^2} \tag{13.15}$$

By using equations (13.13) and (13.14) in (13.15), we obtain the dispersion relationship as:

$$b_0\varsigma^2 + ib_1\varsigma + b_0 = 0 \tag{13.16}$$

$$b_0 = \rho_n \coth\left(kh_n\right) + \rho_f \coth\left(kh_f\right)$$
$$b_1 = 2k^2 \left(\mu_{eff}\left(1 - \frac{\phi_{ag}}{\phi_m}\right)^{-[\delta]\phi_m} \coth\left(kh_n\right) + \mu_f \coth\left(kh_f\right)\right)$$
$$b_2 = \left(\rho_n - \rho_f\right)gk - \sigma k^3$$

To find the real and imaginary values of equation (13.16) as $\varsigma = \varsigma_r + i\varsigma_i$, we get:

$$b_0\left(\varsigma_r^2 - \varsigma_i^2\right) - b_1\varsigma_i + b_2 = 0 \tag{13.17}$$

$$b_0\left(2\varsigma_r\varsigma_i\right) + b_1\varsigma_r = 0 \Rightarrow \varsigma_r = 0 \tag{13.18}$$

Therefore, equation (13.16) takes the form as

$$b_0\varsigma_i^2 + b_1\varsigma_i - b_2 = 0 \tag{13.19}$$

In case of neutral stability, $\varsigma_i = 0$, equation (13.19) gives as

$$b_2 = 0 \Rightarrow \left(\rho_n - \rho_f\right)gk - \sigma k^3 = 0 \tag{13.20}$$

Hence,

$$k = \sqrt{\frac{\left(\rho_n - \rho_f\right)g}{\sigma}}.$$

13.5 NONDIMENSIONAL FORM

If V is the assumed velocity and $H = h_n + h_f$ is the length, the equation (13.19) has the form

$$\left[\coth\left(\hat{k}\hat{h}_n\right) + \rho\coth\left(\hat{k}\hat{h}_f\right)\right]N^2$$
$$+ \left[2 \times \frac{\hat{k}^2}{\mathrm{Re}_n}\left(\left(1 - \frac{\phi_{ag}}{\phi_m}\right)^{-[\delta]\phi_m}\coth\left(\hat{k}\hat{h}_n\right) + \mu\coth\left(\hat{k}\hat{h}_f\right)\right)\right]$$
$$N^2 + (1-\rho)\hat{k}Fr + \frac{\hat{k}^3}{We} = 0 \tag{13.21}$$

where

$$\mathrm{Re}_n = \frac{\rho_n V H}{\mu_{eff}}, \quad We = \frac{\rho_n V^2 H}{\sigma}, \quad \rho = \frac{\rho_f}{\rho_n}, \quad \mu = \frac{\mu_f}{\mu_{eff}}, \quad \hat{h}_n = \frac{h_n}{H},$$
$$\hat{h}_f = \frac{h_f}{H}, \quad N = \frac{\varsigma_r H}{V}, \quad Fr = \frac{gH}{V^2}$$

Here Fr refers to the Froude number, Re_n denotes Reynolds number, and We represents the Weber number.

13.6 RESULTS AND DISCUSSION

In this section, equation (13.21) numerically solved for taking water-alumina nanofluid in the upper region while the lower region contains water. The physical variables to compute the perturbations growth from equation (13.21) is given as follows:

$$\rho_f = 1.2, \mu_f = 0.000018, \rho_{H_2O} = 1000, \mu_{H_2O} = 0.001$$

$$\sigma_{water\&air} = 0.0473, a = b = 15^*10e - 09, g = 9.8$$

The impact of Reynolds number on the growth of perturbations is examined in Figure 13.2. It is noted that the perturbations growth increases on increasing the Reynolds number and this trend confirms that the Reynolds number is destabilizing the nanofluid-viscous fluid interface. As density of the nanofluid contributes directly to the Reynolds number, the density of the nanofluid is destabilizing the interface. At the same time, the Reynolds number

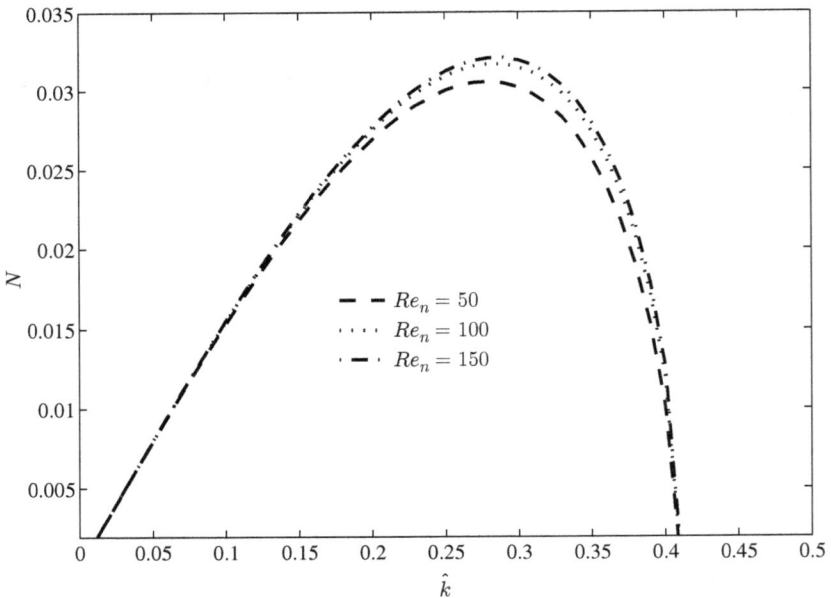

Figure 13.2 Effect of Reynolds number.

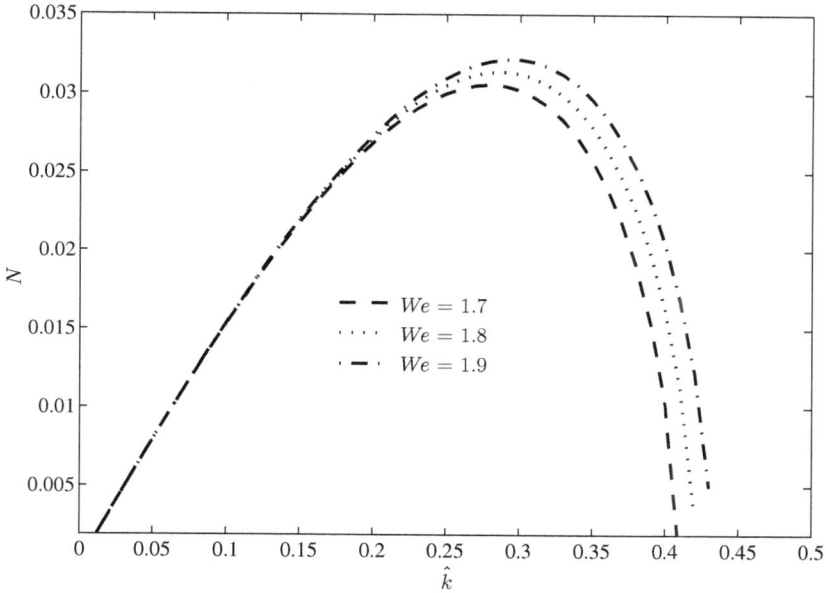

Figure 13.3 Effect of Weber number.

decreases with an increase in effective viscosity of nanofluid. Therefore, the nanofluid viscosity has a stabilizing impact.

Figure 13.3 shows the variation of Weber number on the growth of perturbations. This figure clearly indicates that the perturbations grows with the Weber number, showing the destabilizing impact. As the Weber number increases with the increasing nanofluid density, the nanofluid density destabilizes the interface. It is also observed that the surface tension has a stabilizing role as the surface tension varies inversely with the Weber number.

The Froude number shows the impact of gravitational acceleration. The perturbation growth variation with the Froude number is displayed in Figure 13.4. As the Froude number increases, the perturbations grow faster and showing that the Froude number has a destabilizing role. Therefore, gravitational acceleration plays a destabilizing role.

The role of the density ratio of the viscous fluid to the nanofluid is explained in Figure 13.5. It is found the perturbation growth decreases with an increase in the density ratio confirming that the density ratio of fluids has a stabilizing role. As the density ratio varies directly with the viscous fluid density, the viscous fluid density has a stabilizing impact. At the same time, the nanofluid density destabilizes the interface.

Figure 13.6 illustrates the growth rate concerning varying upper fluid fractions. Elevating the upper fluid fraction leads to an augmentation in the

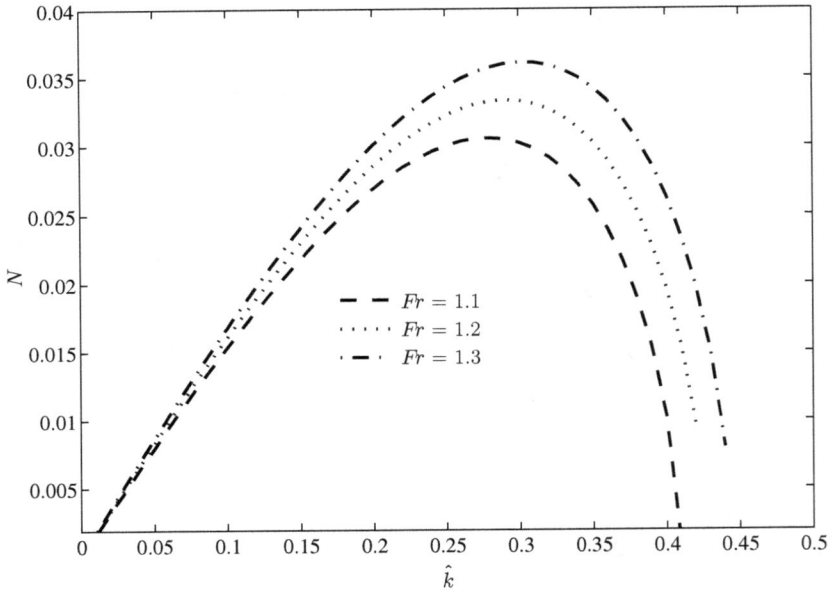

Figure 13.4 Effect of Froude number.

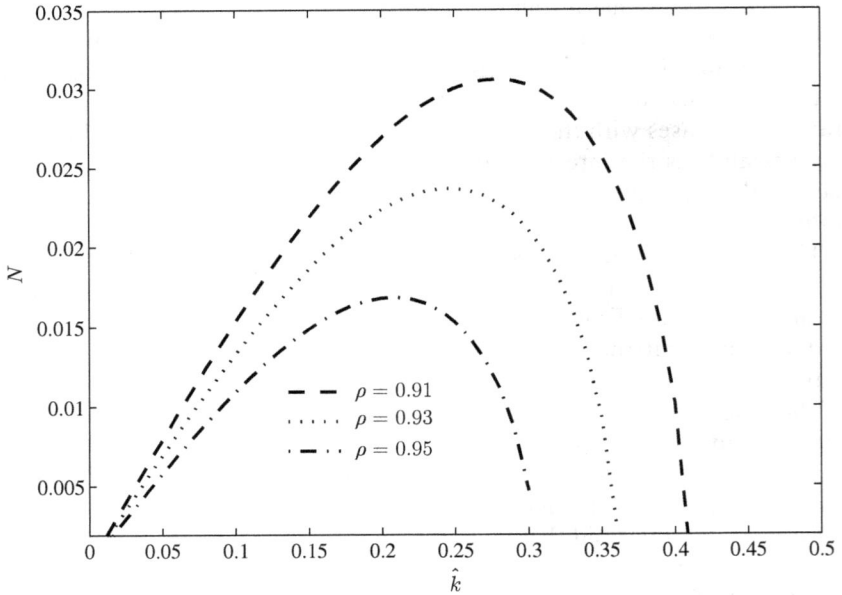

Figure 13.5 Effect of density ratio of the fluids.

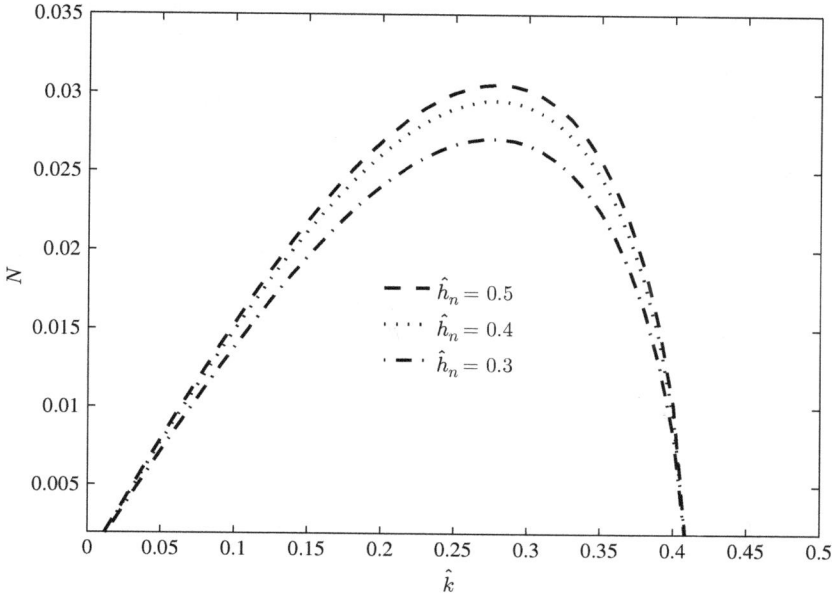

Figure 13.6 Effect of nanofluid thickness.

growth rate, consequently destabilizing the system. The increase in the upper fluid fraction results in a heightened influence of gravity on the interface, causing the gravitational acceleration to further amplify the perturbation growth. The introduction of gravitational acceleration accelerates the perturbations, exacerbating the system's destabilization.

The effect of the viscosity ratio on the perturbation growth is shown in Figure 13.7. The perturbation growth decreases with an increase in the viscosity ratio indicating that viscosity ratio plays a stabilizing role. As nanofluid viscosity varies inversely with the viscosity ratio, the nanofluid viscosity has a destabilizing effect. At the same time, the viscous fluid viscosity has a stabilizing impact.

Figure 13.8 demonstrates the impact of nanoparticle concentration on the growth rate. As the nanoparticle concentration rises, the growth of disturbances decreases. Consequently, the system experiences stabilization due to the increasing nanoparticle concentration. This phenomenon is visually evident in Figure 13.8.

The shape parameter δ accounts for the rotating behavior of nanoparticles about their axes, considering them as prolate spheroids. The effect of shape parameter on the interface is shown in Figure 13.9. As shaper parameter increases, perturbations growth decreases indicating stabilizing behavior of shape parameter.

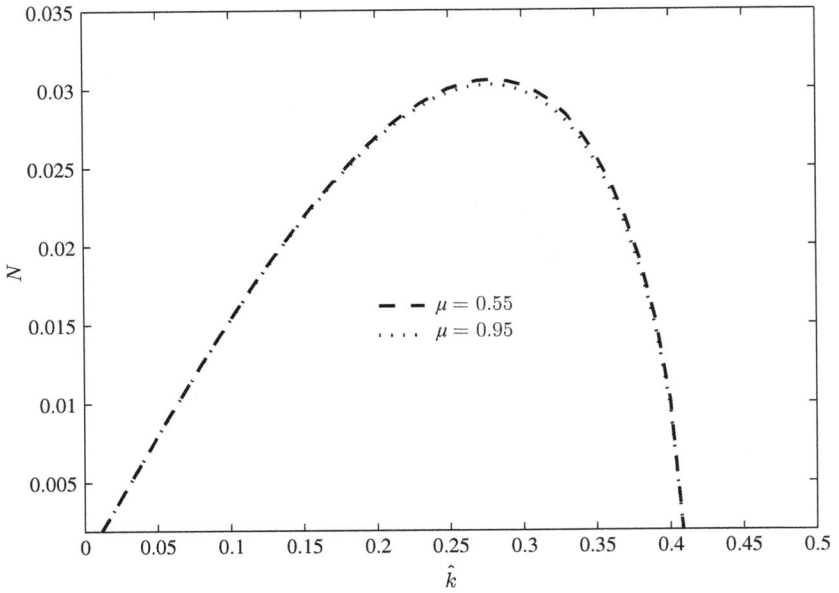

Figure 13.7 Effect of viscosity ratio of fluids.

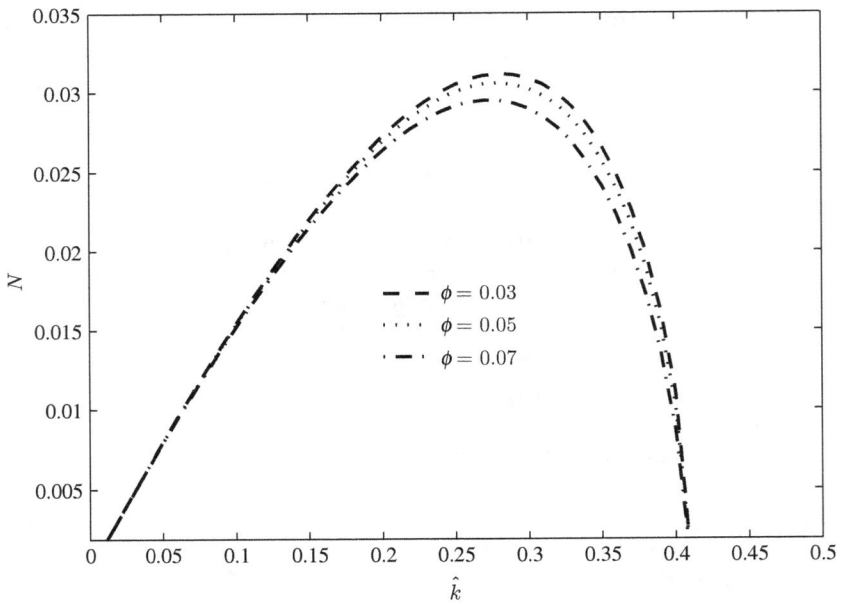

Figure 13.8 Effect of nanoparticle concentrations.

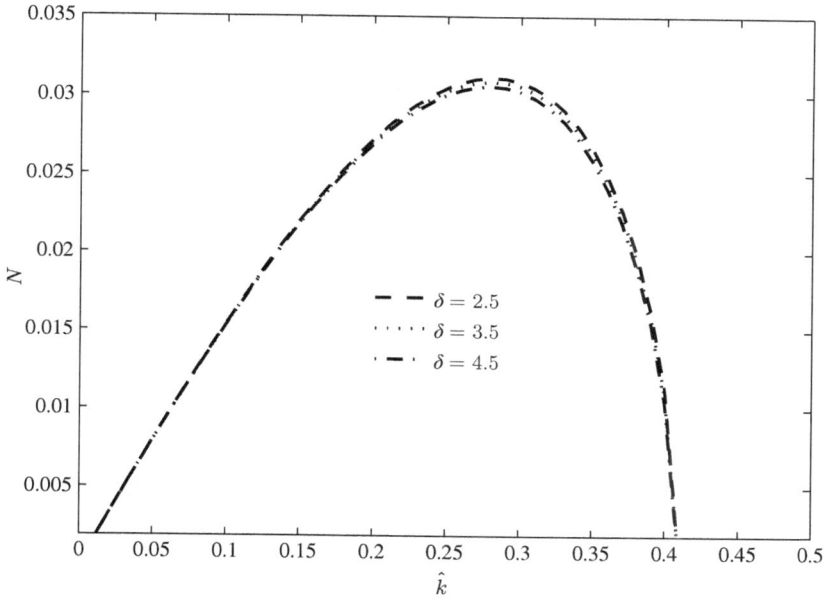

Figure 13.9 Effect of nanofluid shape parameter.

13.7 CONCLUSIONS

This study focuses on investigating the RT instability occurring at the interface between a viscous fluid and a nanofluid in a planar configuration. The lower region contains the viscous fluid, while the upper region consists of the nanofluid. The growth rate's dispersion relationship follows a quadratic pattern. The interfacial stability is influenced positively by increasing the viscosity of the viscous fluid but is destabilized by the density of the nanofluids. Interestingly, the presence of nanoparticles contributes to the stability of the system. The viscosity of the nanofluid supports the growth of perturbations, while the density of the viscous fluid plays a stabilizing role. Additionally, the shape parameter of the nanofluid contributes to stabilization as well.

REFERENCES

1. D. H. Sharp, "An overview of Rayleigh-Taylor instability" *Physica D*, 12, (1984), 3–18.
2. J. W. Jacobs and I. Catton, "Three-dimensional Rayleigh-Taylor instability Part 1. Weakly nonlinear theory" *J. Fluid Mech.* 187 (1988), 329–352.
3. H. J. Kull, "Theory of the Rayleigh-Taylor instability" *Phys. Rep.* 206 (1991) 197–325.

4. K. O. Mikaelian, "Exact, approximate, and hybrid treatments of viscous Rayleigh-Taylor and Richtmyer-Meshkov instabilities" *Phys. Review E* 99 (2019) 023112.

5. D. D. Joseph, J. Belanger and G. S. Beavers "Breakup of a liquid drop suddenly exposed to a high-speed airstream" *Int. J. Multiphase Flow* 25 (1999) 1263–1303.

6. R. Asthana, M. K. Awasthi, and G. S. Agrawal, "Viscous potential flow analysis of Rayleigh-Taylor instability of cylindrical interface" *Appl. Mech. Mat.* 110–116 (2012) 769–775.

7. M. K. Awasthi, R. Asthana, and G. S. Agrawal "Viscous potential flow analysis of nonlinear Rayleigh-Taylor instability with heat and mass transfer" *Microgravity Sci. Technol.* 24 (2012) 351–363.

8. M. K. Awasthi, "Rayleigh–Taylor Instability of Swirling Annular Layer with Mass Transfer" *ASME-J. Fluid Eng.* 141 (2019) 071202 (5 pages).

9. M. K. Awasthi "Nonlinear analysis of Rayleigh-Taylor instability of cylindrical flow with heat and mass transfer" *ASME J. Fluid Eng.* 135 (2013) 061205 (7 pages).

10. M. K. Awasthi, Z. Uddin, and R. Asthana, "Temporal instability of a power-law viscoelastic nanofluid layer" *Eur. Phys. J. ST* 230 (2021) 1427–1434.

11. M. K. Awasthi, Dharamendra, and D. Yadav, "Temporal instability of nanofluid layer in a circular cylindrical cavity" *Eur. Phys. J. ST* 231 (2022) 2773–2779.

12. S. Agarwal, M. K. Awasthi, and A. K. Shukla, "Stability analysis of water-alumina nanofluid film at the spherical interface" *Proc. InstMech Eng.: Part E J. Proc. Mech. Eng. Sci.* (In Press).

13. G. M. Moatimid, and M. A. Hassan, "Convection instability of non-Newtonian Walter's nanofluid along a vertical layer" *J. Egypt. Math. Soc.* 25 (2017) 220–229.

14. G. M. Moatimid, and M. Gaber, "Linear instability of Water–Oil electrohydrodynamic nanofluid layers: analytical and numerical study" *J. Comp. Theo. Nanosci.* 15 (2018) 1495–1510.

15. G. M. Moatimid, M. Gaber, and M. A. A. Mohamed, "Temporal instability of a confined nano-liquid film with the Marangoni convection effect: viscous potential theory" *Microsyst. Technol.* 26 (2020) 2123–2136.

16. N. Dutt, A. J. Hedau, A. Kumar, M. K. Awasthi, V. P. Singh, and G. Dwivedi, "Thermo-hydraulic performance of solar air heater having discrete D-shaped ribs as artificial roughness" *Environ. Sci. Pollut. Res.* (2023). https://doi.org/10.1007/s11356-023-28247-9

17. M. K. Awasthi, N. Dutt, A. Kumar, and S. Kumar, "Electrohydrodynamic capillary instability of Rivlin–Ericksen viscoelastic fluid film with mass and heat transfer" *Heat Transf.* (2023) 1–19. https://doi.org/10.1002/htj.22944

Chapter 14

Shock wave effects on hydrodynamic instability in elliptical bubbles

Satyvir Singh
RWTH Aachen University, Germany

Bidesh Sengupta
Nanyang Technological University, Singapore

14.1 INTRODUCTION

A study on the hydrodynamic instability in fluid dynamics establishes whether a flow is stable or unstable, and if unstable, how these instabilities produce turbulent mixing. When an originally perturbed interface separated by different fluid characteristics is driven by an incident shock wave, the Richtmyer-Meshkov (RM) instability, a shock-accelerated hydrodynamic instability, arises in conjunction with the Kelvin-Helmholtz (KH) instability (Richtmyer, 1960; Meshkov, 1969). The Rayleigh-Taylor (RT) instability, whose basic disturbances eventually erupt into a turbulent fluid mixing as the uniform gravitational acceleration rises, can be thought of as the impulsive limit of the RM instability.

Baroclinic vorticity production, which results from the disruption of the pressure gradient connected to the planar shock wave and the density gradient of the surface, can lead to the RM instability. As soon as the amplitudes and wavelengths of these perturbations coincide, the perturbed interface changes into a turbulent mixing layer. The importance of RM instability in fields including astrophysics, inertial confinement fusion, combustion ramjet, and compressible turbulent flows has drawn considerable attention in recent years. The occurrence and applicability of hydrodynamic instabilities have been thoroughly examined by Brouillette, Zhou, and co-authors (2002; 2017a; 2017b).

Over the years, one of the most basic research subjects for describing the physical architecture of the RM instability has been the study of shock-accelerated bubble. Much research on the shock-accelerated bubbles of different configurations, containing different gases, has been conducted from theoretical, experimental, and numerical viewpoints to explore the development of RM instability. Various experimental results and a greater understanding of shock-bubble interaction were accomplished by Haas and Sturtevant (1987), Jacobs (1992), Layes et al. (2009), Ranjan et al. (2007),

DOI: 10.1201/9781003465171-14

Zhai et al. (2011), as well as many others, using progressively improved experimental methodologies, after the important research work of Markstein (1957). Numerous outstanding numerical approaches have been developed to provide a clearer picture of the intricate shock-bubble interactions by Rudinger and Somers (1960), Quirk and Karni (1996), Zabusky and Zeng (1998), Bagabir and Drikakis (2001) Giordano and Burtschell (2006), Niederhaus et al. (2008), and many more. On the other hand, Picone and Boris (1988), Yang et al. (1988), Samtaney and Zabusky (1994), and Li et al. (2019) developed distinct circulation models based on theoretical predictions of the velocity circulation to be used in various conditions.

In the recent research on the shock-accelerated hydrodynamic instability, the Atwood number effect on the growth of RM instability induced by the interaction of a shock wave with a square bubble was explored numerically by Singh (2020). The dynamics of a shock-heavy cylindrical bubble interaction in both diatomic and polyatomic gases were studied numerically by Singh and Battiato (2021). Subsequently, Singh et al. (2021) examined the bulk viscosity effects on the flow field of a shock-light cylindrical bubble interaction. Further, this study was extended to the interaction of a shock wave and light helium square bubble by Singh (2021a). Recent numerical simulations of the RM instability of the heavy SF_6 square bubble in diatomic and polyatomic gases were published by Singh and Battiato (2022). The shock Mach number is a significant regulating parameter for the exploration of flow-mixing events in RM instability studies. Bagabir and Drikakis (2001) used a numerical model to examine the impacts of shock Mach numbers on the shock-light bubble interaction. When Zhu et al. (2018) used numerical simulations to examine the impact of different incoming shock Mach numbers on the flow fields of a shock-SF_6 bubble interaction, they discovered that the gas bubble deforms differently depending on the shock Mach number. Further, Singh (2021b) looked at the impact of the shock Mach number on the development of the RM instability brought on by the shock-helium square bubble interaction. Singh and Battiato (2023) expanded on this research to examine the impact of shock Mach number on the convergent RM instability in a square heavy gas cylinder. Further, Singh (2023), and Singh and Torrilhon (2023) presented a comparative numerical investigation of the hydrodynamic instability development on the shock-driven square and rectangular light gas bubbles. Recently, Singh and Jalleli (2023) investigated numerically the coupling effect on the evolution of RM instability at double heavy square bubbles.

The strength of the incident shock wave is well known as a critical controlling parameter for the investigation of compressible flows, such as compressible hydrodynamic instability and turbulence mixing. Therefore, the effects of different Mach numbers (M_s = 1.12, 1.25, and 1.5) on the evolution of the RM instability in a shock-accelerated square light bubble are examined numerically in the current study. Utilizing numerical simulations based on an explicit mixed-type modal discontinuous Galerkin method, the effects of Mach numbers on the wave patterns, bubble deformation, vortex creation, vorticity generation, evolution of enstrophy, and dissipation rate, and interface features are discussed. The remainder of this chapter is organized as follows: Section 14.2 outlines the computational model including the problem setup and employed numerical method. Section 14.3 presents the grid refinement analysis and the validation of the numerical method. Section 14.4 discusses in detail the Mach number effect on the shock-accelerated square light bubble in terms of flow evolution, vorticity generation, and their quantitative analysis. Section 14.5 draws some concluding remarks with further development in this topic.

14.2 PROBLEM SETUP

Figure 14.1 displays the computational setup of the shock-accelerated hydrodynamic instability problem. A rectangle with dimensions of 150×50 mm^2 is chosen for the computational domain. A stationary oblate elliptical bubble collides with a shock wave driving in the $x-$direction from left to right sides. For simulations, three shock wave strengths with $M_s = 1.12$, 1.25, and 1.5 are selected, which represent the strength of the propagating IS wave. From the left-hand edge of the computational domain, the shock wave is initially positioned at $x = 20$ mm. The major and minor axes of the elliptical bubble are chosen as $a = 20$ mm and $b = 10$ mm, respectively. Around the elliptical bubble, the initial pressure and temperature are taken to be $P_0 = 101, 325$ Pa, and $T_0 = 273$ K, respectively. Nitrogen gas surrounds the light helium (He) gas that fills the elliptical bubble. The left boundary is designated as the inflow, and the upper, bottom, and right borders serve as the outflow. We use the ambient state on the right-hand side of the incident shock wave to start the numerical simulation. The Rankine-Hugoniot conditions are used to determine primitive variables at the shock wave's left side.

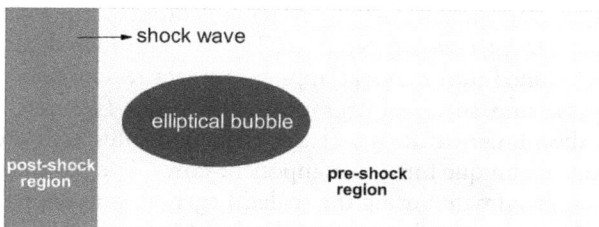

shock wave

elliptical bubble

post-shock region

pre-shock region

Figure 14.1 Schematic diagram of the problem setup.

14.3 NUMERICAL PROCEDURE

In this study, a two-dimensional system of compressible Euler equations for an ideal binary gas mixture is adopted, which can be written in the conservative form as

$$\frac{\partial U}{\partial t} + \frac{\partial F(U)}{\partial x} + \frac{\partial G(U)}{\partial y} = 0, \tag{14.1}$$

with

$$U = \begin{bmatrix} \rho \\ \rho u \\ \rho v \\ \rho E \\ \rho\phi \end{bmatrix}, \quad F(U) = \begin{bmatrix} \rho u \\ \rho u^2 + p \\ \rho uv \\ (\rho E + p)u \\ \rho u\phi \end{bmatrix}, \quad G(U) = \begin{bmatrix} \rho v \\ \rho uv \\ \rho v^2 + p \\ (\rho E + p)v \\ \rho v\phi \end{bmatrix}.$$

In the above expression, ρ denotes the density, u and v represent the velocity components in x- and y- directions, respectively, E signifies the total energy, ϕ indicates the mass fraction, and p refers to the static pressure determined by the ideal gas law

$$p = (\gamma_{mix} - 1)\left(\rho E - \frac{1}{2}(u^2 + v^2)\right), \tag{14.2}$$

where γ_{mix} represents the specific heat ratio of a gas mixture, which is evaluated from the following mathematical expression as

$$\gamma_{mix} = \frac{C_{p1}\phi + C_{p2}(1-\phi)}{C_{v1}\phi + C_{v2}(1-\phi)}. \tag{14.3}$$

Here, subscripts 1 and 2 are meant for bubble and ambient gas, respectively. C_p and C_v represents the specific heat coefficients at constant pressure and volume, respectively.

A high-order explicit modal discontinuous Galerkin method is applied for solving the two-dimensional, unstable system of compressible Euler equations for multi-component gas flows (Singh, 2023). The considered computational domain is partitioned into nonoverlapping uniform rectangular meshes. For both the volume and surface integrations, the Gauss-Legendre quadrature rule is used, the numerical fluxes at the elemental interfaces are estimated using the HLLC technique for two-component flows. The finite element space solutions are approximated using the scaled Legendre polynomial expansion with third-order precision. An explicit third-order precise SSP Runge-Kutta scheme is applied to conduct the time discretization.

Experimental 3D results

Numerical 2D results

Experimental 3D results

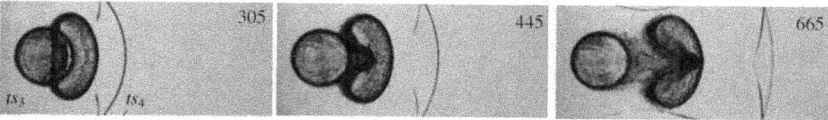

Numerical 2D results

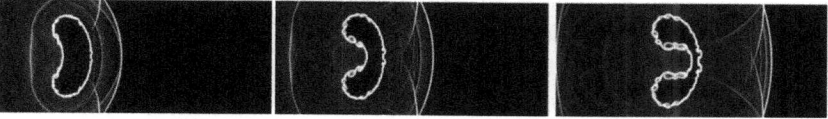

Figure 14.2 Comparison of numerical shadowgraphs between the experimental results (upper) and present simulation results (bottom) at different time instants.

An excellent comparison between the experimental results of Ding et al. (2017) and the current results are shown in Figure 14.2 to demonstrate the validation of the numerical DG solver. In this validation, a moving shock with the strength of Ma = 1.28 interacts with a stationary 3D convex N_2 gas bubble, which is surrounded by SF_6. The numerical findings demonstrate that the cylindrical bubble is continuously distorted by shock waves that are propagating and being reflected, and they also demonstrate the identical beginning condition, resolution, shock wave evolution, and thickness of the diffusion layer. The experimental and current results exhibit good qualitative agreement, demonstrating the dependability of the current numerical solver.

Furthermore, the time variations of the interfacial characteristic scales, i.e., the length and the height of the evolving interface for the cylindrical bubble, are also illustrated in Figure 14.3. It can be seen from the plot that the present results, including the general trend of the interfacial characteristic scales changing with time, are found very close to the experimental results of Ding et al. (2017).

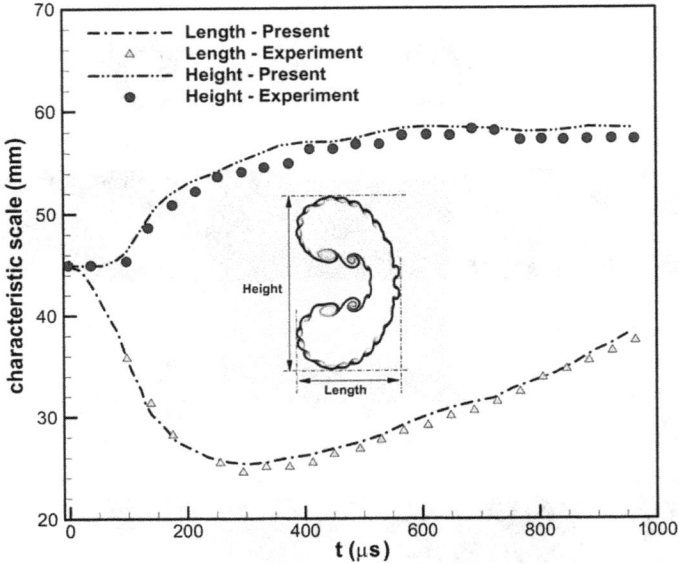

Figure 14.3 Comparison of evolving interfaces between the experimental results and present numerical results for a shock-accelerated cylindrical light gas bubble.

14.4 NUMERICAL RESULTS AND DISCUSSION

In this section, we present the numerical results of shock wave effects on the hydrodynamic-instability at the elliptical light gas bubble. The effects of an initial interface disruption on the gas bubble shape, wave evolutions, vorticity generation, and time evolution of enstrophy as well as kinetic energy are highlighted. Three different shock Mach numbers (M_s = 1.12, 1.25, and 1.50) are chosen for the numerical simulations in order to examine the impact of shock wave strength Mach numbers on the hydrodynamic instability. All the computational simulations are done with grid points 1200 × 600 to ensure numerical accuracy.

Visual representation of the flow when a shock wave and an elliptical gas bubble interact, evolution is thought to be the most fascinating event. To understand the impact of shock wave strengths on the hydrodynamic instability, we conducted an extensive investigation of the time evolution of flow morphology for the shock-accelerated elliptical He bubble. Figure 14.4 illustrates the shock Mach number effects (M_s = 1.12, 1.25, 1.5) on the time-dependent flow morphology of the shock-accelerated elliptical He bubble through density gradient contours. When the shock wave reaches the bubble interface, the bubble begins to compress. In addition, a transmitted shock wave propagating downstream inside the bubble is generated, while a reflected shock wave simultaneously travels upstream. The propagation

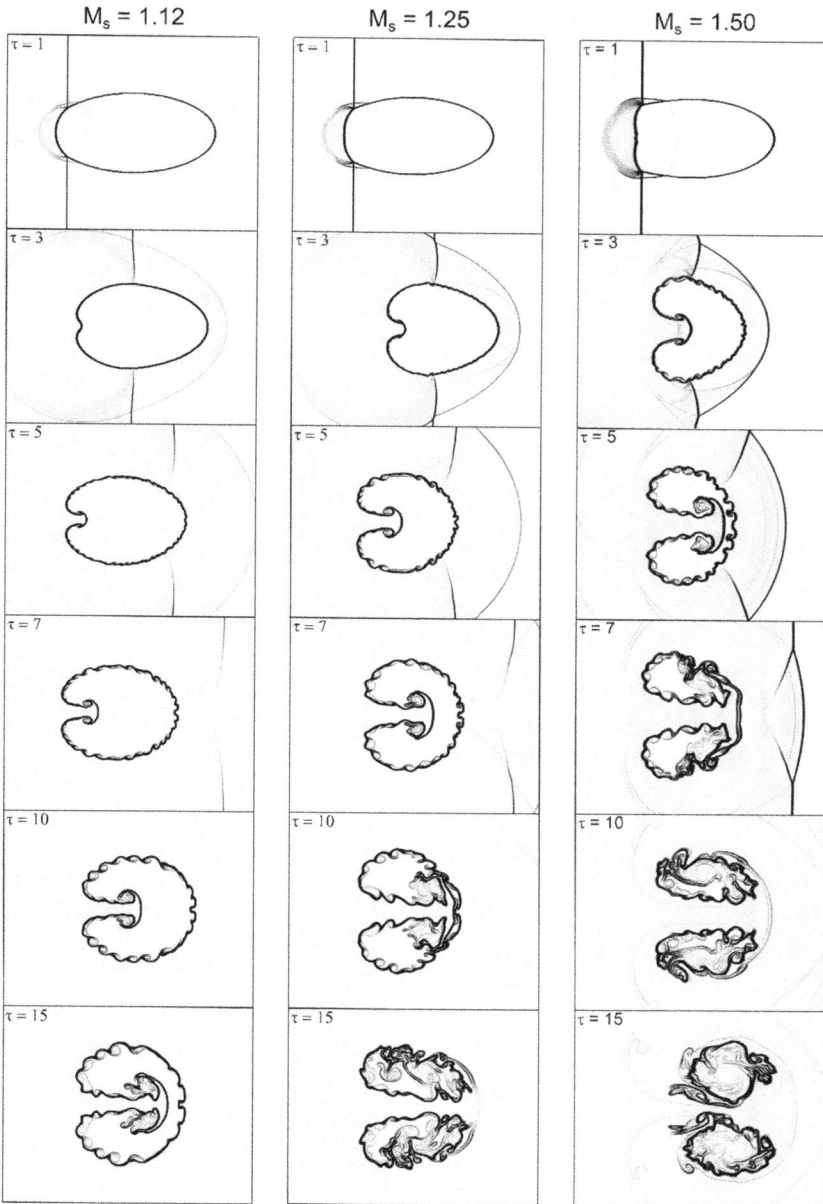

Figure 14.4 Impact of shock wave strength on the time evolution of numerical Schlieren images in shock-accelerated elliptical helium gas bubble.

speed of the shock wave inside the bubble is smaller than that in the surrounding gas due to the small acoustic impedance. Therefore, the generated transmitted shock wave inside the bubble travels faster behind the shock wave. As the interaction develops, the shock wave and transmitted wave form a quadruple shock in the gas, which reveals irregular refraction. As a consequence, a Mach reflection configuration is generated with a Mach stem, triple point, and slip surface. Later, a secondary transmitted shock wave is observed near the downstream interface. As time proceeds, another reflected shock wave within the bubble moves upward and then hits the upstream interface, generating a third transmitted shock wave. During the interaction, the bubble is gradually deformed. The density inhomogeneity is accelerated at the very beginning, and the upstream bubble interface is flat. The evolving bubble interface then starts to transform into a mushroom shape, and a re-entrant gas jet head is subsequently generated near the center of the bubble. As time proceeds, the jet catches up with the downstream bubble interface, and then, a pair of vortex rings connected with a bridge emerges and grows almost symmetrically. Eventually, the flow field is completely controlled by the vortex ring. It can be observed that the high shock Mach number causes a stronger interaction between the shock and bubble. The height of generated jet increases with the increasing the strength of the incident shock wave due to the larger expansion of the bubble interface upwards. Furthermore, as the Mach number increases, the bubble deforms significantly. As a result, the generated wave patterns become more complex, and the size of the bubbles decreases noticeably. The re-entrant jet structure at $M_s = 1.5$ is observed as the longest among three Mach numbers. Additionally, the size and strength of the rolled-up vortices increase significantly at high Mach numbers, and these vortices are conspicuous at the interface between the bubble and surrounding gas due to the baroclinic vorticity deposition.

By attentively examining the terms in the vorticity transport equation, it is possible to get a better understanding of the mechanisms of vortex dynamics. The vorticity transport equation for compressible inviscid flows can be written as

$$\frac{D\omega}{Dt} = (\omega.\nabla)u - \omega(\nabla.u) + \frac{1}{\rho^2}\nabla\rho \times \nabla p \qquad (14.4)$$

Here, the left-hand side term denotes the material derivatives expressed as the sum of unsteady. On the right-hand side, the first term represents the vorticity stretching due to the flow velocity gradients, which is zero for two-dimensional flows. The second term expresses the stretching of vorticity due to flow compressibility. Finally, the third term represents the baroclinic vorticity production term, which is responsible for the generation of small-scale rolled-up vortices at the bubble interface. Notable, this term is most prominent in shock-driven interface problems due to the mismatch of the density and pressure gradients.

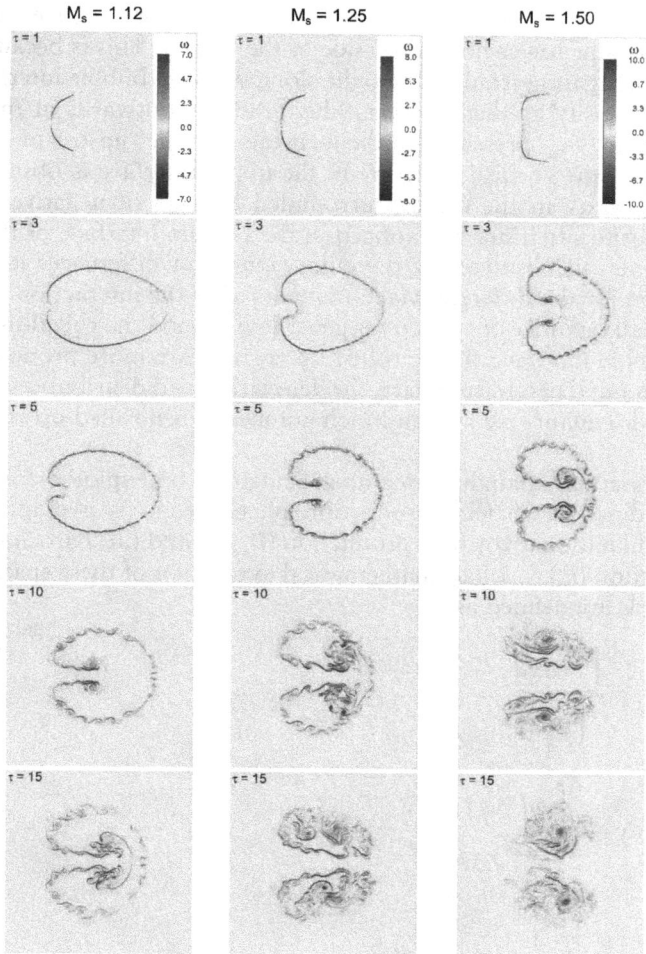

Figure 14.5 Impact of shock wave strength on the time evolution of vorticity contours in shock-accelerated elliptical helium gas bubble.

Figure 14.5 illustrates the effect of Mach number on the vorticity distribution of the shock-accelerated square helium bubble at different time instants. Initially, the vorticity is equal to zero everywhere. When the IS wave passes across the bubble, the baroclinic vorticity is mainly deposited locally on the bubble interface in the early stage, where the discontinuity between helium gas and the ambient gas exists. At the top and bottom locations of the bubble where the density and pressure gradients are orthogonal, the magnitude of the vorticity is maximum and it is zero at the interface along the axis of the bubble where the density and pressure gradients are collinear. A significant quantity of positive vorticity is generated on the upper horizontal side of the bubble interface, while a significant quantity of negative vorticity is generated on the lower horizontal side of the bubble. This is because of the IS wave propagating from left to right along with the bubble interface. As a result, the density gradient is everywhere radially outwards at the bubble interface and the pressure gradient is across the upstream IS wave. Furthermore, the vortical structure in the upper interface is observed with positive vorticity in the center, surrounded by tails of negative vorticity, while opposite situations are noticed in the bottom interface of the square bubble. One can observe that there are significant differences in vorticity distribution for the different Mach numbers after the interaction. For M_s = 1.12, a small quantity of vorticity is generated around the rolled-up vortices on the bubble interface. These rolled-up vortices are more pronounced for high Mach numbers. In summary, the generation and distribution of vorticity play a dominant role at high Mach numbers when rolled-up vortices are formed.

Three essential spatially integrated fields can be explored for a better understanding of vorticity production phenomena: (i) average vorticity (ω_{av}), (ii) dilatational vorticity production ($P_{\omega, dil}$), and (iii) baroclinic vorticity production ($P_{\omega, baro}$). The mathematical expression of these spatially integrated fields are defined as

$$\omega_{av}(\tau) = \frac{\int |\omega| \, dxdy}{\int dxdy}, \tag{14.5}$$

$$P_{\omega, dil}(\tau) = -\frac{\int |\omega(\nabla . u)| \, dxdy}{\int dxdy}, \tag{14.6}$$

$$P_{\omega, baro}(\tau) = \frac{\int \left| \frac{1}{\rho^2} \nabla\rho \times \nabla p \right| dxdy}{\int dxdy}. \tag{14.7}$$

Figure 14.6 illustrates the effect of Mach number on the spatially integrated fields of average vorticity, absolute dilatational vorticity, and absolute baroclinic vorticity in the shock-accelerated elliptical helium bubble. It can be observed that these spatially integrated fields in case of M_s = 1.12 is the smallest among the three cases when the incident and reflected shock waves collide with the bubble. The spatially integrated fields are substantially enhanced in the case of M_s = 1.25 For all three Mach numbers, the spatially integrated fields increase with time, which implies that the ambient gas is increasingly entrained into the distorted elliptical helium bubble. The vortices produced by the shock wave-bubble interaction encourage the mixing of ambient gas with the elliptical helium bubble. When the reflected

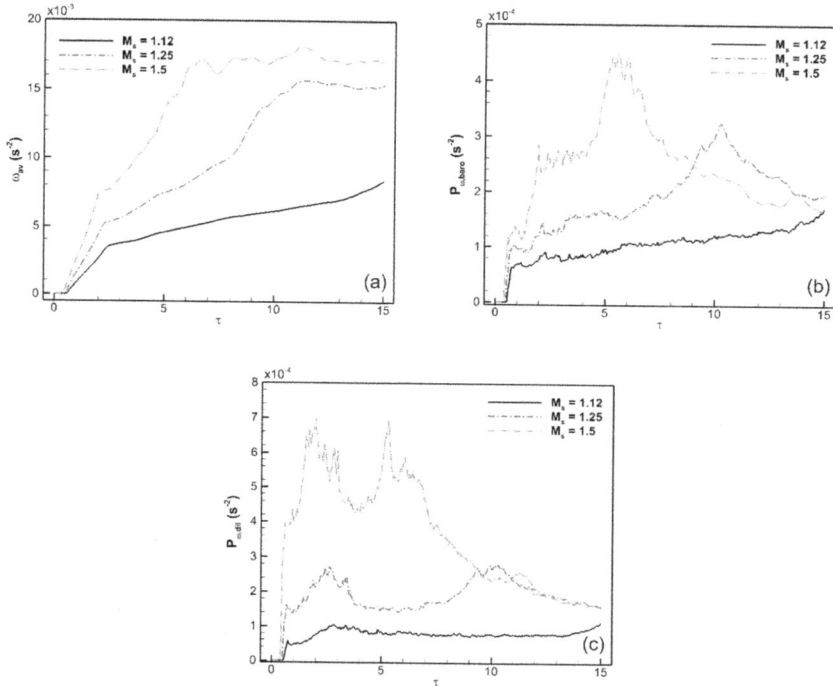

Figure 14.6 Impact of shock wave strength on the spatially integrated fields of (a) average vorticity, (b) dilatational vorticity production, and (c) baroclinic vorticity production in the shock-accelerated elliptical helium bubble.

shock waves impinge on the distorted helium bubble again, the spatially integrated fields exhibit their greatest growth rate, which indicates that the vorticities are significantly enhanced during this period.

The enstrophy and kinetic energy are two fundamental fields in the shock-accelerated interface problems. These fields are addressed here to better understand the effects of shock wave strengths on the interaction process. The time evolution of enstrophy and kinetic energy (K.E.) are defined as

$$\Omega(\tau) = \frac{1}{2} \int \omega^2 dx dy, \tag{14.8}$$

$$\text{K.E.}(\tau) = \frac{1}{2} \int u^2 dx dy. \tag{14.9}$$

Figure 14.7 investigates the Mach number effects on the spatially integrated fields of the enstrophy and kinetic energy over time. The evolution of enstrophy over time is shown in Figure 14.7a. The enstrophy is zero until the shock wave reaches the upstream pole of the bubble. Baroclinic vorticity production leads to an increase during the shock wave passage. A first local

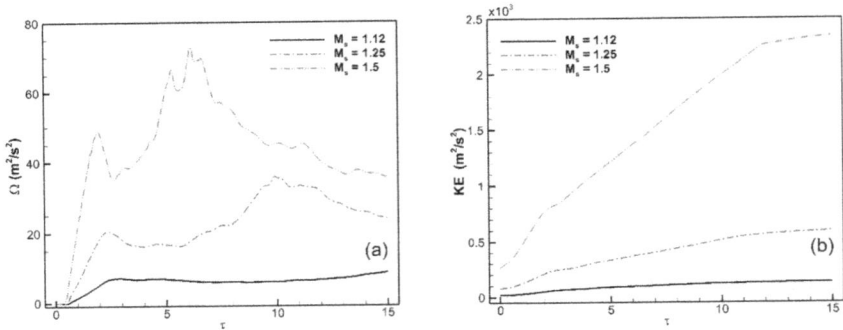

Figure 14.7 Shock wave strengths on the time evolution of (a) enstrophy (Ω) and (b) kinetic energy (KE) in the shock-accelerated elliptical helium bubble.

maximum in enstrophy is reached after the shock has passed half of the bubble, an effect that can be observed for all simulations. Thereafter, a slight decay is visible, followed by another increase due to shock transmission and shock reflections at the interface. Subsequently, the enhanced vorticity promotes the mixing of gases inside and outside the gas bubble, and thus accelerates the transfer and consumption of vorticity energy, which gradually weakens the enstrophy intensity in the bubble region. Further, the evolution of the kinetic energy is plotted in Figure 14.7b, which is different and dependent on shock Mach number. It is obvious that the kinetic energy is highly raised with the increase in the shock Mach numbers.

14.5 CONCLUSION

The hydrodynamic instability has long been a fascinating issue due to its fundamental importance in scientific study as well as its critical function in engineering applications. In this research, the impact of different shock wave strengths (M_s = 1.12, 1.25, and 1.5) on the evolution of the RM instability in a shock-accelerated elliptical light bubble is investigated numerically. A two-dimensional system of unsteady compressible Navier-Stokes-Fourier equations is solved by using an explicit mixed-type modal discontinuous Galerkin method with uniform meshes. The numerical results reveal that the shock Mach numbers play a significant role in describing the RM instability during the interaction between a planar shock wave and a light bubble. The effects of Mach numbers result in a substantial change in the flow morphology with complex wave patterns, vortex creation, vorticity generation, and bubble deformation. In contrast to low Mach numbers, high Mach numbers produce larger rolled-up vortex chains, larger inward jet formation, and a stronger mixing zone with larger expansion. At high Mach numbers, the bubble deforms differently and the reflected shock wave promotes a more complicated deformation of the bubble. Additionally, larger distortions of the bubble occur at the early time instants for higher Mach numbers. A detailed study of the Mach numbers effects is investigated through vorticity generation. It is interesting to observe that vorticity plays a significant role to describe essential features in the study of the shock-accelerated bubble. It is found that the vorticity in the bubble region is enhanced with the increment of shock Mach number, especially for the period of the incident and reflected shock waves impinging on the bubbles. Finally, a significant increment in kinetic energy and enstrophy is also found with increasing the shock Mach numbers.

REFERENCES

Bagabir, A., & Drikakis, D. (2001). Mach number effects on shock-bubble interaction. *Shock Waves*, *11*, 209–218.

Brouillette, M. (2002). The Richtmyer-Meshkov instability. *Annual Review of Fluid Mechanics*, *34*, 445–468.

Ding, J., Si, T., Chen, M. Zhai, Z., Lu, X., Luo, X. (2017). On the interaction of a planar shock with a three-dimensional light gas cylinder. *The Journal of Fluid Mechanics*, *828*, 289–317.

Giordano, J., & Burtschell, Y. (2006). Richtmyer-Meshkov instability induced by shock-bubble interaction: Numerical and analytical studies with experimental validation. *Physics Fluids*, *18*(3), 036102.

Haas, J.F., & B. Sturtevant, B. (1987). Interaction of weak shock waves with cylindrical and spherical gas inhomogeneities. *Journal of Fluid Mechanics*, *181*, 41.

Jacobs, J.W. (1992). Shock-induced mixing of a light-gas cylinder. *Journal of Fluid Mechanics*, *234*, 629–649.

Layes, G., Jourdan, G., & Houas, L. (2009). Experimental study on a plane shock wave accelerating a gas bubble. *Physics of Fluids*, *21*, 074102.

Li, D., Wang, W., & Guan, B. (2019). On the circulation prediction of shock-accelerated elliptical heavy gas cylinders. *Physics Fluids*, *31*, 056104.

Markstein, G.H. (1957). A shock-tube study of flame front-pressure wave interaction. *Symposium on Combustion*, *6(1)*, 387–398.

Meshkov, E.E. (1969). Instability of the interface of two gases accelerated by a shock wave. *Fluid Dynamics*, *4(5)*, 101–104.

Niederhaus, J.J., Greenough, J. A., Oakley, J.G., Ranjan, D., Anderson, M.H., & Bonazza, R. (2008). A computational parameter study for the three-dimensional shock-bubble interaction. *Journal of Fluid Mechanics*, *594*, 85.

Picone, J.M., & Boris, J.P. (1988). Vorticity generation by shock propagation through bubbles in a gas. *Journal of Fluid Mechanics*, *189*, 23–51.

Quirk, J.J., & Karni, S. (1996). On the dynamics of a shock–bubble interaction. *Journal of Fluid Mechanics*, *318*, 129–163.

Ranjan, D., Niederhaus, J.H.J., Motl, B., Anderson, M.H., Oakley, J., & Bonazza, R. (2007). Experimental investigation of primary and secondary features in high Mach-number shock-bubble interaction. *Physical Review Letters*, *98*, 024502.

Richtmyer, R.D. (1960). Taylor instability in shock acceleration of compressible fluids. *Communications on Pure and Applied Mathematics*, *13*, 297.

Rudinger, G., & Somers. (1960). Behaviour of small regions of different gases carried in accelerated gas flows. *Journal of Fluid Mechanics, 7*(2), 161–176.

Samtaney, R., & Zabusky, N.J. (1994). Circulation deposition on shock-accelerated planar and curved density-stratified interfaces: Models and scaling laws. *Journal of Fluid Mechanics, 269,* 45–78.

Singh, S. (2020). Role of Atwood number on flow morphology of a planar shock-accelerated square bubble: A numerical study. *Physics of Fluids, 32,* 126112.

Singh, S. (2021a). Numerical investigation of thermal non-equilibrium effects of diatomic and polyatomic gases on the shock-accelerated square light bubble using a mixed-type modal discontinuous Galerkin method. *International Journal of Heat and Mass Transfer, 169,* 121708.

Singh, S. (2021b). Contribution of Mach number on the evolution of Richtmyer-Meshkov instability induced by a shock-accelerated square light bubble. *Physical Review Fluids, 6,* 044001.

Singh, S. (2023). Investigation of aspect ratio effects on flow characteristics and vorticity generation in shock-induced rectangular bubble. *European Journal of Mechanics/B Fluids, 101,* 131–148.

Singh, S., & Battiato, M. (2021). Behavior of a shock-accelerated heavy cylindrical bubble under nonequilibrium conditions of diatomic and polyatomic gases. *Physical Review Fluids, 6,* 044001.

Singh, S., & Battiato, M. (2022). Numerical simulations of Richtmyer-Meshkov instability of SF6 square bubble in diatomic and polyatomic gases. *Computers and Fluids, 242,* 105502.

Singh, S., & Battiato, M. (2023). Numerical investigation of shock Mach number effects on convergent Richtmyer-Meshkov instability in a heavy square bubble. *Physica D: Nonlinear Phenomena, 453,* 133844.

Singh, S., Battiato, M. & Myong, R.S. (2021). Impact of bulk viscosity on flow morphology of shock-accelerated cylindrical light bubble in diatomic and polyatomic gases. *Physics of Fluids, 33,* 066103.

Singh, S., & Jalleli, D.T. (2024). Investigation of coupling effect on the evolution of Richtmyer-Meshkov instability at double heavy square bubbles. *SCIENCE CHINA Physics, Mechanics & Astronomy, 67,* 214711.

Singh, S., & Torrilhon, M. (2023). On the shock-driven hydrodynamic instability in square and rectangular light gas bubbles: a comparative study from numerical simulations. *Physics of Fluids, 35,* 012117.

Yang, J., Kubota, T., & Zukoski, E.E. (1988). A model for characterization of a vortex pair formed by shock passage over a light-gas inhomogeneity. *Journal of Fluid Mechanics, 258,* 217–244.

Zabusky, N.J., & Zeng, S.M. (1998). Shock cavity implosion morphologies and vortical projectile generation in axisymmetric shock-spherical fast/slow bubble interactions. *Journal of Fluid Mechanics, 362,* 327–346.

Zhai, Z., Si, T., Luo, X., & Yang, J. (2011). On the evolution of spherical gas interfaces accelerated by a planar shock wave. *Physics Fluids, 23,* 084104.

Zhou, Y. (2017a). Rayleigh-Taylor and Richtmyer-Meshkov instability induced flow, turbulence, and mixing. I. *Physics Reports, 720,* 1–136.

Zhou, Y. (2017b). Rayleigh-Taylor and Richtmyer-Meshkov instability induced flow, turbulence, and mixing. II. *Physics Reports, 723,* 1–60.

Zhu, Y., Yang, Z., Pan, Z., Zhang, P., & Pan, J. (2018). Numerical investigation of shock-SF6 bubble interaction with different Mach numbers. *Computers & Fluids, 177,* 78.

Index

For Product Safety Concerns and Information please contact our EU
representative GPSR@taylorandfrancis.com
Taylor & Francis Verlag GmbH, Kaufingerstraße 24, 80331 München, Germany